[redacted]

[redacted]

Totally Random

[redacted]

[redacted]

[redacted]

[redacted]

[redacted]

[redacted]

[redacted]

[redacted]

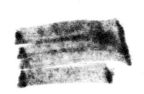

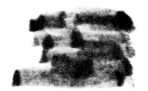

TOTALLY
RANDOM

WHY NOBODY UNDERSTANDS
QUANTUM
MECHANICS
(A SERIOUS COMIC ON ENTANGLEMENT)

TANYA BUB AND JEFFREY BUB

Princeton University Press
Princeton and Oxford

Published by Princeton University Press
41 William Street, Princeton, New Jersey 08540
6 Oxford Street, Woodstock, Oxfordshire OX20 1TR

press.princeton.edu

Library of Congress Control Number: 2017963960

ISBN (pbk.) 978-0-691-17695-6

British Library Cataloging-in-Publication Data is available

Editorial: Eric Henney and Arthur Werneck
Production Editorial: Mark Bellis
Cover Credit: artwork by Tanya Bub
Production: Erin Suydam
Publicity: Sara Henning-Stout (US) and Katie Lewis (UK)

This book has been composed in Adventure, Agency, Arial, Badger,
BethHand, Garamond, Sitka, and VTC Letterer

Printed on acid-free paper. ∞

Printed in Canada

1 3 5 7 9 10 8 6 4 2

For
ANOUK
and
ARLO.

Dramatis

PART I - A Curious Correlation
Meta-characters (you can't see them but they're there)

J (black on white): A superego alter ego of author Jeffrey Bub, serious, professorial, anally retentive and devastatingly handsome, J drives the book's content & direction.

T (white on black): Author Tanya Bub's id, T, the never-seen but nonetheless super-hot illustrator decides what actually goes in the book. She's not sold on J's ideas.

The Narrator: T's creation, this surrogate unhinged tell-us-what-you-really-think voice follows the letter of J's direction while simultaneously trying to derail the book.

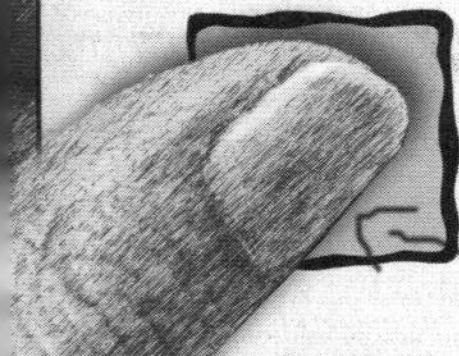

Reader: You, (sort of) also quite hot, represented by illustrated hands, have been "drawn" into the book to do the physical & mental heavy lifting. Feel the burn!

John Stuart Bell: Arguably the most important never-seen character. Sense his presence as the insidious implications of his deceptively simple proof unfold!

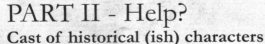

PART II - Help?
Cast of historical (ish) characters

Einstein-ish: Deliverer of goods and reason (and good reason), Einstein-ish attempts to apply common sense to the quoins (and fails).

Schrödinger-ish: A classical type, Schrödinger-ish uses a machine to see what an ambivalent state of reality would look like—get this—in a cat!

Everett-ish: This carny huckster will try to sell you on a branching universe with many worlds — but look out, you may just find he has a point.

Personæ
A.K.A. who's in it

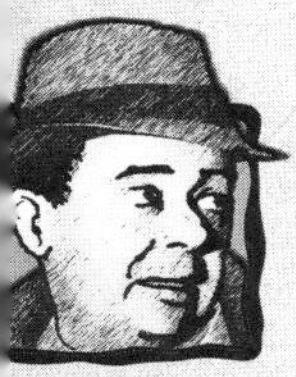

von Neumann-ish: The rigorous math genius who brought consciousness into the story. If anyone can sleuth out who collapsed reality to a definite state, it's him.

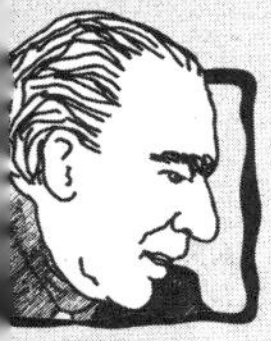

Bohr-ish: Do you suffer from uncontrollable urges to picture an underlying reality? Dr. Bohr, ringing in from the Copenhagen interpretation, will help you let it go!

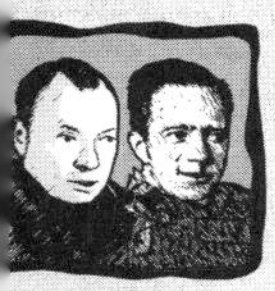

Pauli-ish and Heisenberg-ish: Physicists by day, sign painters by night, Bohr's brilliant and loyal henchies.

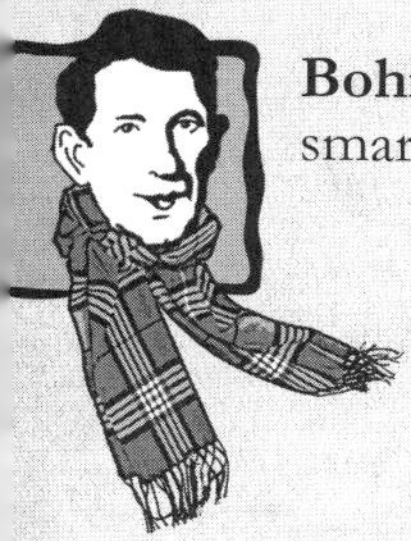

Bohm-ish: Hidden variables and pilot waves are his thing. Is this guy smart enough to take on Einstein-ish? You be the judge.

PART III - Beyond the great debate
Mostly extras and clip art (budgeting problems)

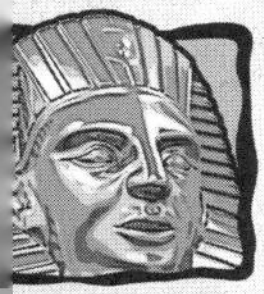

E.V.E: Get an unhackable, uncrackable encrypted secret message past her all-seeing, all-knowing eye or suffer the consequences!

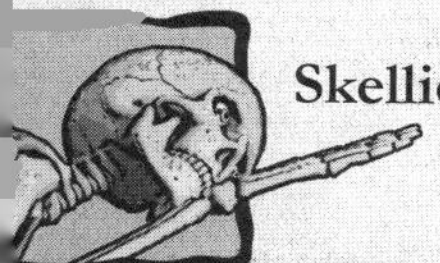

Skellies: Tried the above but failed. Sad.

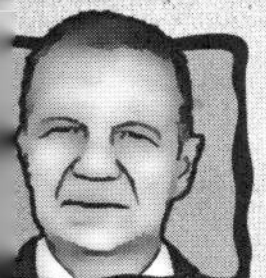

Quasino Croupier: Actually Jeffrey Bub (no relation to J), this guy gets around (budgeting problems). Compute a solution to beat his odds & take the big prize!

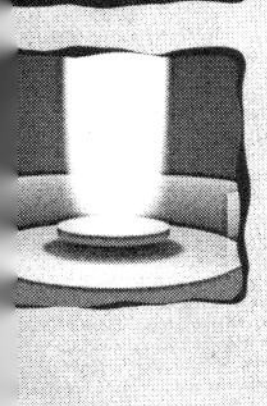

Teleporter: Find out why teleportation is really a question of information. The ending that isn't an ending because it's hard to come up with a real ending.

Hey T.

Yes, J.

The book? How are we starting it?

How about...

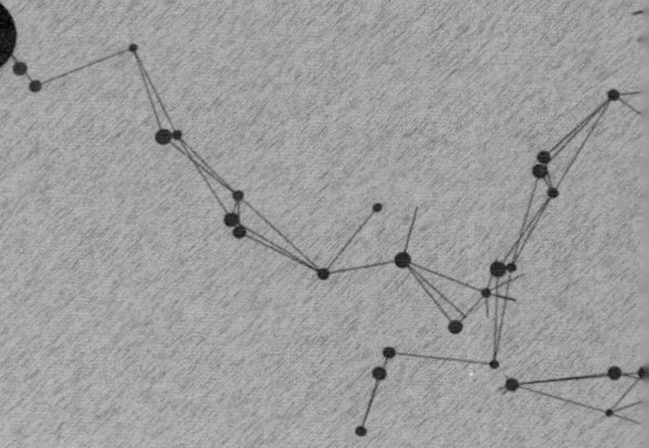

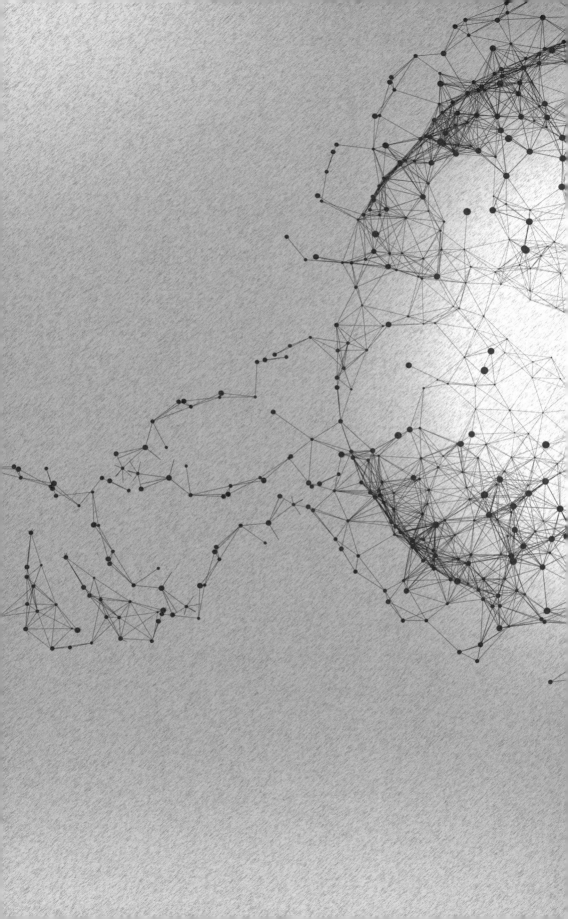

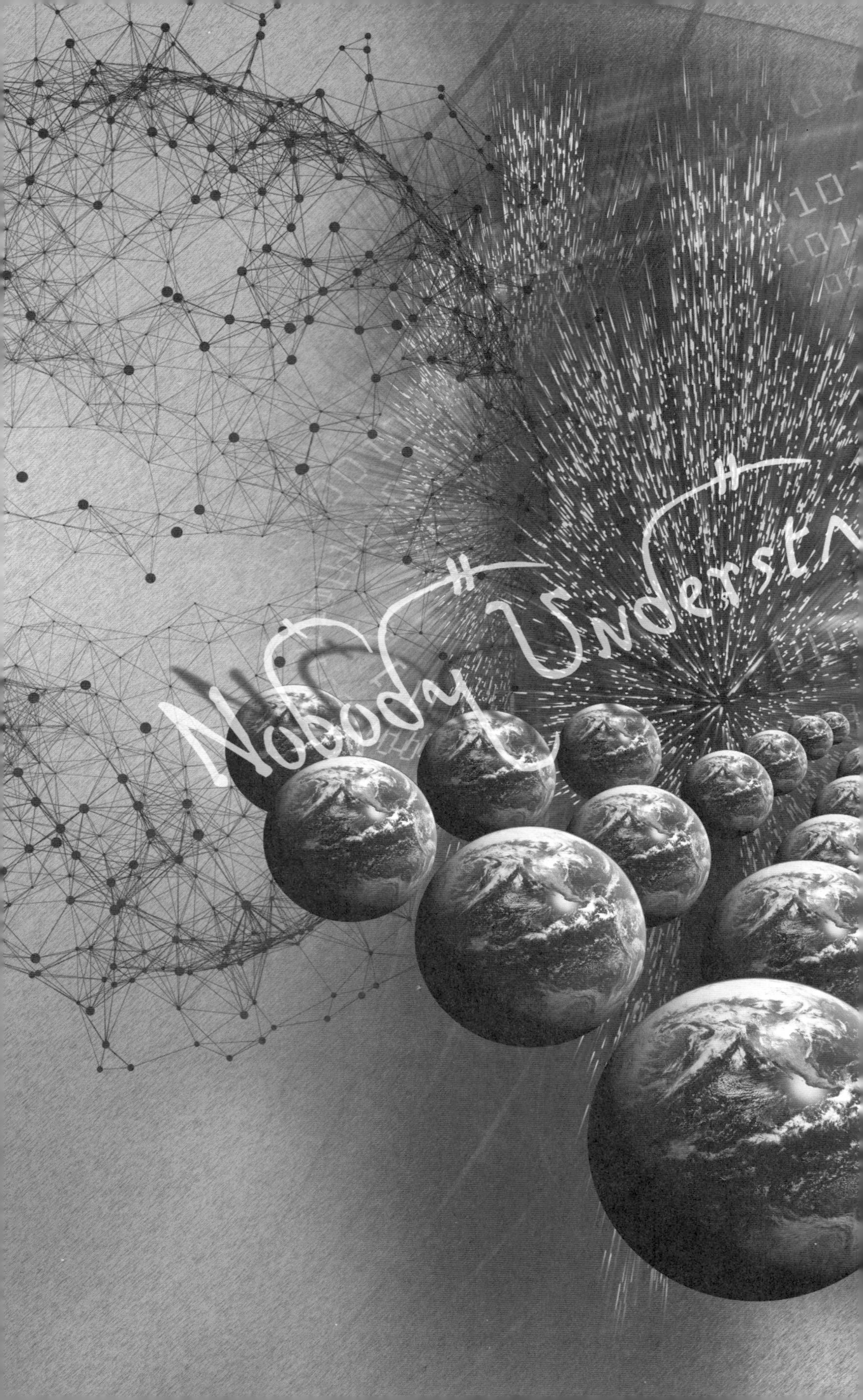
Nobody Underst

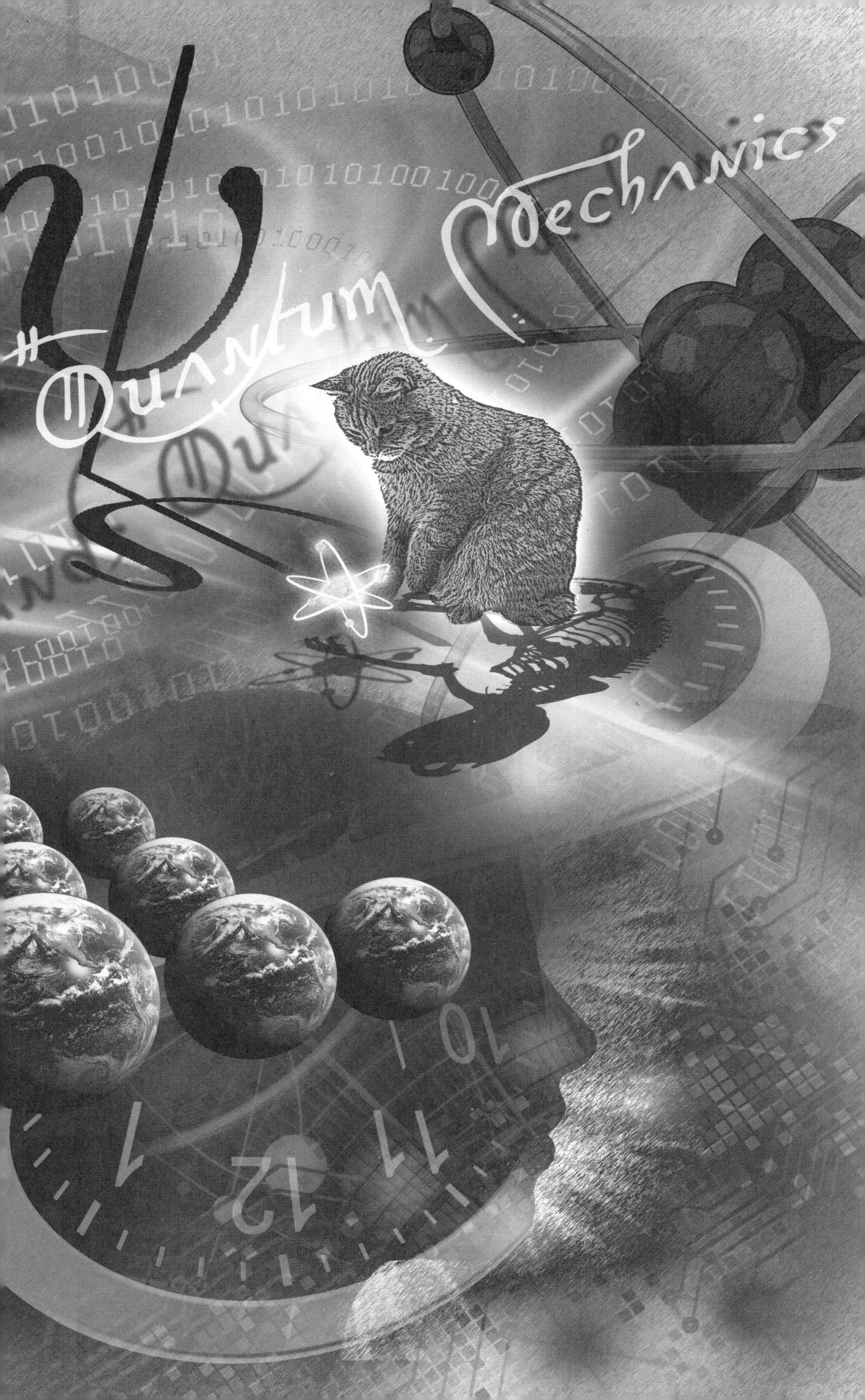
Quantum Mechanics

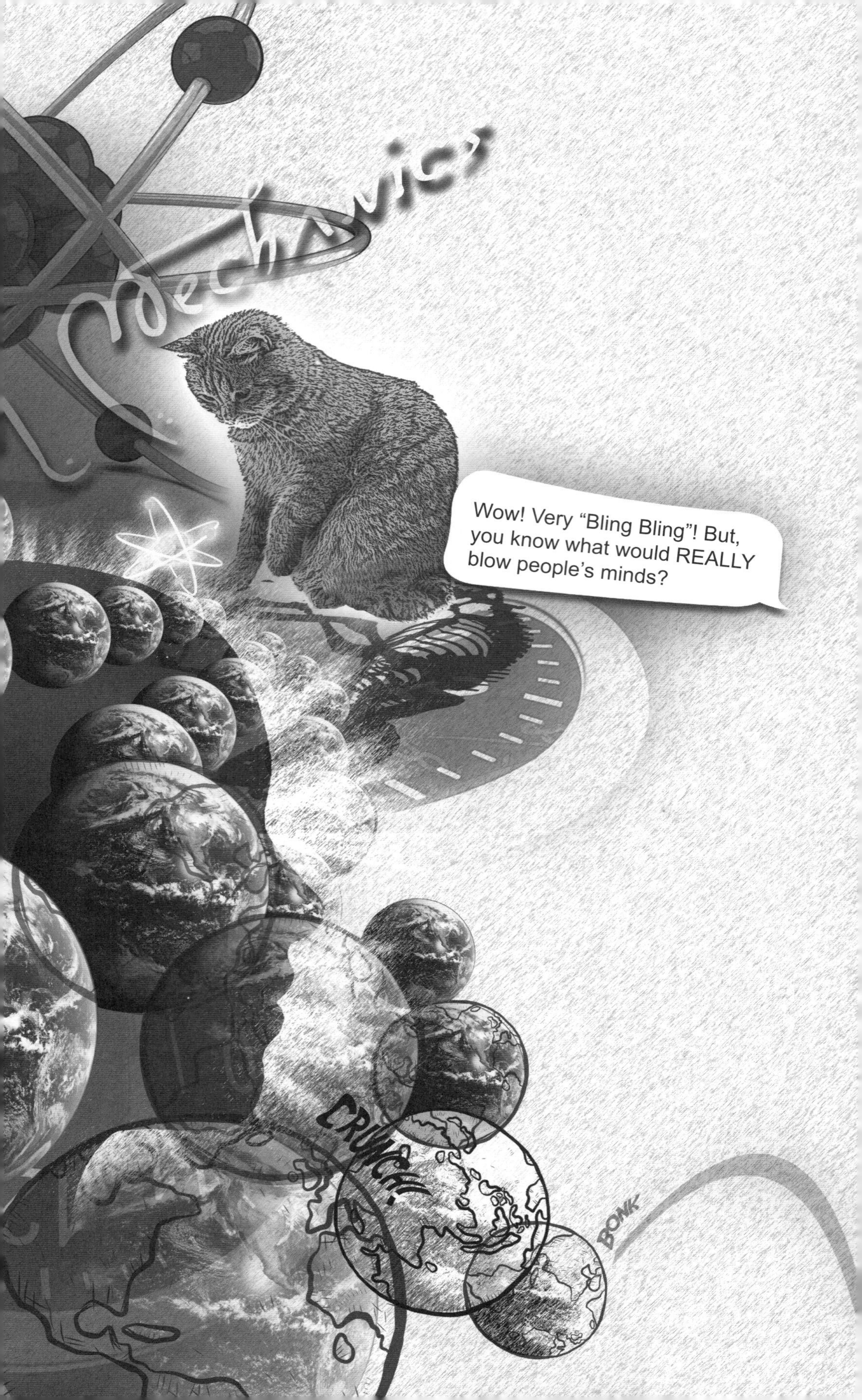
Wow! Very "Bling Bling"! But, you know what would REALLY blow people's minds?
CRUNCH!!
BONK

PART I

We show them that behind it all there's just a...

A CURIOUS CORRELATION

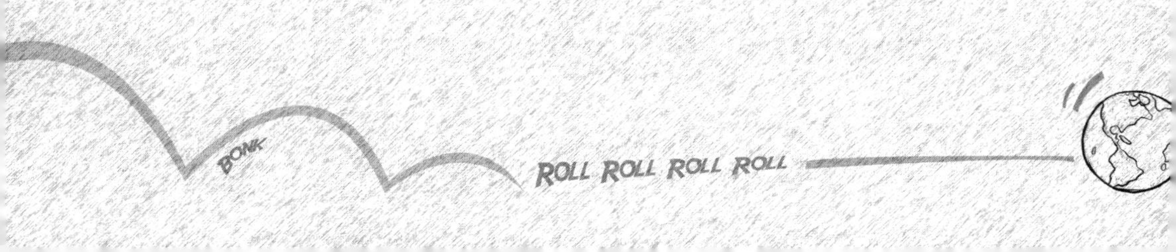
BONK
ROLL ROLL ROLL ROLL

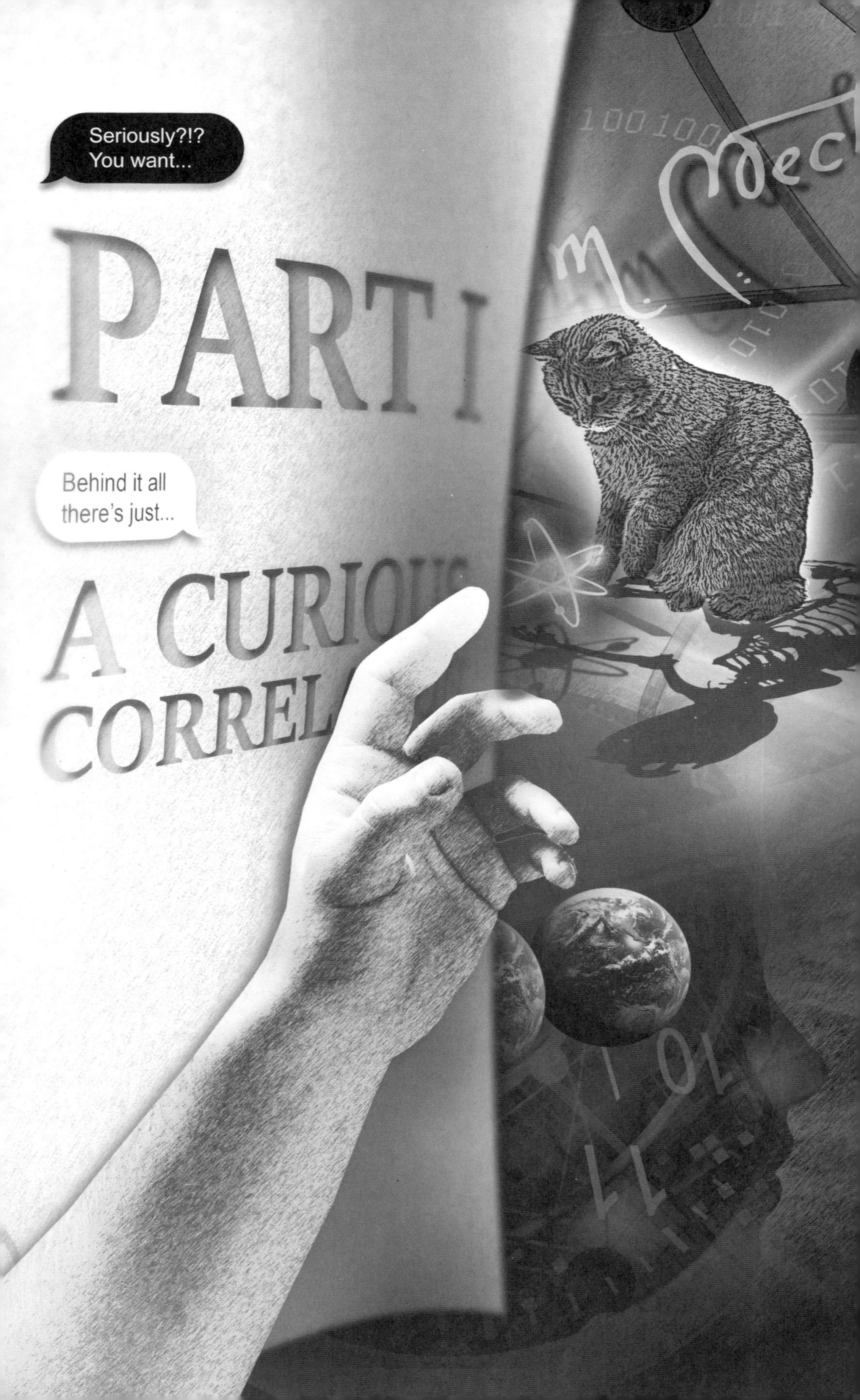
Seriously?!? You want...
PART I
Behind it all there's just...
A CURIOU
CORREL

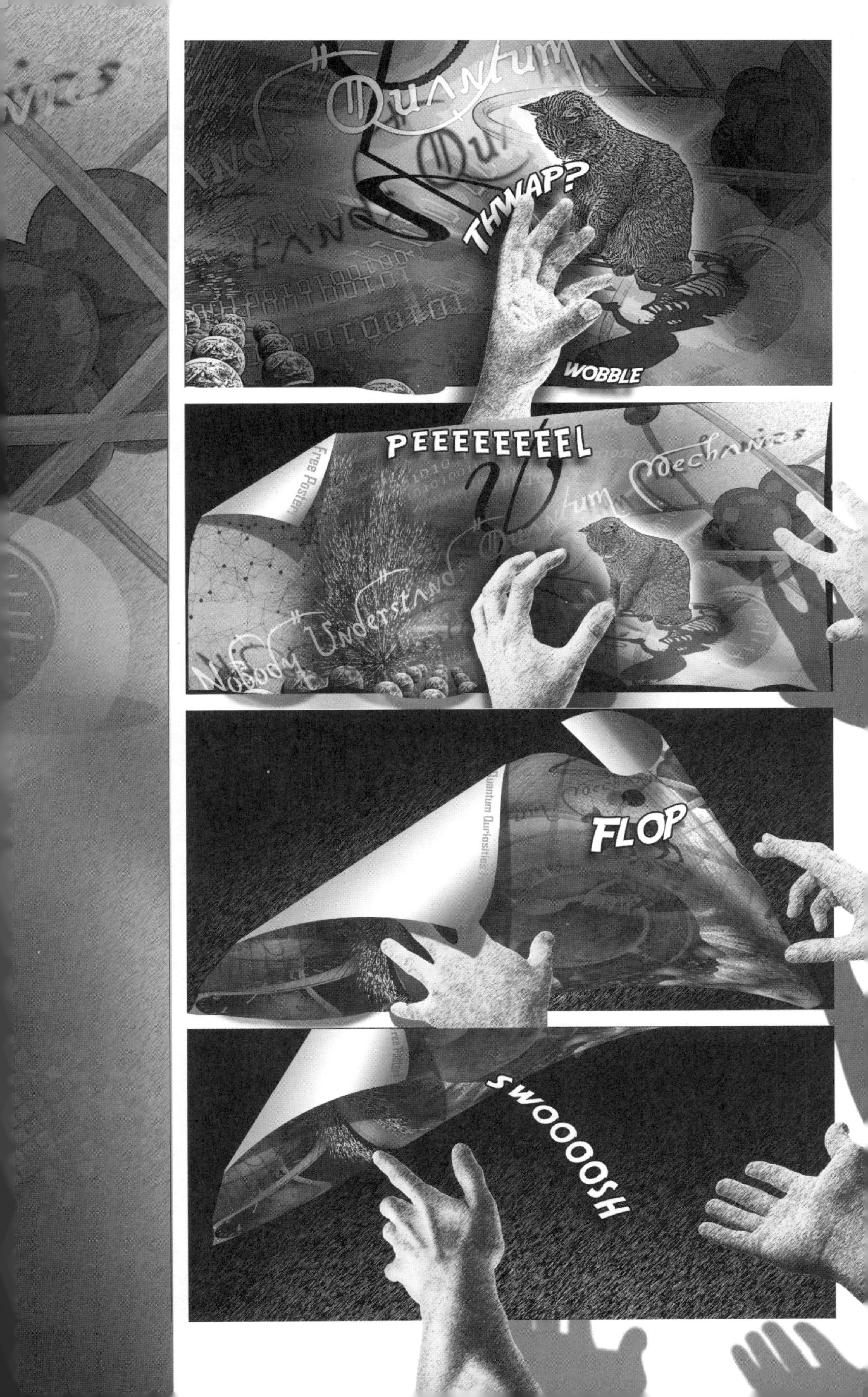
Quantum
THWAP?
WOBBLE
PEEEEEEEEL
Nobody Understands Quantum Mechanics
Free Poster!
FLOP
SWOOOOSH

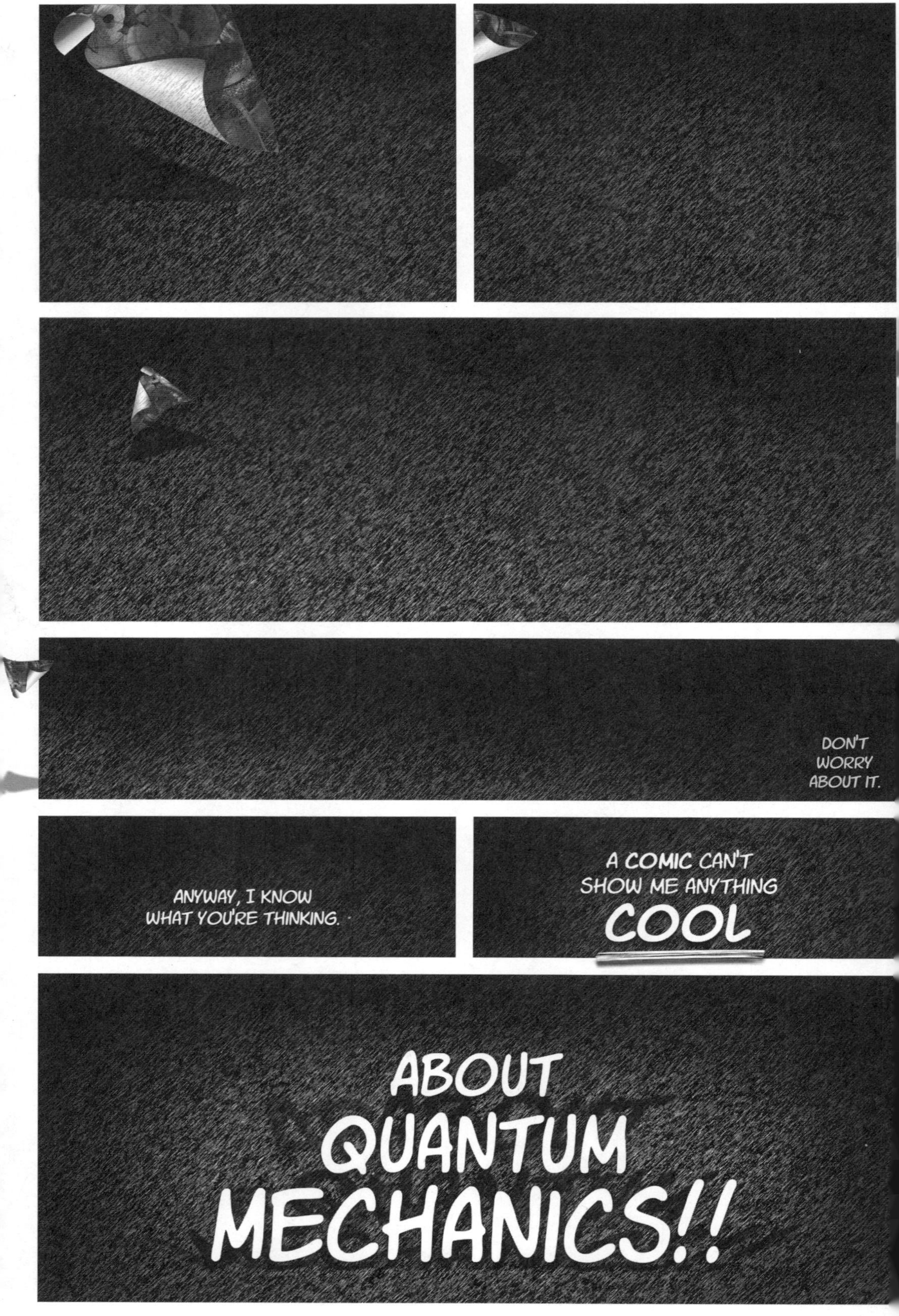
DON'T WORRY ABOUT IT.
ANYWAY, I KNOW WHAT YOU'RE THINKING.
A COMIC CAN'T SHOW ME ANYTHING COOL
ABOUT QUANTUM MECHANICS!!

STUFF LIKE SPOOKY ACTION AT A DISTANCE!
MATH THAT MAKES YOU SAD JUST TO LOOK AT IT!
CATS THAT ARE BOTH ALIVE AND DEAD AT THE SAME TIME!

PUH-LEEZE!!

$$\hat{H} = \sum_{n=1}^{N} \frac{\hat{p}_n^2}{2m_n} + V(\ldots x_n)$$

$$= -\frac{\hbar^2}{2} \sum_{n=1}^{N} \frac{1}{m_n} \frac{\partial^2}{\partial x_n^2} + V(x$$

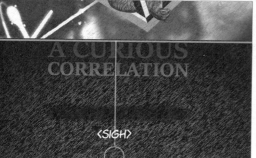

A CURIOUS
CORRELATION

<SIGH>

BUT MAYBE
— JUST MAYBE —
A COMIC CAN SHOW YOU A
CORRELATION SO TOTALLY
UM, **CURIOUS**

IT WILL BLOW YOUR MIND!

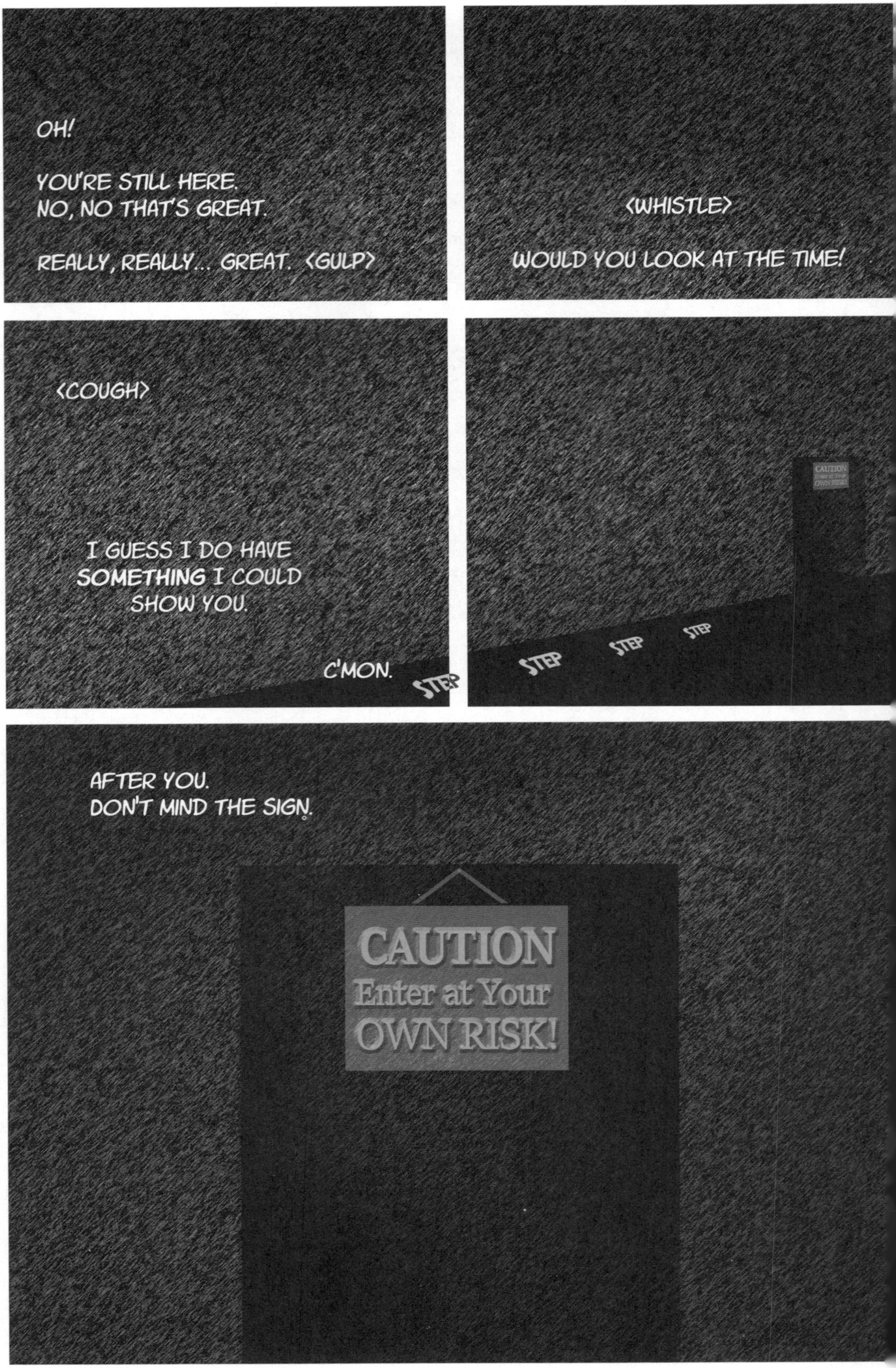
OH!
YOU'RE STILL HERE.
NO, NO THAT'S GREAT.
REALLY, REALLY... GREAT. <GULP>
<WHISTLE>
WOULD YOU LOOK AT THE TIME!
<COUGH>
I GUESS I DO HAVE **SOMETHING** I COULD SHOW YOU.
C'MON.
STEP
STEP
STEP
STEP
AFTER YOU.
DON'T MIND THE SIGN.
CAUTION
Enter at Your
OWN RISK!

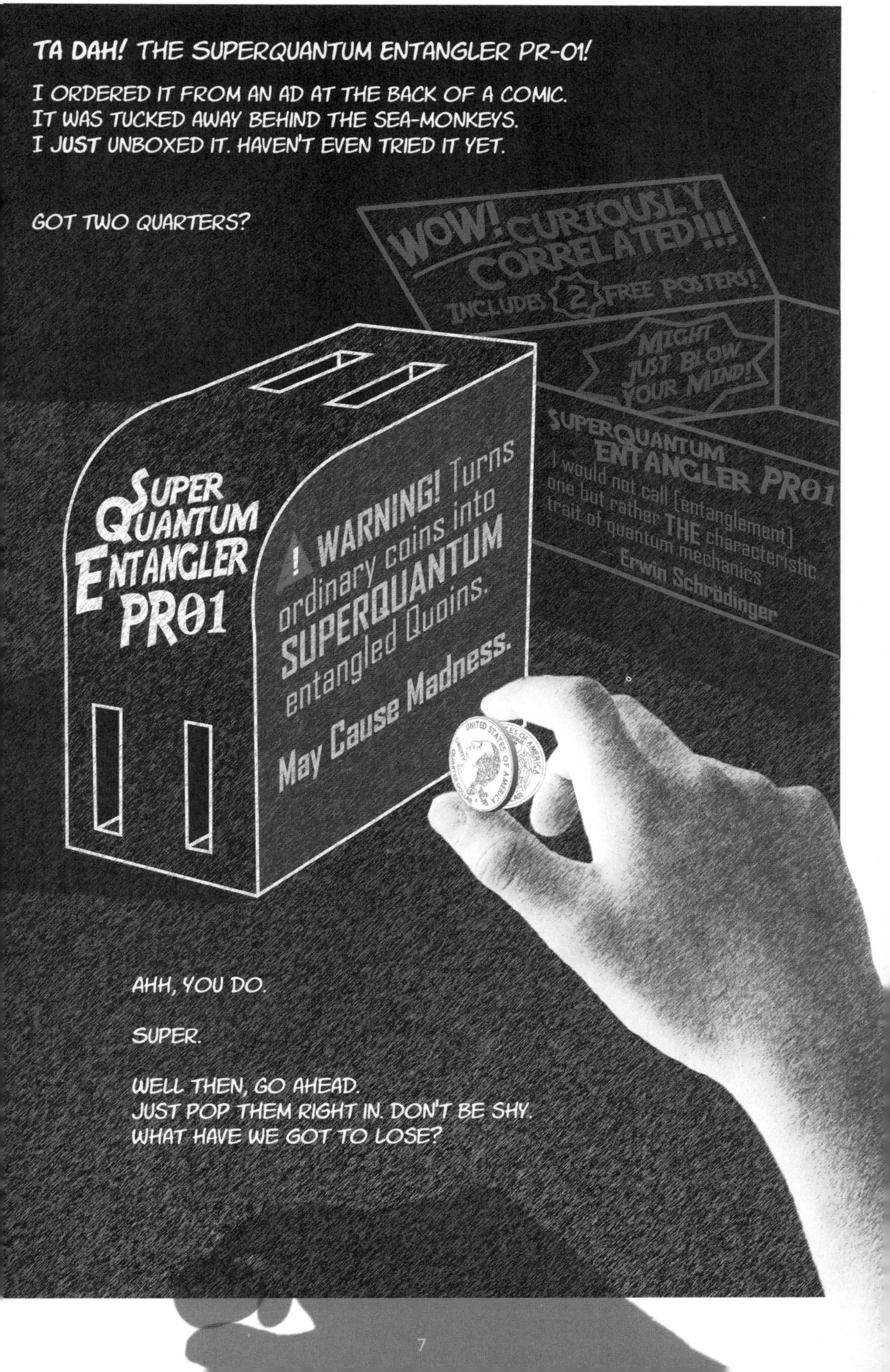
TA DAH! THE SUPERQUANTUM ENTANGLER PR-01!
I ORDERED IT FROM AN AD AT THE BACK OF A COMIC.
IT WAS TUCKED AWAY BEHIND THE SEA-MONKEYS.
I **JUST** UNBOXED IT. HAVEN'T EVEN TRIED IT YET.
GOT TWO QUARTERS?
WOW! CURIOUSLY CORRELATED!!!
INCLUDES 2 FREE POSTERS!
MIGHT JUST BLOW YOUR MIND!
SUPERQUANTUM ENTANGLER PR01
I would not call [entanglement] one but rather THE characteristic trait of quantum mechanics
— Erwin Schrödinger
SUPER QUANTUM ENTANGLER PR01
! WARNING! Turns ordinary coins into SUPERQUANTUM entangled Quoins.
May Cause Madness.
AHH, YOU DO.
SUPER.
WELL THEN, GO AHEAD.
JUST POP THEM RIGHT IN. DON'T BE SHY.
WHAT HAVE WE GOT TO LOSE?

Ok, I did it.

Great. Thanks.

I took out all the cool stuff that people might actually find interesting and put in your Super Quantum Entangler thing.

I saw that.

Because the world needs a comic about a weird relationship between two quarters. We can call it 50 Cents in Gray.

Ha ha. It's not about quarters. It's about entanglement, the bizarre link between quantum particles. The comic will give readers a gut feeling for entanglement through our superquantum correlated coins.

SUPERquantum?!? 😒

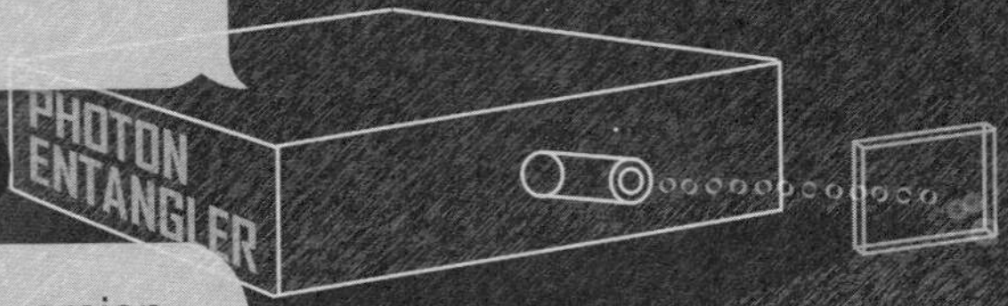
PHOTON ENTANGLER

It's a slightly souped-up version of the photon quantum correlation that highlights precisely what's so strange about entanglement. You need to GET entanglement to really GET the other "cool stuff." That's what "the world" needs :)

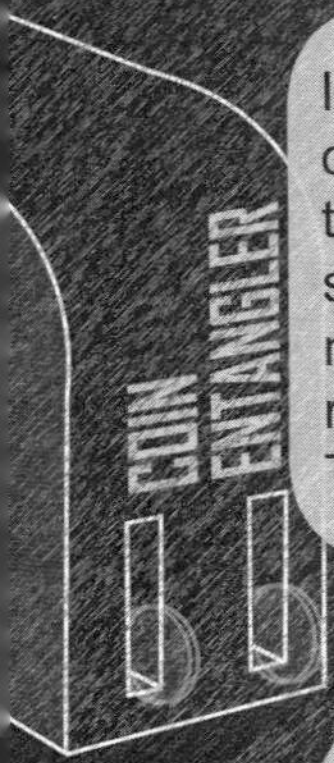
COIN ENTANGLER

"Ok"

Also, can that Schrödinger quote stand out more? It could even go on its own page.

How's this?

Perfect. Last thing. Does the Entangler look like a toaster, or is that just me?

Just you.

Entanglement

I would not call [entanglement] ***one*** but rather ***the*** characteristic trait of quantum mechanics, the one that enforces its entire departure from classical lines of thought.

—Erwin Schrödinger

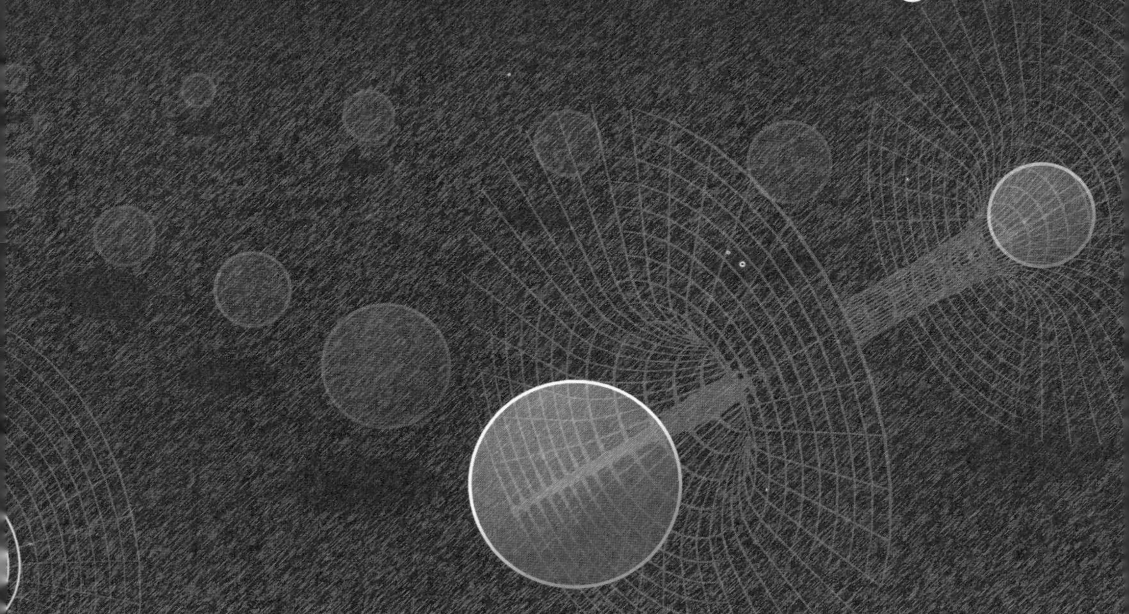

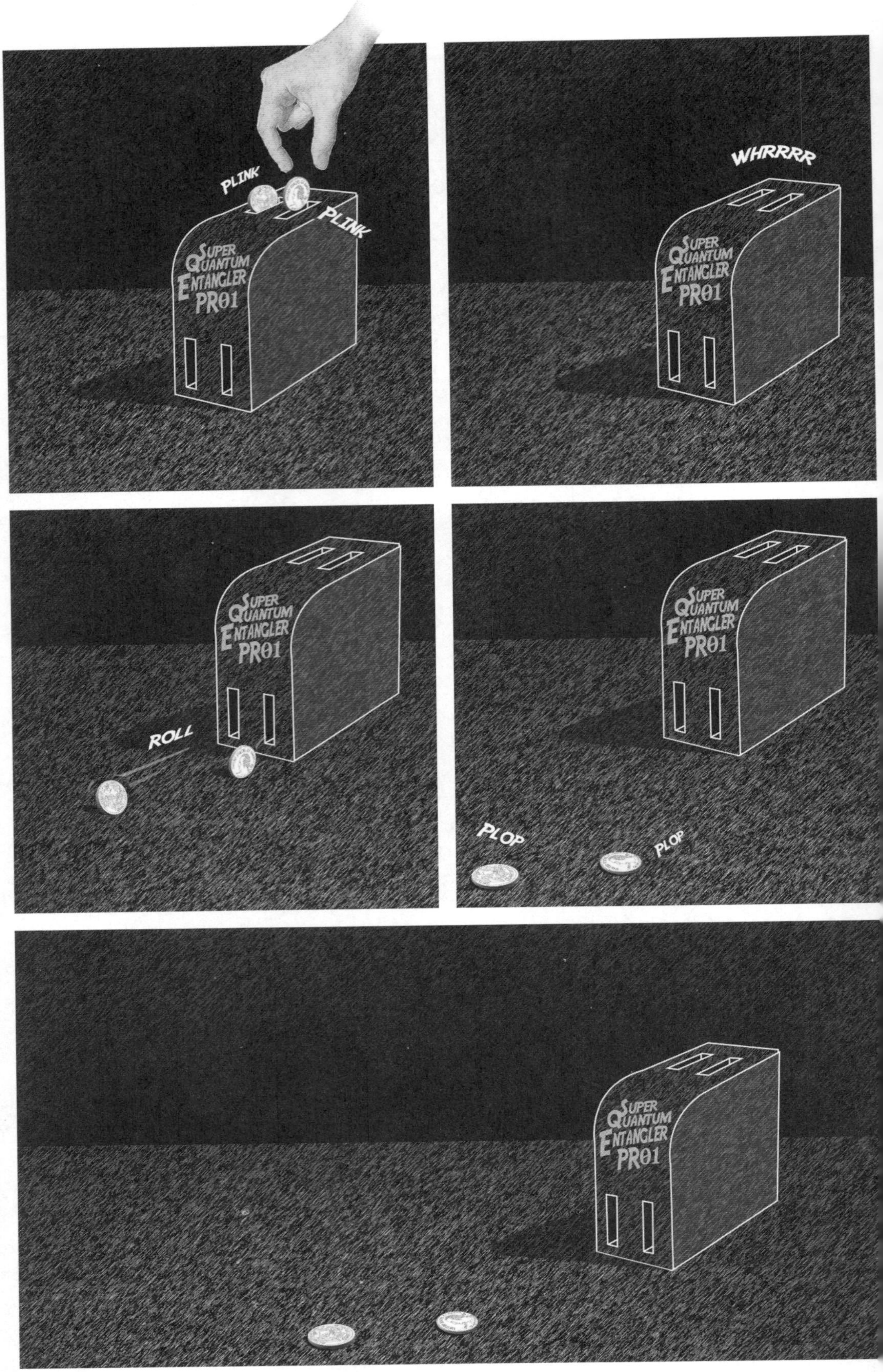
PLINK
PLINK
SUPER QUANTUM ENTANGLER PR01
WHRRRR
SUPER QUANTUM ENTANGLER PR01
SUPER QUANTUM ENTANGLER PR01
ROLL
SUPER QUANTUM ENTANGLER PR01
PLOP
PLOP
SUPER QUANTUM ENTANGLER PR01

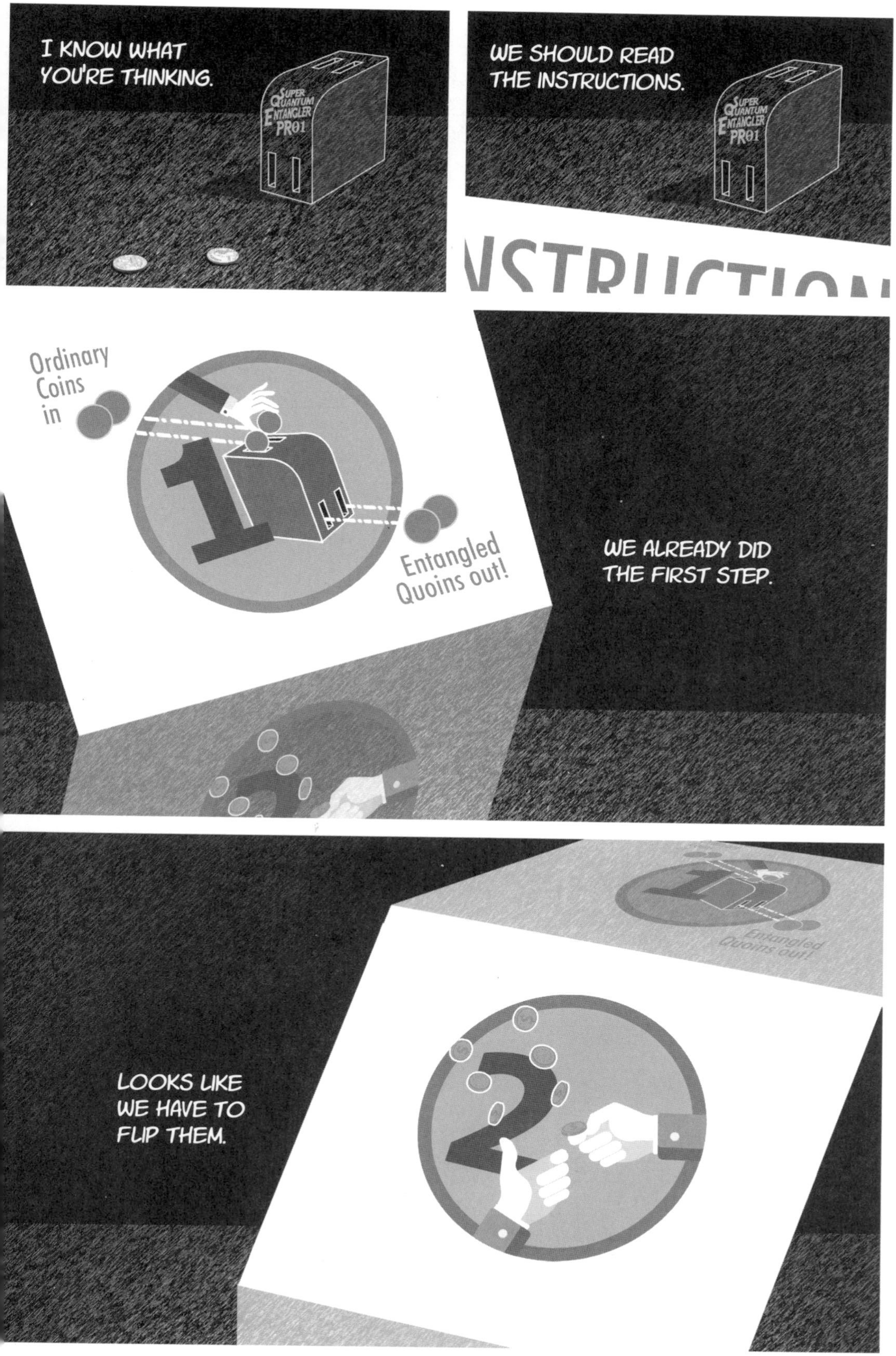
I KNOW WHAT YOU'RE THINKING.
SUPER QUANTUM ENTANGLER PR01
WE SHOULD READ THE INSTRUCTIONS.
SUPER QUANTUM ENTANGLER PR01
Ordinary Coins in
Entangled Quoins out!
WE ALREADY DID THE FIRST STEP.
Entangled Quoins out!
LOOKS LIKE WE HAVE TO FLIP THEM.

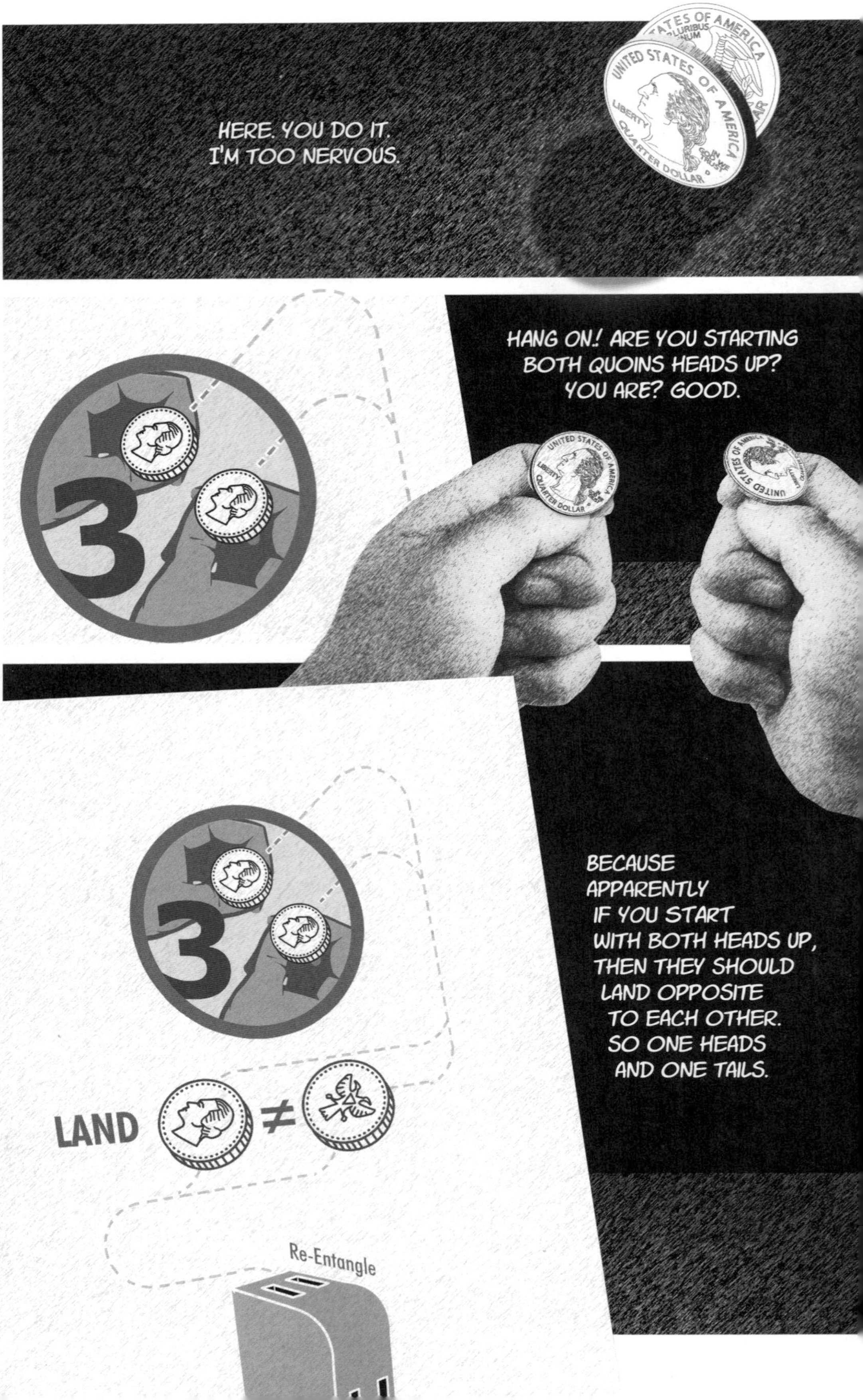
HERE. YOU DO IT.
I'M TOO NERVOUS.
UNITED STATES OF AMERICA
LIBERTY
QUARTER DOLLAR
HANG ON! ARE YOU STARTING BOTH QUOINS HEADS UP? YOU ARE? GOOD.
3
BECAUSE APPARENTLY IF YOU START WITH BOTH HEADS UP, THEN THEY SHOULD LAND OPPOSITE TO EACH OTHER. SO ONE HEADS AND ONE TAILS.
3
LAND ≠
Re-Entangle

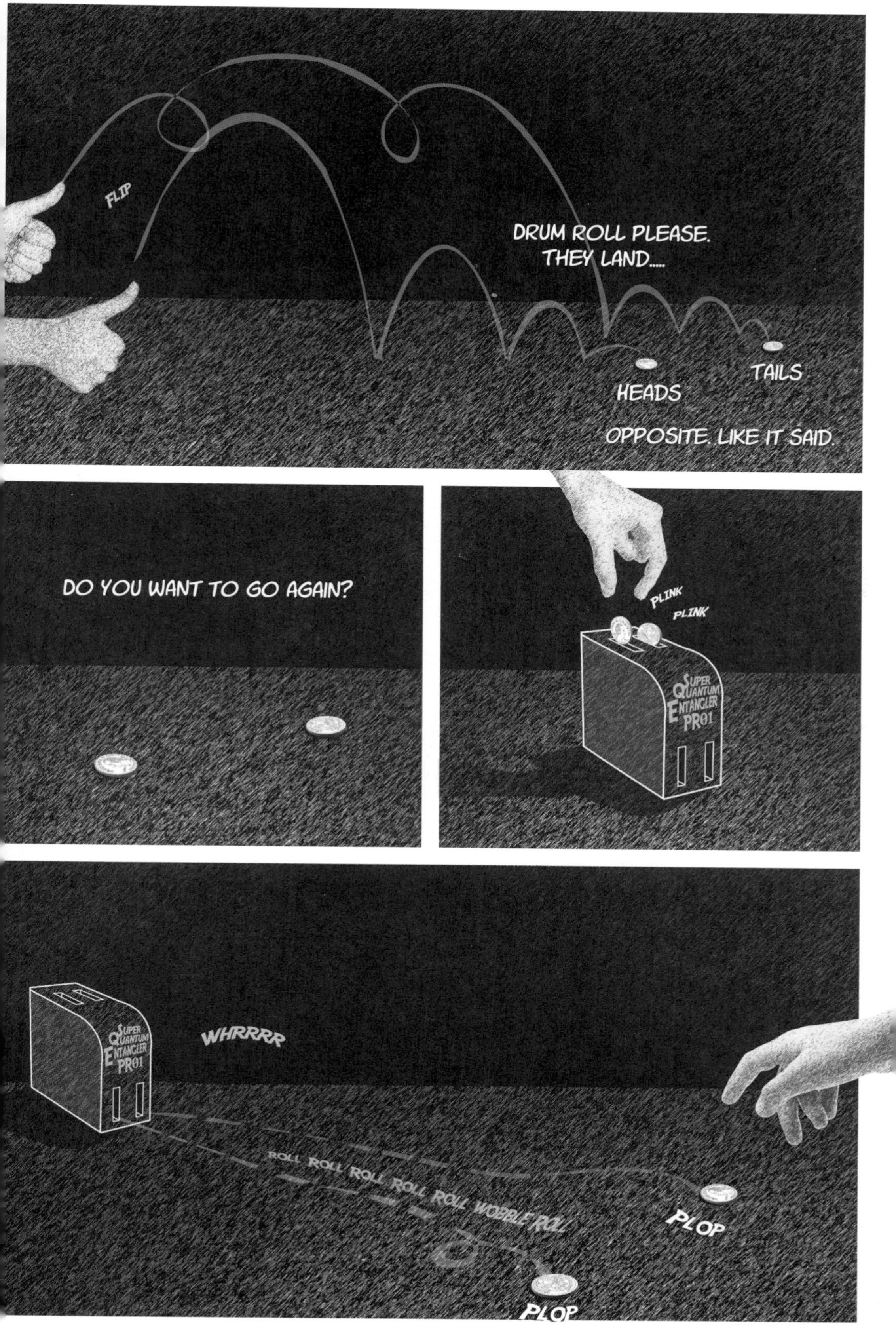
FLIP
DRUM ROLL PLEASE.
THEY LAND.....
HEADS
TAILS
OPPOSITE. LIKE IT SAID.
DO YOU WANT TO GO AGAIN?
PLINK
PLINK
SUPER QUANTUM ENTANGLER PRO1
SUPER QUANTUM ENTANGLER PRO1
WHRRRR
ROLL ROLL ROLL ROLL ROLL WOBBLE ROLL
PLOP
PLOP

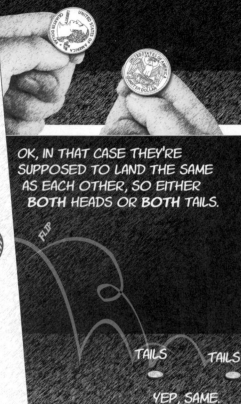

YOU'VE PUT ONE HEADS UP AND ONE TAILS UP THIS TIME, I SEE.

LAND =

OR

Re-Entangle

OK, IN THAT CASE THEY'RE SUPPOSED TO LAND THE SAME AS EACH OTHER, SO EITHER **BOTH** HEADS OR **BOTH** TAILS.

FLIP

TAILS   TAILS

YEP, SAME.

AGAIN? YOU WANT TO GO **AGAIN**?

PLINK
PLINK

WHRRRR

YOU DON'T GET OUT MUCH, DO YOU?

PLOP

PLOP

STARTING BOTH TAILS UP, HUH? WE'RE LIVING ON THE EDGE NOW.

THEY SHOULD LAND THE SAME, JUST LIKE LAST TIME.

FLIP

PLINK

PLONK

HEADS.

HEADS.
YUP, SAME.

YAWN!
               SORRY.
I WAS UP LATE LAST NIGHT.

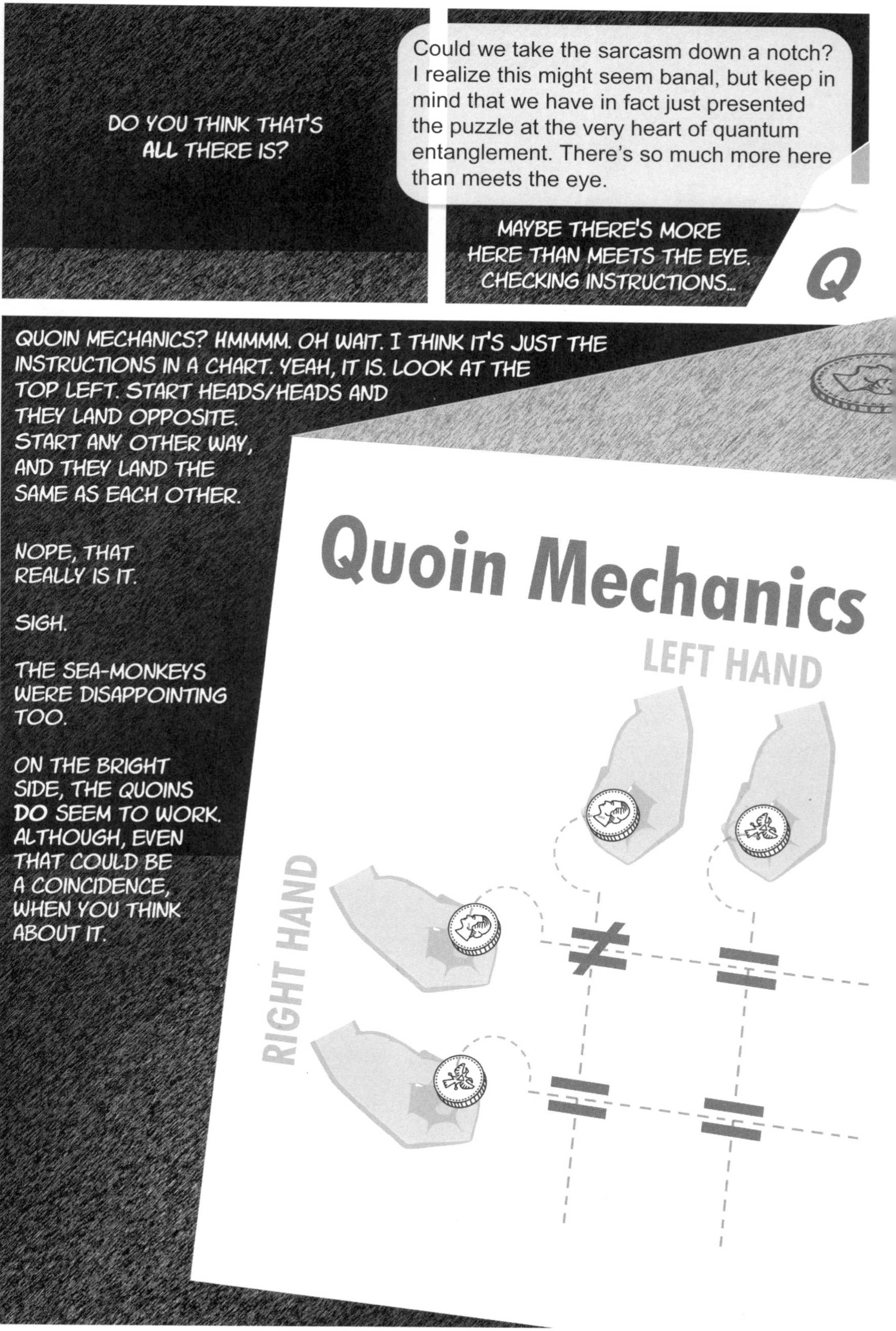

DO YOU THINK THAT'S ALL THERE IS?

Could we take the sarcasm down a notch? I realize this might seem banal, but keep in mind that we have in fact just presented the puzzle at the very heart of quantum entanglement. There's so much more here than meets the eye.

MAYBE THERE'S MORE HERE THAN MEETS THE EYE. CHECKING INSTRUCTIONS...

Q

QUOIN MECHANICS? HMMMM. OH WAIT. I THINK IT'S JUST THE INSTRUCTIONS IN A CHART. YEAH, IT IS. LOOK AT THE TOP LEFT. START HEADS/HEADS AND THEY LAND OPPOSITE. START ANY OTHER WAY, AND THEY LAND THE SAME AS EACH OTHER.

NOPE, THAT REALLY IS IT.

SIGH.

THE SEA-MONKEYS WERE DISAPPOINTING TOO.

ON THE BRIGHT SIDE, THE QUOINS DO SEEM TO WORK. ALTHOUGH, EVEN THAT COULD BE A COINCIDENCE, WHEN YOU THINK ABOUT IT.

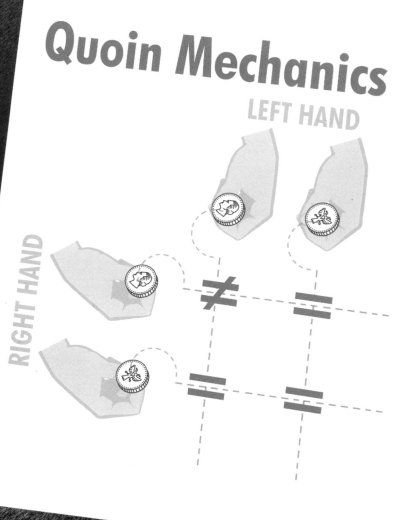

Quoin Mechanics
LEFT HAND
RIGHT HAND

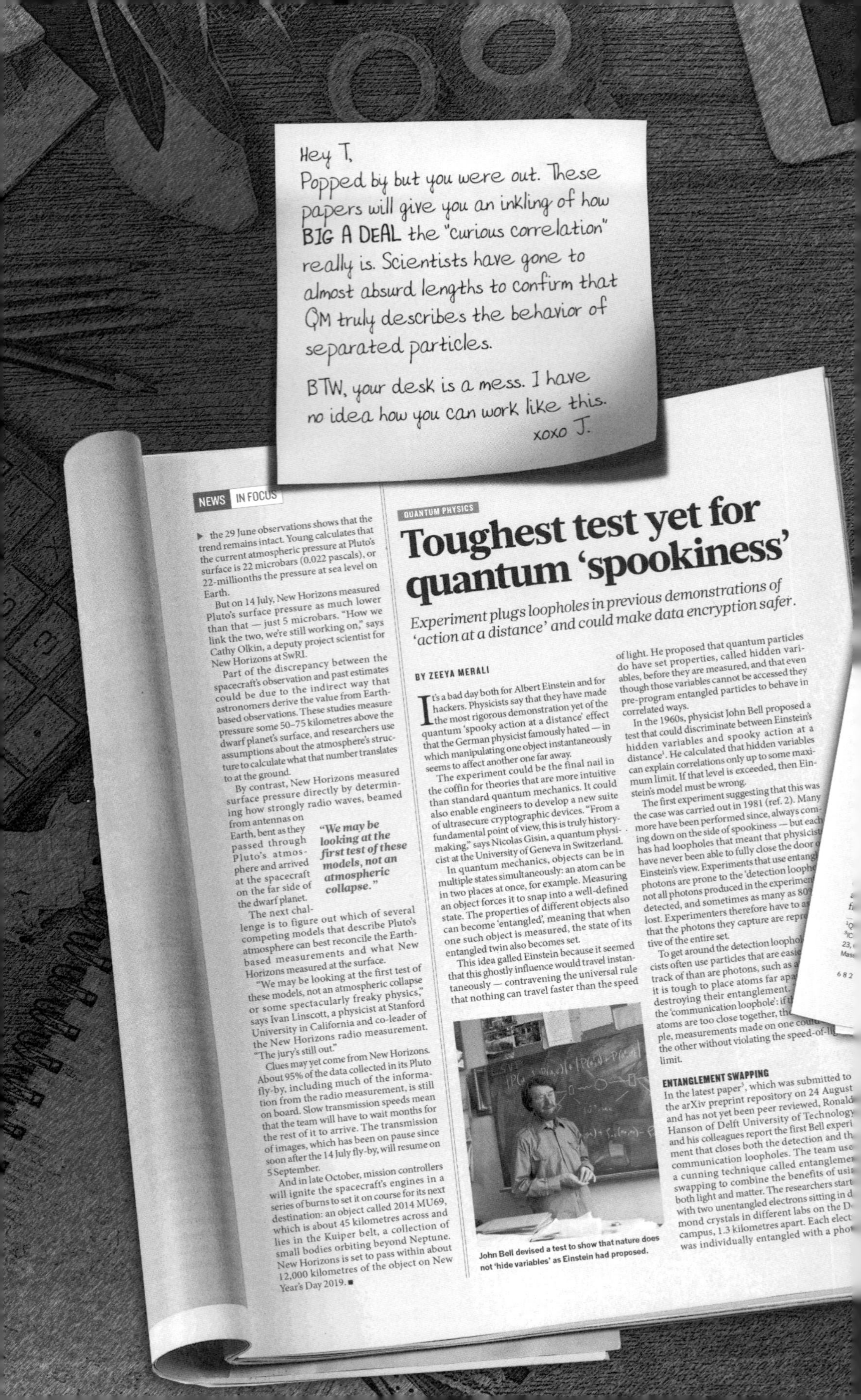

Hey T,
Popped by but you were out. These papers will give you an inkling of how **BIG A DEAL** the "curious correlation" really is. Scientists have gone to almost absurd lengths to confirm that QM truly describes the behavior of separated particles.

BTW, your desk is a mess. I have no idea how you can work like this.

xoxo J.

NEWS IN FOCUS

▶ the 29 June observations shows that the trend remains intact. Young calculates that the current atmospheric pressure at Pluto's surface is 22 microbars (0.022 pascals), or 22-millionths the pressure at sea level on Earth.

But on 14 July, New Horizons measured Pluto's surface pressure as much lower than that — just 5 microbars. "How we link the two, we're still working on," says Cathy Olkin, a deputy project scientist for New Horizons at SwRI.

Part of the discrepancy between the spacecraft's observation and past estimates could be due to the indirect way that astronomers derive the value from Earth-based observations. These studies measure pressure some 50–75 kilometres above the dwarf planet's surface, and researchers use assumptions about the atmosphere's structure to calculate what that number translates to at the ground.

By contrast, New Horizons measured surface pressure directly by determining how strongly radio waves, beamed from antennas on Earth, bent as they passed through Pluto's atmosphere and arrived at the spacecraft on the far side of the dwarf planet.

"We may be looking at the first test of these models, not an atmospheric collapse."

The next challenge is to figure out which of several competing models that describe Pluto's atmosphere can best reconcile the Earth-based measurements and what New Horizons measured at the surface.

"We may be looking at the first test of these models, not an atmospheric collapse or some spectacularly freaky physics," says Ivan Linscott, a physicist at Stanford University in California and co-leader of the New Horizons radio measurement. "The jury's still out."

Clues may yet come from New Horizons. About 95% of the data collected in its Pluto fly-by, including much of the information from the radio measurement, is still on board. Slow transmission speeds mean that the team will have to wait months for the rest of it to arrive. The transmission of images, which has been on pause since soon after the 14 July fly-by, will resume on 5 September.

And in late October, mission controllers will ignite the spacecraft's engines in a series of burns to set it on course for its next destination: an object called 2014 MU69, which is about 45 kilometres across and lies in the Kuiper belt, a collection of small bodies orbiting beyond Neptune. New Horizons is set to pass within about 12,000 kilometres of the object on New Year's Day 2019. ■

QUANTUM PHYSICS

Toughest test yet for quantum 'spookiness'

Experiment plugs loopholes in previous demonstrations of 'action at a distance' and could make data encryption safer.

BY ZEEYA MERALI

It's a bad day both for Albert Einstein and for hackers. Physicists say that they have made the most rigorous demonstration yet of the quantum 'spooky action at a distance' effect that the German physicist famously hated — in which manipulating one object instantaneously seems to affect another one far away.

The experiment could be the final nail in the coffin for theories that are more intuitive than standard quantum mechanics. It could also enable engineers to develop a new suite of ultrasecure cryptographic devices. "From a fundamental point of view, this is truly history-making," says Nicolas Gisin, a quantum physicist at the University of Geneva in Switzerland.

In quantum mechanics, objects can be in multiple states simultaneously: an atom can be in two places at once, for example. Measuring an object forces it to snap into a well-defined state. The properties of different objects also can become 'entangled', meaning that when one such object is measured, the state of its entangled twin also becomes set.

This idea galled Einstein because it seemed that this ghostly influence would travel instantaneously — contravening the universal rule that nothing can travel faster than the speed of light. He proposed that quantum particles do have set properties, called hidden variables, before they are measured, and that even though those variables cannot be accessed they pre-program entangled particles to behave in correlated ways.

In the 1960s, physicist John Bell proposed a test that could discriminate between Einstein's hidden variables and spooky action at a distance[1]. He calculated that hidden variables can explain correlations only up to some maximum limit. If that level is exceeded, then Einstein's model must be wrong.

The first experiment suggesting that this was the case was carried out in 1981 (ref. 2). Many more have been performed since, always coming down on the side of spookiness — but each has had loopholes that meant that physicist[s] have never been able to fully close the door o[n] Einstein's view. Experiments that use entang[led] photons are prone to the 'detection loopho[le']: not all photons produced in the experimen[t are] detected, and sometimes as many as 80[%] lost. Experimenters therefore have to as[sume] that the photons they capture are repre[sentative] of the entire set.

To get around the detection loophol[e, physi]cists often use particles that are easie[r to keep] track of than are photons, such as a[toms. But] it is tough to place atoms far apa[rt without] destroying their entanglement. [...] the 'communication loophole': if t[he] atoms are too close together, th[...] ple, measurements made on one cou[ld ...] the other without violating the speed-of-li[ght] limit.

John Bell devised a test to show that nature does not 'hide variables' as Einstein had proposed.

ENTANGLEMENT SWAPPING

In the latest paper[3], which was submitted to the arXiv preprint repository on 24 August and has not yet been peer reviewed, Ronald Hanson of Delft University of Technology and his colleagues report the first Bell experi[ment] that closes both the detection and th[e] communication loopholes. The team use[d] a cunning technique called entanglemen[t] swapping to combine the benefits of usi[ng] both light and matter. The researchers start[ed] with two unentangled electrons sitting in d[ia]mond crystals in different labs on the D[elft] campus, 1.3 kilometres apart. Each elect[ron] was individually entangled with a pho[ton]

682

LETTER

doi:10.1038/nature15759

Loophole-free Bell inequality violation using electron spins separated by 1.3 kilometres

.Hensen[1,2], H. Bernien[1,2]†, A. E. Dréau[1,2], A. Reiserer[1,2], N. Kalb[1,2], M. S. Blok[1,2] J. Ruitenberg[1,2], R. F. L. Vermeulen[1,2],
N. Schouten[1,2], C. Abellán[3], W. Amaya[3], V. Pruneri[3,4], M. W. Mitchell[3,4], M. Markham[5], D. J. Twitchen[5], D. Elkouss[1],
.ehner[1], T. H. Taminiau[1,2] & R. Hanson[1,2]

than 50 years ago[1], John Bell proved that no theory of nature
beys locality and realism[2] can reproduce all the predictions of
um theory: in any local-realist theory, the correlations
n outcomes of measurements on distant particles satisfy
uality that can be violated if the particles are entangled.
us Bell inequality tests have been reported[3-13], however,
riments reported so far required additional assump-
obtain a contradiction with local realism, resulting in
[13-16]. Here we report a Bell experiment that is free of
dditional assumption and thus directly tests the principles
Bell's inequality. We use an event-ready scheme[17-19] that
generation of robust entanglement between distant
ns (estimated state fidelity of 0.92 ± 0.03). Efficient
ut avoids the fair-sampling assumption (detection
, while the use of fast random-basis selection and spin
 combined with a spatial separation of 1.3 kilometres
quired locality conditions[13]. We performed 245 trials
the CHSH–Bell inequality[20] $S \leq 2$ and found
(where S quantifies the correlation between mea-
omes). A null-hypothesis test yields a probability
.039 that a local-realist model for space-like sepa-
produce data with a violation at least as large as
when allowing for memory[16,21] in the devices.
imply statistically significant rejection of the
e experiments; for instance, reaching a value of
quire approximately 700 trials for an observed
rovements, our experiment could be used for
ional theories, and for implementing device-
um-secure communication[22] and randomness

test in the form proposed by Clauser, Horne,
HSH)[20] (Fig. 1a). The test involves two boxes
box accepts a binary input (0 or 1) and subse-
y output (+1 or −1). In each trial of the Bell
t is generated on each side and input to the
dom input bit triggers the box to produce an
ded. The test concerns correlations between
x and y for boxes A and B, respectively) and
and b for A and B, respectively) ge

Bell is that in any theory of physic
ces do not propagate faster than lig
are defined before, and independe
ns are bounded more strongly tha
icular, if the input bits can be cons
tion of 'free will') and the box

Box 5046, 2600 GA Delft, The Netherlands. [2]
arcelona Institute of Science and Technology, 0
novation. Fermi Avenue, Harwell Oxford, Did

sufficiently separated such that locality prevents communication
between the boxes during a trial, then the following inequality holds
under local realism:

$$S = \left| \langle x \cdot y \rangle_{(0,0)} + \langle x \cdot y \rangle_{(0,1)} + \langle x \cdot y \rangle_{(1,0)} - \langle x \cdot y \rangle_{(1,1)} \right| \leq 2 \quad (1)$$

where $\langle x \cdot y \rangle_{(a,b)}$ denotes the expectation value of the product of x and y
for input bits a and b. (A mathematical formulation of the concepts
underlying Bell's inequality is found in, for example, ref. 25.)
Quantum theory predicts that the Bell inequality can be significantly
violated in the following setting. We add one particle, for example an
electron, to each box. The spin degree of freedom of the electron forms
a two-level system with eigenstates $|\uparrow\rangle$ and $|\downarrow\rangle$. For each trial, the two
spins are prepared into the entangled state $|\psi^-\rangle = (|\uparrow\downarrow\rangle - |\downarrow\uparrow\rangle)/\sqrt{2}$.
The spin in box A is then measured along direction Z (for input bit
$a = 0$) or X (for $a = 1$) and the spin in box B is measured along
$(-Z + X)/\sqrt{2}$ (for $b = 0$) or $(-Z - X)/\sqrt{2}$ (for $b = 1$). The measurement outcomes are used as outputs of the boxes, then quantum
theory predicts a value of $S = 2\sqrt{2}$, which shows that the combina-
tions of locality and realism is fundamentally incompatible with the predictions of quantum mechanics.
Bell's inequality provides a powerful recipe for probing fundamental
properties of nature: all local-realist theories that specify where and
when the free random input bits and the output values are generated
can be experimentally tested against it.
Violating Bell's inequality with entangled particles poses two main
challenges: excluding any possible communication between the boxes
(locality loophole[13]) and guaranteeing efficient measurements (detection loophole[14,15]). First, if communication is possible, a box can in
principle respond using knowledge of both input settings, rendering
the Bell inequality invalid. The locality conditions thus require boxes A
and B and their respective free-input-bit generations to be separated in
such a way that signals travelling at the speed of light (the maximum
allowed under special relativity) cannot comm
setting of box A to box B
record
does
allow
samp
subs

PRL **115**, 250402 (2015)

Selected for a Viewpoint in *Physics*

PHYSICAL REVIEW LETTERS

week ending
18 DECEMBER 2015

PHYSICAL REVIEW LETTERS

VOLUME 81   7 DECEMBER 1998   NUMBER 23

Violation of Bell's Inequality under Strict Einstein Locality Conditions

Gregor Weihs, Thomas Jennewein, Christoph Simon, Harald Weinfurter, and Anton Zeilinger

Institut für Experimentalphysik, Universität Innsbruck, Technikerstraße 25, A-6020 Innsbruck, Austria
(Received 6 August 1998)

We observe strong violation of Bell's inequality in an Einstein-Podolsky-Rosen-type experiment with
independent observers. Our experiment definitely implements the ideas behind the well-known work
by Aspect *et al.* We for the first time fully enforce the condition of locality, a central assumption in
the derivation of Bell's theorem. The necessary spacelike separation of the observations is achieved
by sufficient physical distance between the measurement stations, by ultrafast and random setting of the
analyzers, and by completely independent data registration. [S0031-9007(98)07901-0]

PACS numbers: 03.65.Bz

The stronger-than-classical correlations between entangled quantum systems, as first discovered by Einstein, Podolsky, and Rosen (EPR) in 1935 [1], have
ever since occupied a central position in the discussions
of the foundations of quantum mechanics. After Bell's
discovery [2] that EPR's implication to explain the correlations using hidden parameters would contradict the predictions of quantum physics, a number of experimental
tests have been performed [3–5]. All recent experiments
confirm the predictions of quantum mechanics. Yet, from
a strictly logical point of view, they don't succeed in rul-
ing out a local realistic explanation completely, because of
two essential loopholes. The first loophole builds on the

the directions of polarization analysis were switched after
the photons left the source. Aspect *et al.*, however, used
periodic sinusoidal switching, which is predictable into
the future. Thus communication slower than the speed
of light, or even at the speed of light [8], could in
principle explain the results obtained. Therefore this
second loophole is still open.
The assumption of locality in the derivation of Bell's
theorem requires that the individual measurement processes of the two observers are spacelike separated
(Fig. 1). We define an individual measurement to last
from the first point in time which can influence the choice
of the analyzer setting

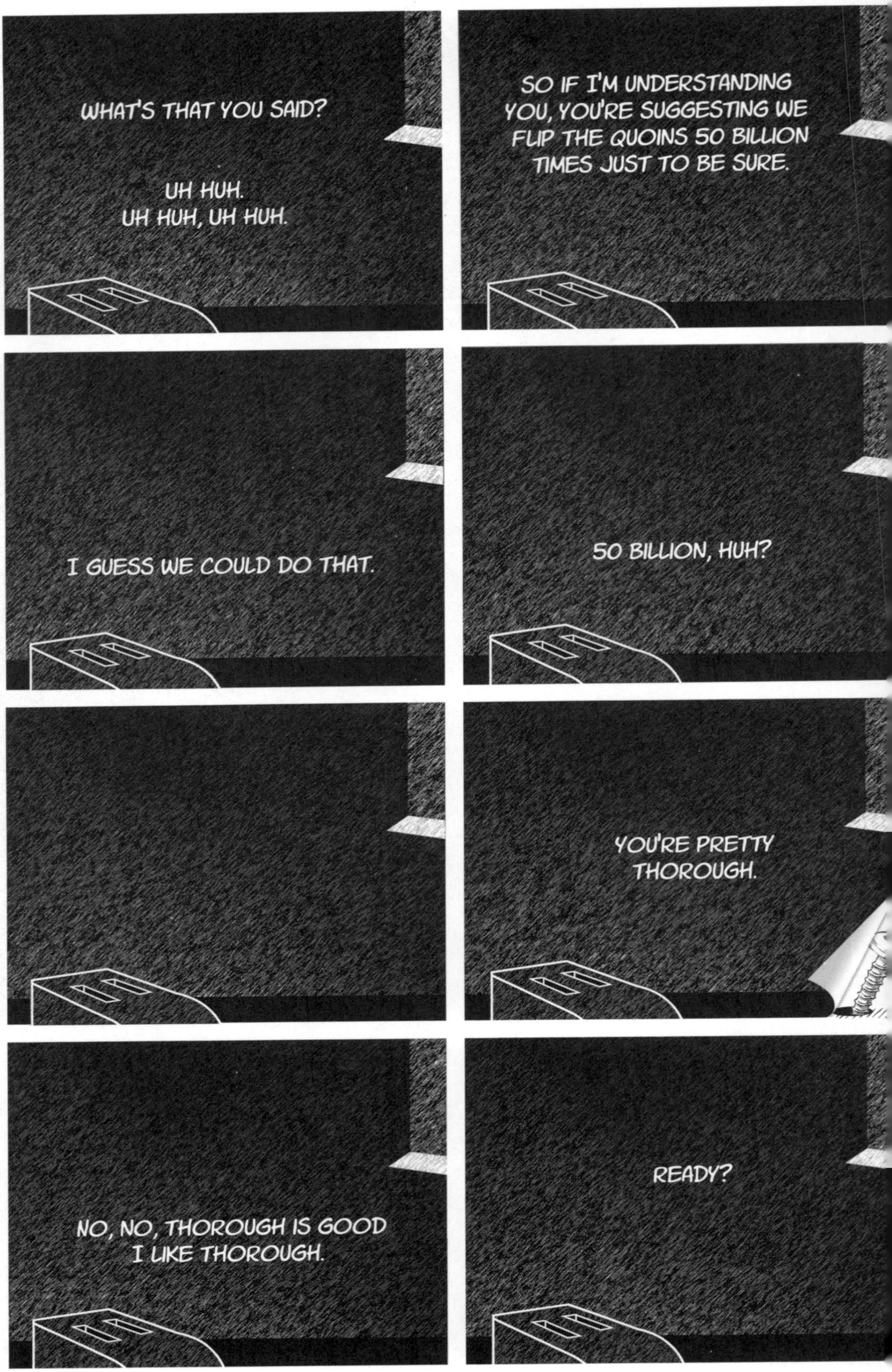
WHAT'S THAT YOU SAID?
UH HUH.
UH HUH, UH HUH.
SO IF I'M UNDERSTANDING YOU, YOU'RE SUGGESTING WE FLIP THE QUOINS 50 BILLION TIMES JUST TO BE SURE.
I GUESS WE COULD DO THAT.
50 BILLION, HUH?
YOU'RE PRETTY THOROUGH.
NO, NO, THOROUGH IS GOOD
I LIKE THOROUGH.
READY?

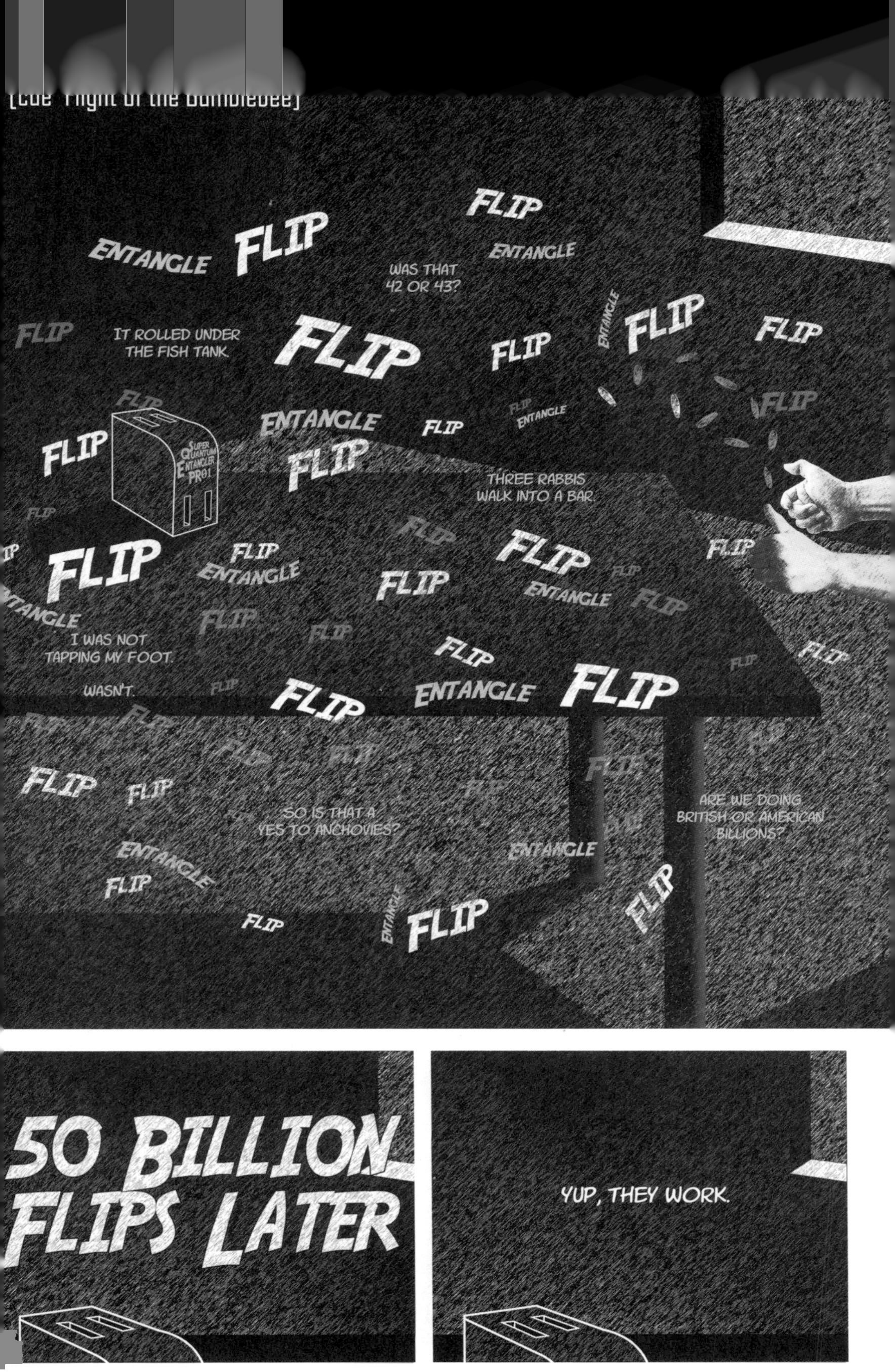
[Cue "Flight of the Bumblebee"]
ENTANGLE
FLIP
WAS THAT 42 OR 43?
IT ROLLED UNDER THE FISH TANK.
SUPER QUANTUM ENTANGLER PRO1
THREE RABBIS WALK INTO A BAR.
I WAS NOT TAPPING MY FOOT.
WASN'T.
SO IS THAT A YES TO ANCHOVIES?
ARE WE DOING BRITISH OR AMERICAN BILLIONS?
50 BILLION FLIPS LATER
YUP, THEY WORK.

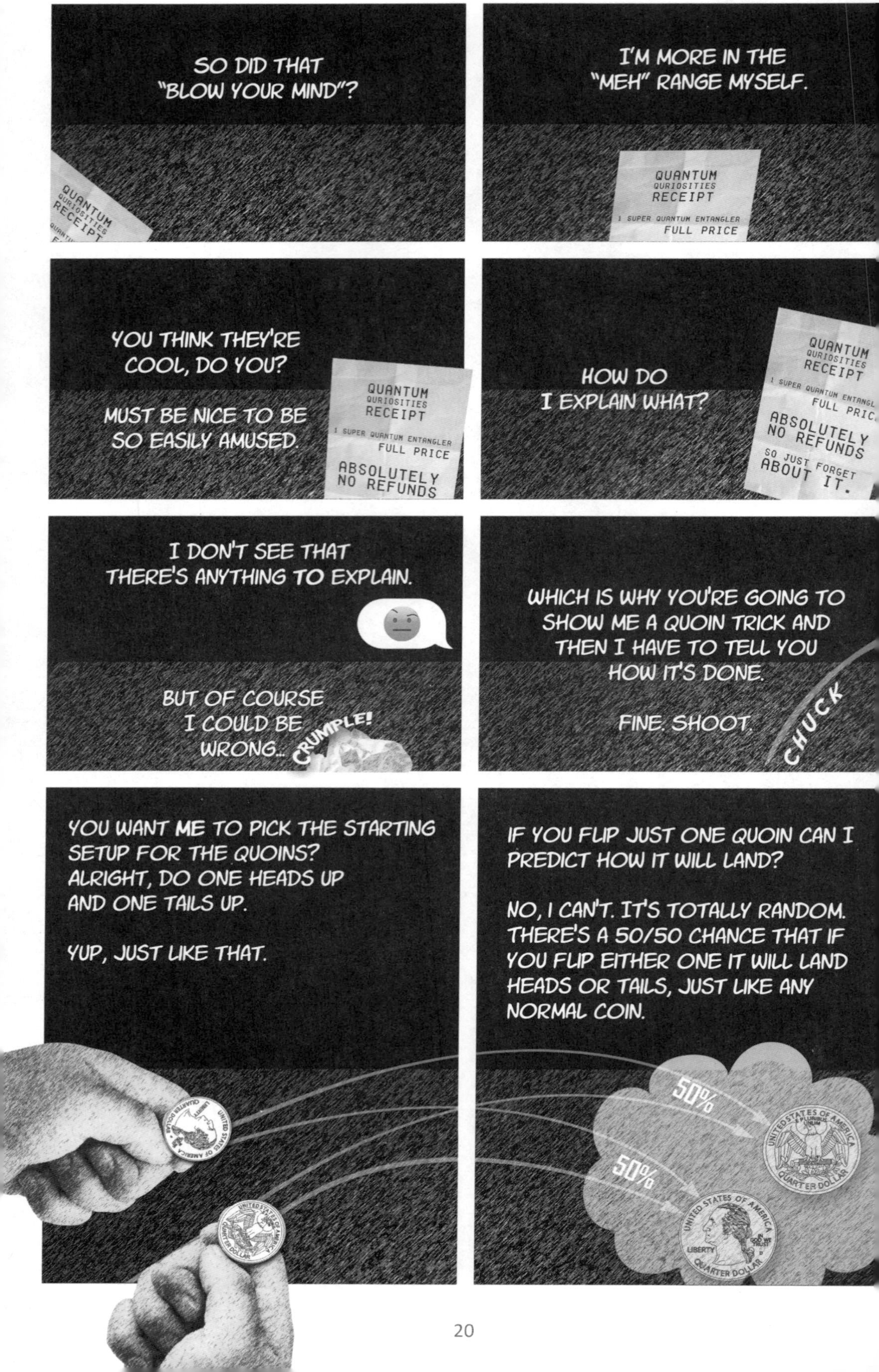
SO DID THAT "BLOW YOUR MIND"?
QUANTUM QURIOSITIES RECEIPT
I'M MORE IN THE "MEH" RANGE MYSELF.
QUANTUM QURIOSITIES RECEIPT
1 SUPER QUANTUM ENTANGLER
FULL PRICE
YOU THINK THEY'RE COOL, DO YOU?
MUST BE NICE TO BE SO EASILY AMUSED.
QUANTUM QURIOSITIES RECEIPT
1 SUPER QUANTUM ENTANGLER
FULL PRICE
ABSOLUTELY NO REFUNDS
HOW DO I EXPLAIN WHAT?
QUANTUM QURIOSITIES RECEIPT
ABSOLUTELY NO REFUNDS
SO JUST FORGET ABOUT IT.
I DON'T SEE THAT THERE'S ANYTHING TO EXPLAIN.
BUT OF COURSE I COULD BE WRONG...
CRUMPLE!
WHICH IS WHY YOU'RE GOING TO SHOW ME A QUOIN TRICK AND THEN I HAVE TO TELL YOU HOW IT'S DONE.
FINE. SHOOT.
CHUCK
YOU WANT ME TO PICK THE STARTING SETUP FOR THE QUOINS? ALRIGHT, DO ONE HEADS UP AND ONE TAILS UP.
YUP, JUST LIKE THAT.
IF YOU FLIP JUST ONE QUOIN CAN I PREDICT HOW IT WILL LAND?
NO, I CAN'T. IT'S TOTALLY RANDOM. THERE'S A 50/50 CHANCE THAT IF YOU FLIP EITHER ONE IT WILL LAND HEADS OR TAILS, JUST LIKE ANY NORMAL COIN.
50%
50%

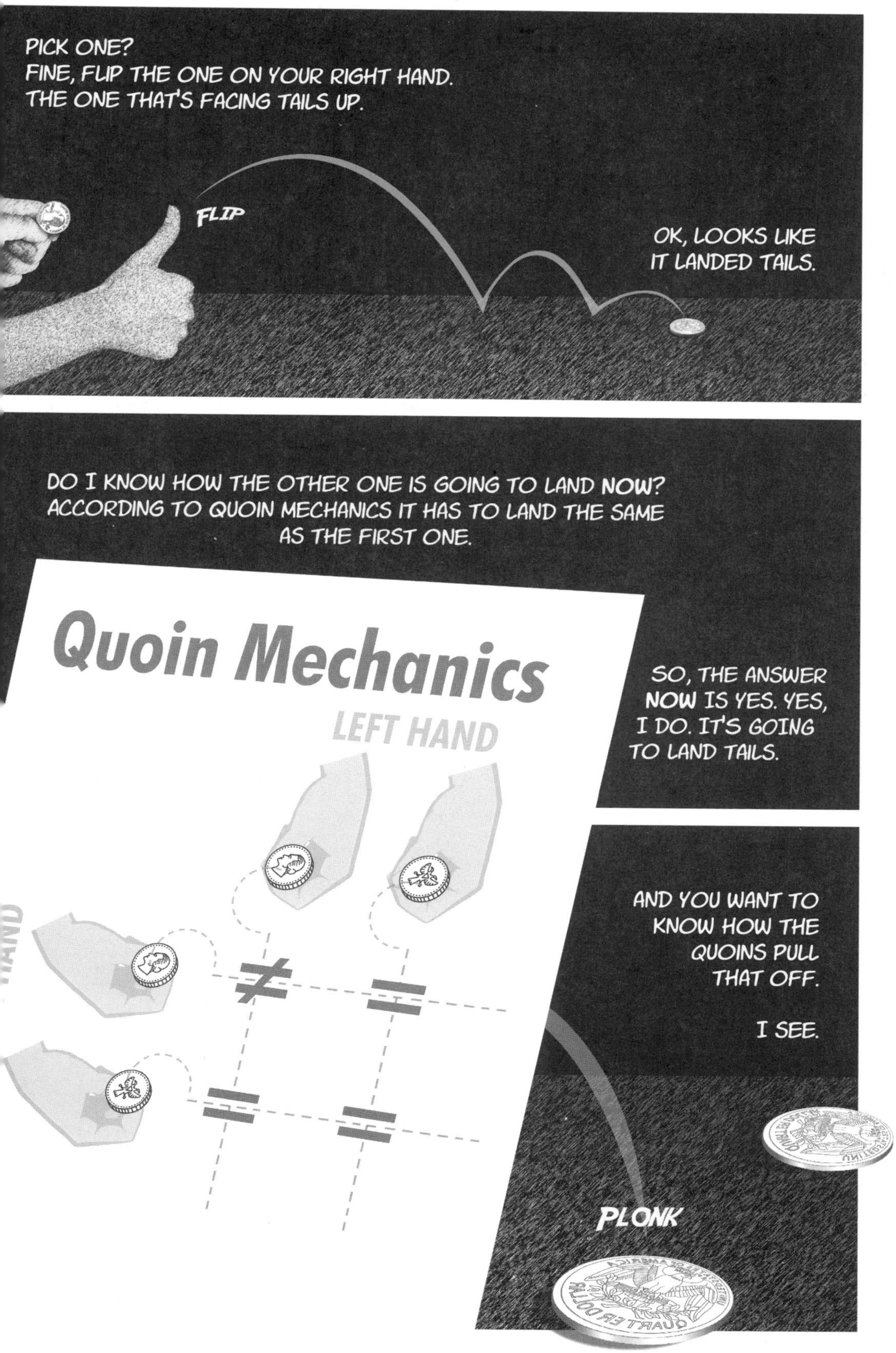
PICK ONE?
FINE, FLIP THE ONE ON YOUR RIGHT HAND.
THE ONE THAT'S FACING TAILS UP.
FLIP
OK, LOOKS LIKE IT LANDED TAILS.
DO I KNOW HOW THE OTHER ONE IS GOING TO LAND NOW? ACCORDING TO QUOIN MECHANICS IT HAS TO LAND THE SAME AS THE FIRST ONE.
Quoin Mechanics
LEFT HAND
SO, THE ANSWER NOW IS YES. YES, I DO. IT'S GOING TO LAND TAILS.
AND YOU WANT TO KNOW HOW THE QUOINS PULL THAT OFF.
I SEE.
PLONK
QUARTER DOLLAR

A Question of Explanation

The scientific attitude is that correlations cry out for explanation.

—John Stewart Bell

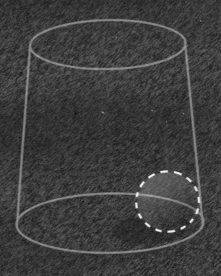

What possible explanation could there be for the quoin correlation? The physicist John Bell asked this question in 1964 about a similarly puzzling quantum correlation and proved that there can't be a causal explanation! This astonishing result has been called "one of the most remarkable papers in the history of physics" by the physicist Alain Aspect and 'the most profound discovery of science' by the science writer John Horgan. We take it for granted that correlations between events in different places should be explained on the basis of a common cause that makes both events happen together, or on the basis of a causal influence that moves from one event to the other. Bell showed that there can't be a common cause that makes the curious correlation happen in more than three out of four cases. But somehow, a pair of photons or electrons, each responding to measurements in a totally random way, can beat this limit.

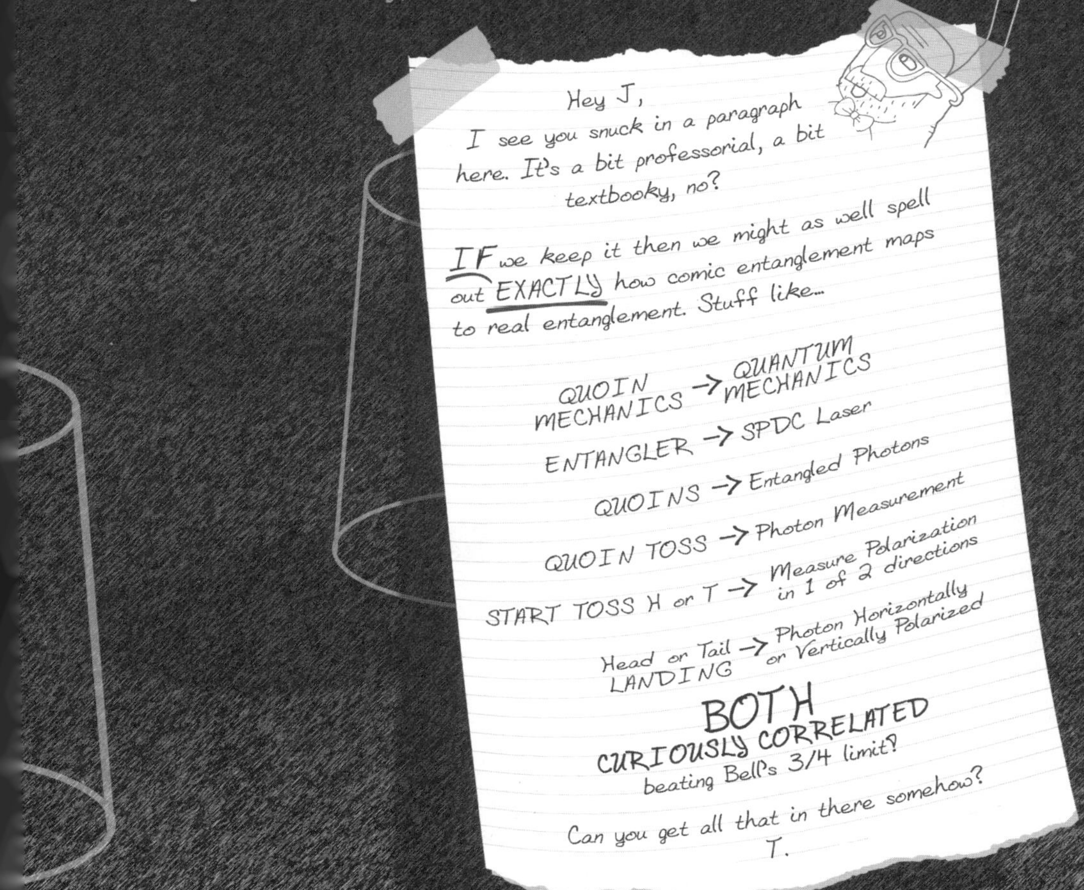
Hey J,
I see you snuck in a paragraph here. It's a bit professorial, a bit textbooky, no?
IF we keep it then we might as well spell out EXACTLY how comic entanglement maps to real entanglement. Stuff like...
QUOIN MECHANICS -> QUANTUM MECHANICS
ENTANGLER -> SPDC Laser
QUOINS -> Entangled Photons
QUOIN TOSS -> Photon Measurement
START TOSS H or T -> Measure Polarization in 1 of 2 directions
Head or Tail LANDING -> Photon Horizontally or Vertically Polarized
BOTH CURIOUSLY CORRELATED beating Bell's 3/4 limit?
Can you get all that in there somehow?
T.

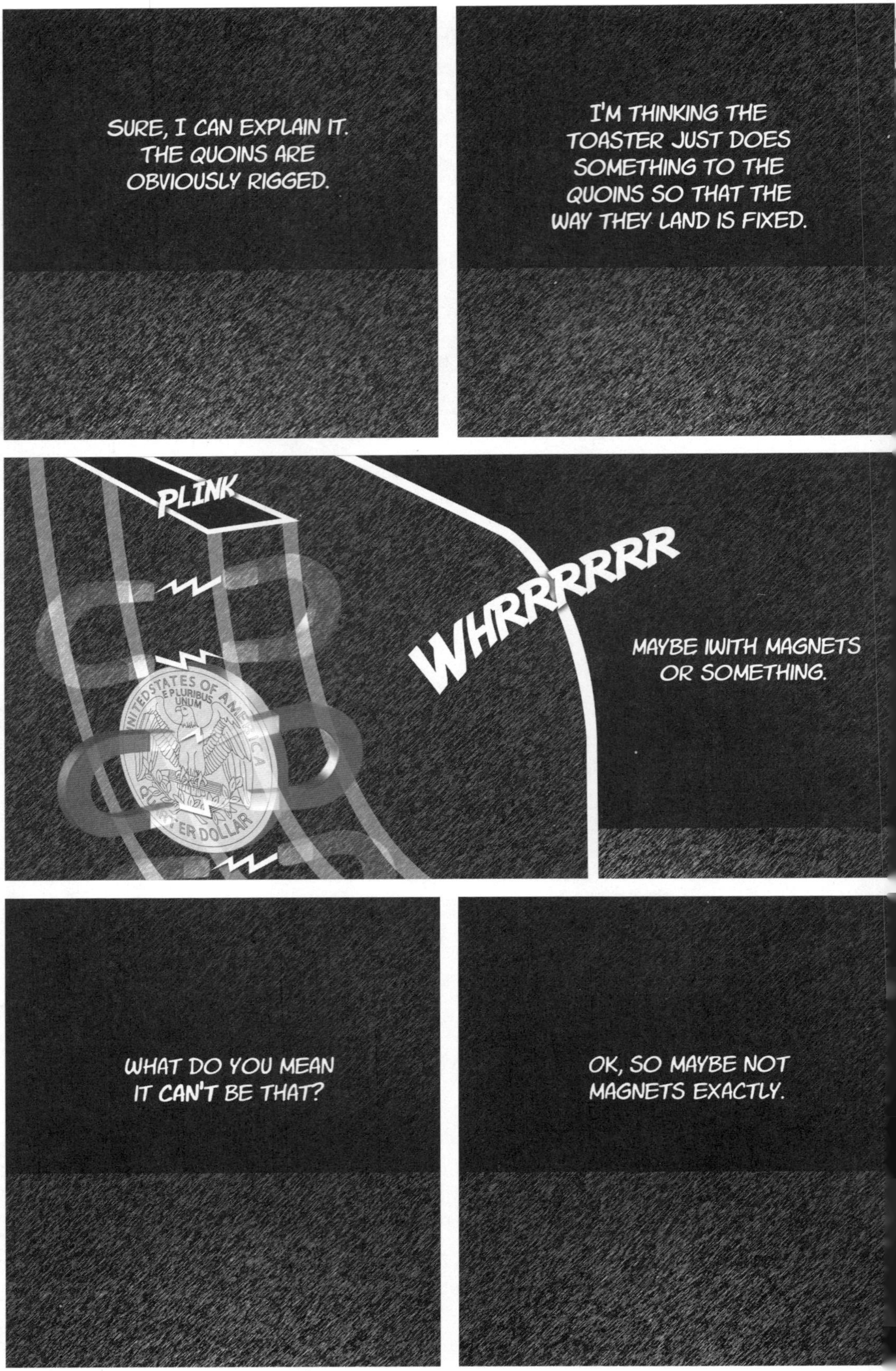
SURE, I CAN EXPLAIN IT. THE QUOINS ARE OBVIOUSLY RIGGED.
I'M THINKING THE TOASTER JUST DOES SOMETHING TO THE QUOINS SO THAT THE WAY THEY LAND IS FIXED.
PLINK
WHRRRRRR
MAYBE IWITH MAGNETS OR SOMETHING.
UNITED STATES OF AMERICA
E PLURIBUS UNUM
QUARTER DOLLAR
WHAT DO YOU MEAN IT **CAN'T** BE THAT?
OK, SO MAYBE NOT MAGNETS EXACTLY.

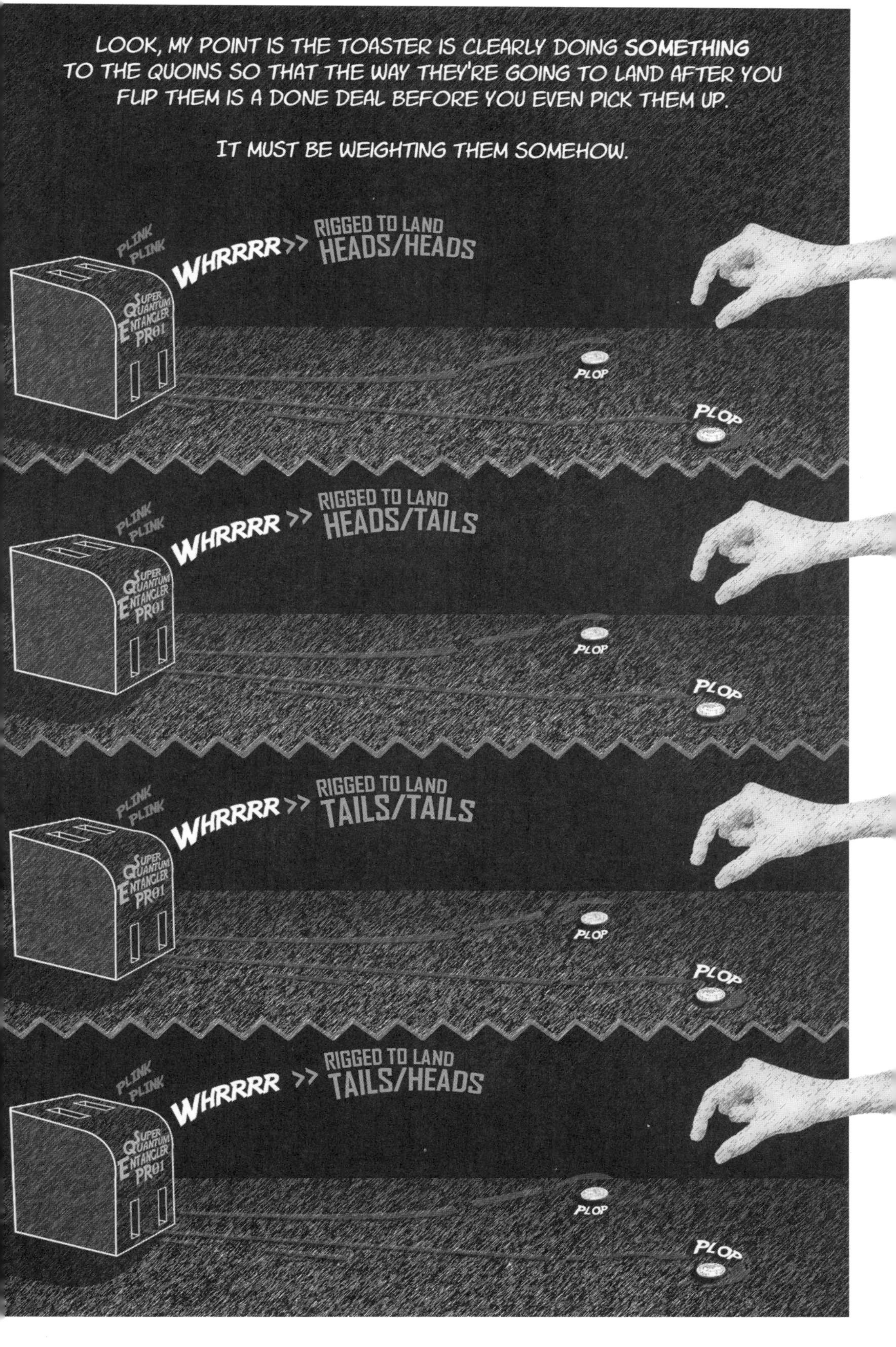
LOOK, MY POINT IS THE TOASTER IS CLEARLY DOING **SOMETHING** TO THE QUOINS SO THAT THE WAY THEY'RE GOING TO LAND AFTER YOU FLIP THEM IS A DONE DEAL BEFORE YOU EVEN PICK THEM UP.
IT MUST BE WEIGHTING THEM SOMEHOW.
PLINK PLINK
SUPER QUANTUM ENTANGLER PRO1
WHRRRR>>
RIGGED TO LAND HEADS/HEADS
PLOP
PLOP
PLINK PLINK
SUPER QUANTUM ENTANGLER PRO1
WHRRRR>>
RIGGED TO LAND HEADS/TAILS
PLOP
PLOP
PLINK PLINK
SUPER QUANTUM ENTANGLER PRO1
WHRRRR>>
RIGGED TO LAND TAILS/TAILS
PLOP
PLOP
PLINK PLINK
SUPER QUANTUM ENTANGLER PRO1
WHRRRR>>
RIGGED TO LAND TAILS/HEADS
PLOP
PLOP

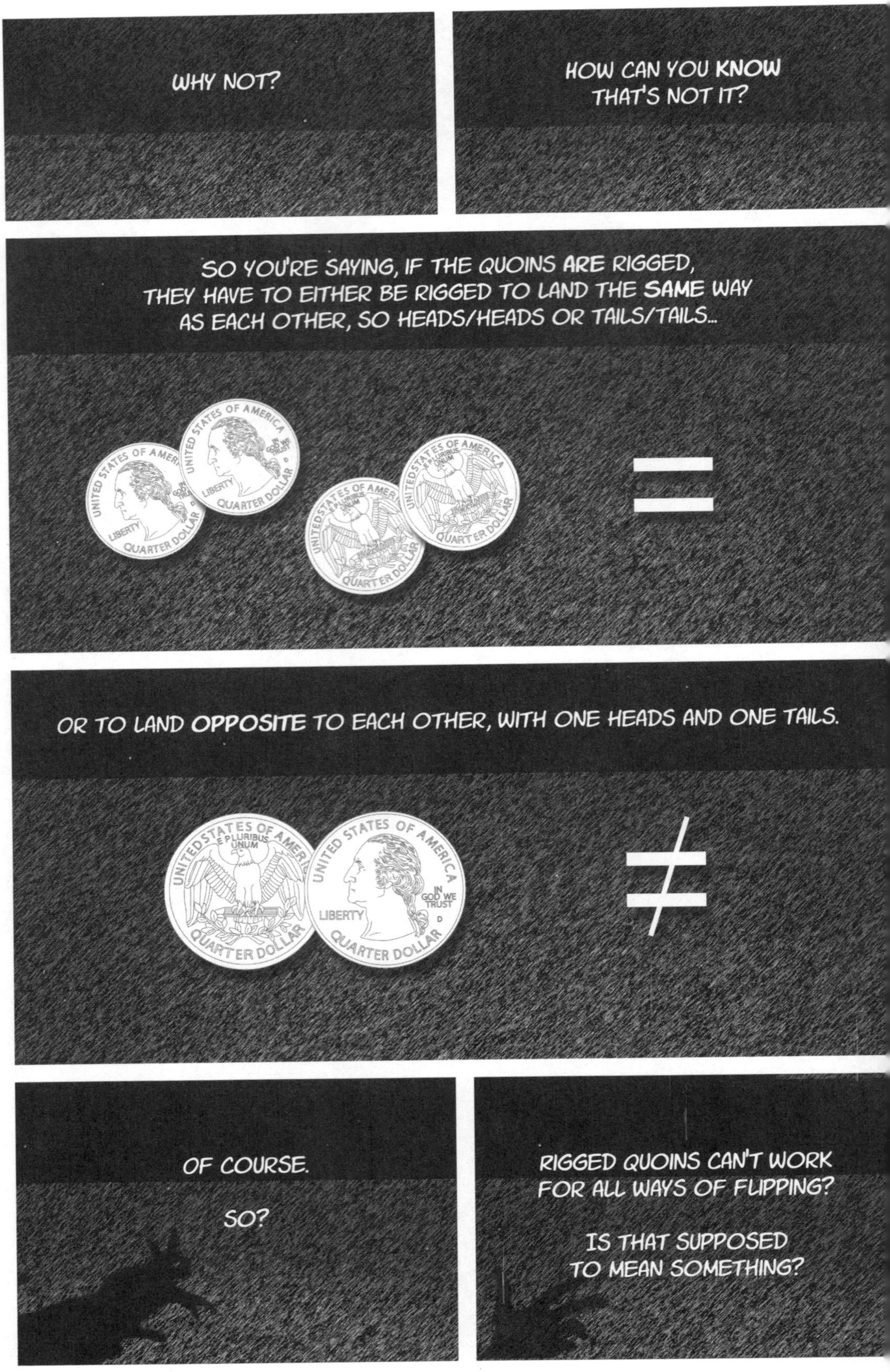
WHY NOT?
HOW CAN YOU KNOW THAT'S NOT IT?
SO YOU'RE SAYING, IF THE QUOINS ARE RIGGED, THEY HAVE TO EITHER BE RIGGED TO LAND THE SAME WAY AS EACH OTHER, SO HEADS/HEADS OR TAILS/TAILS...
OR TO LAND OPPOSITE TO EACH OTHER, WITH ONE HEADS AND ONE TAILS.
OF COURSE.
SO?
RIGGED QUOINS CAN'T WORK FOR ALL WAYS OF FLIPPING?
IS THAT SUPPOSED TO MEAN SOMETHING?

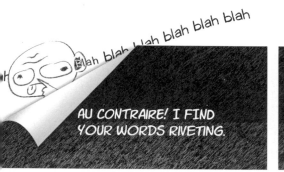

Blah blah blah blah blah blah
AU CONTRAIRE! I FIND YOUR WORDS RIVETING.
WAIT A SECOND,
I JUST THOUGHT OF SOMETHING..
IF THE QUOINS ARE RIGGED TO LAND THE SAME WAY AS EACH OTHER, AND THE PERSON FLIPPING DECIDES TO START THEM BOTH FACING HEADS UP, THEN THEY WON'T FOLLOW THE RULES OF QUOIN MECHANICS..
Quoin Mechanics
LEFT HAND
RIGHT HAND
UNITED STATES OF AMERICA
LIBERTY
QUARTER DOLLAR
AND IF THE QUOINS ARE RIGGED TO LAND OPPOSITE TO EACH OTHER AND THE PERSON FLIPPING STARTS THEM ANY OTHER WAY, LIKE TAILS/TAILS, FOR EXAMPLE, THEN THEY ALSO WON'T BE ABLE TO PULL OFF A QUOIN MECHANICS LANDING.
Quoin Mechanics
LEFT HAND
RIGHT HAND

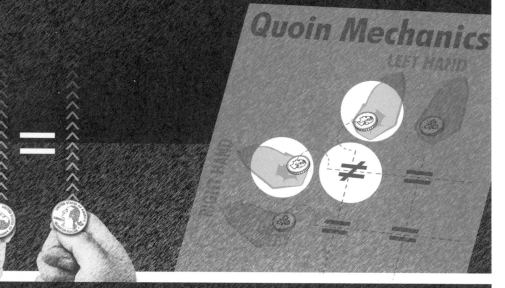

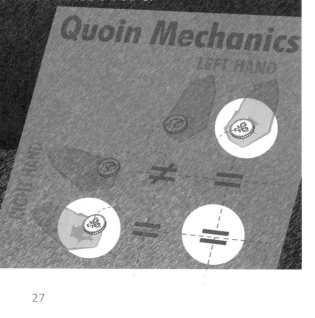

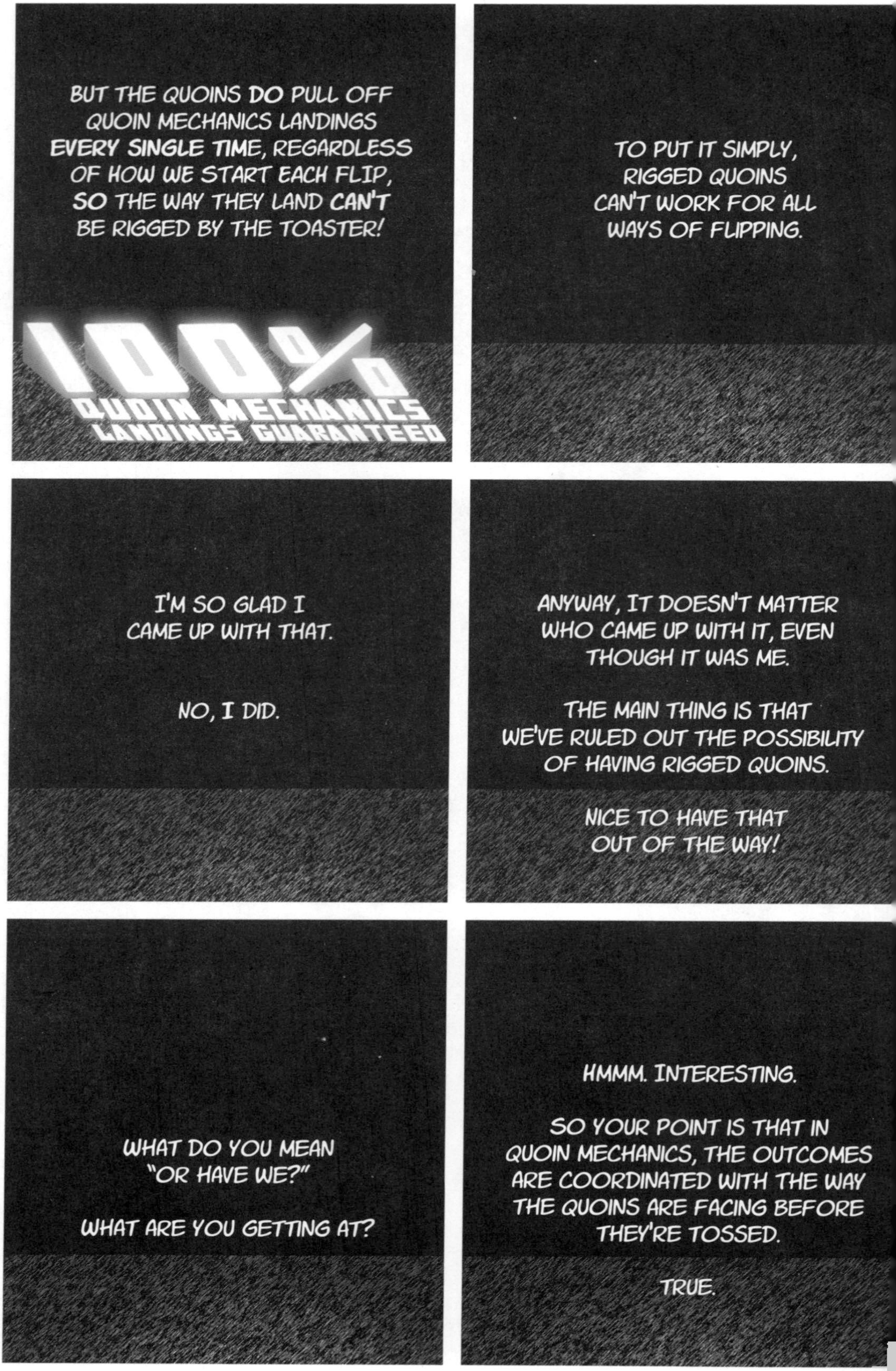
BUT THE QUOINS DO PULL OFF QUOIN MECHANICS LANDINGS EVERY SINGLE TIME, REGARDLESS OF HOW WE START EACH FLIP, SO THE WAY THEY LAND CAN'T BE RIGGED BY THE TOASTER!
100%
QUOIN MECHANICS
LANDINGS GUARANTEED
TO PUT IT SIMPLY, RIGGED QUOINS CAN'T WORK FOR ALL WAYS OF FLIPPING.
I'M SO GLAD I CAME UP WITH THAT.
NO, I DID.
ANYWAY, IT DOESN'T MATTER WHO CAME UP WITH IT, EVEN THOUGH IT WAS ME.
THE MAIN THING IS THAT WE'VE RULED OUT THE POSSIBILITY OF HAVING RIGGED QUOINS.
NICE TO HAVE THAT OUT OF THE WAY!
WHAT DO YOU MEAN "OR HAVE WE?"
WHAT ARE YOU GETTING AT?
HMMM. INTERESTING.
SO YOUR POINT IS THAT IN QUOIN MECHANICS, THE OUTCOMES ARE COORDINATED WITH THE WAY THE QUOINS ARE FACING BEFORE THEY'RE TOSSED.
TRUE.

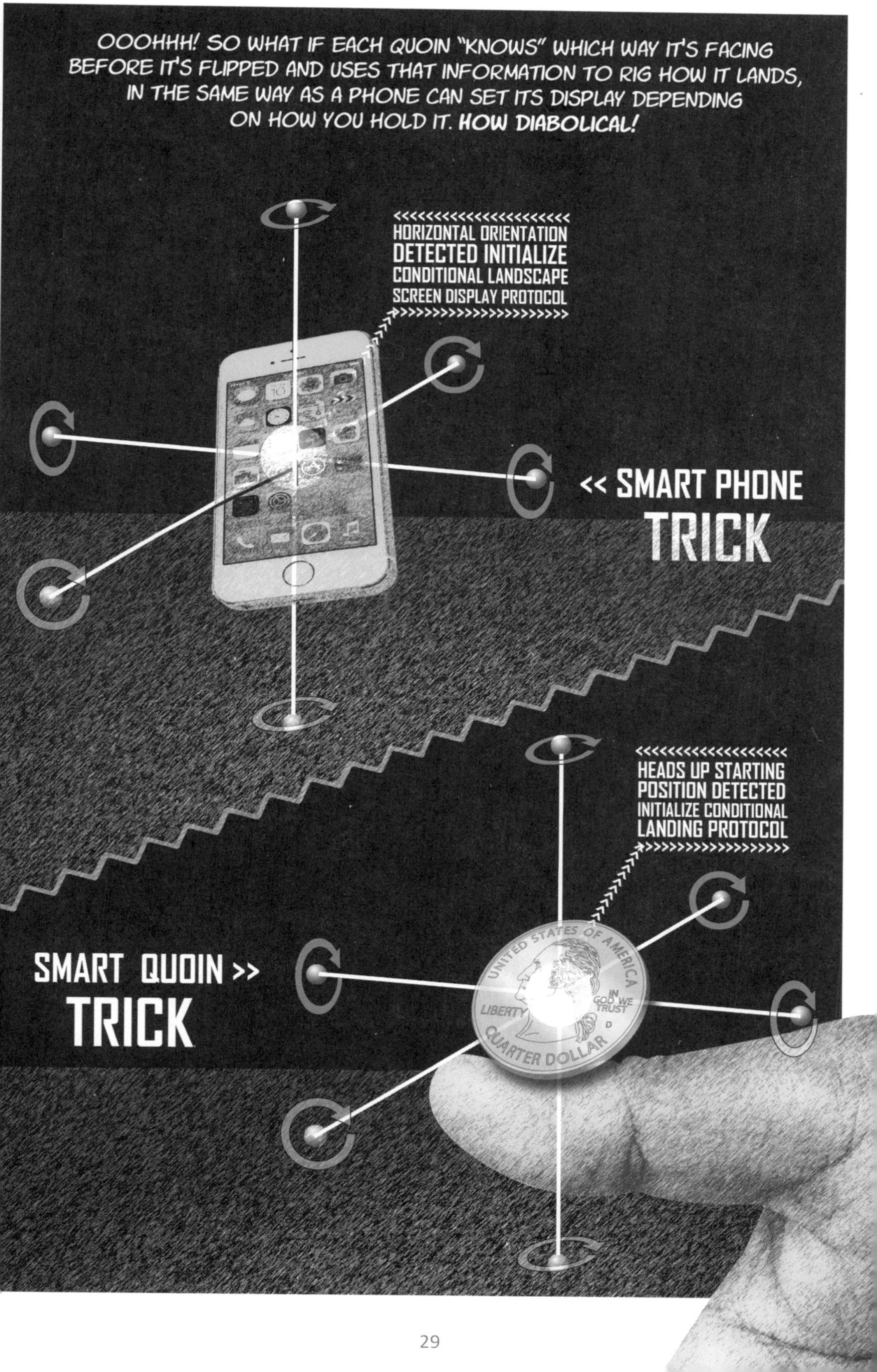
OOOHHH! SO WHAT IF EACH QUOIN "KNOWS" WHICH WAY IT'S FACING BEFORE IT'S FLIPPED AND USES THAT INFORMATION TO RIG HOW IT LANDS, IN THE SAME WAY AS A PHONE CAN SET ITS DISPLAY DEPENDING ON HOW YOU HOLD IT. HOW DIABOLICAL!
HORIZONTAL ORIENTATION DETECTED INITIALIZE CONDITIONAL LANDSCAPE SCREEN DISPLAY PROTOCOL
<< SMART PHONE TRICK
HEADS UP STARTING POSITION DETECTED INITIALIZE CONDITIONAL LANDING PROTOCOL
UNITED STATES OF AMERICA
LIBERTY
IN GOD WE TRUST
D
QUARTER DOLLAR
SMART QUOIN >> TRICK

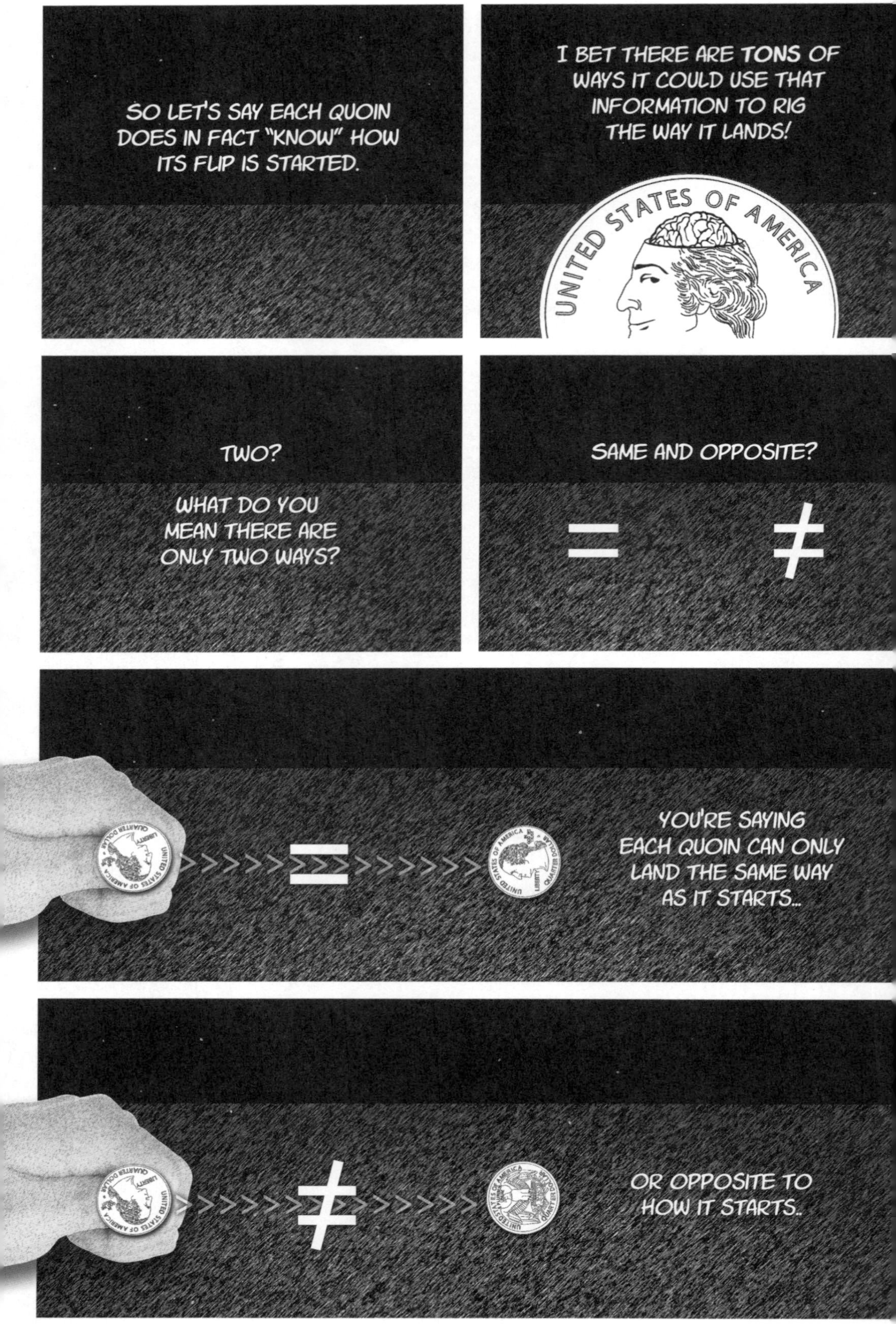
SO LET'S SAY EACH QUOIN DOES IN FACT "KNOW" HOW ITS FLIP IS STARTED.
I BET THERE ARE **TONS** OF WAYS IT COULD USE THAT INFORMATION TO RIG THE WAY IT LANDS!
UNITED STATES OF AMERICA
TWO?
WHAT DO YOU MEAN THERE ARE ONLY TWO WAYS?
SAME AND OPPOSITE?
YOU'RE SAYING EACH QUOIN CAN ONLY LAND THE SAME WAY AS IT STARTS...
OR OPPOSITE TO HOW IT STARTS.

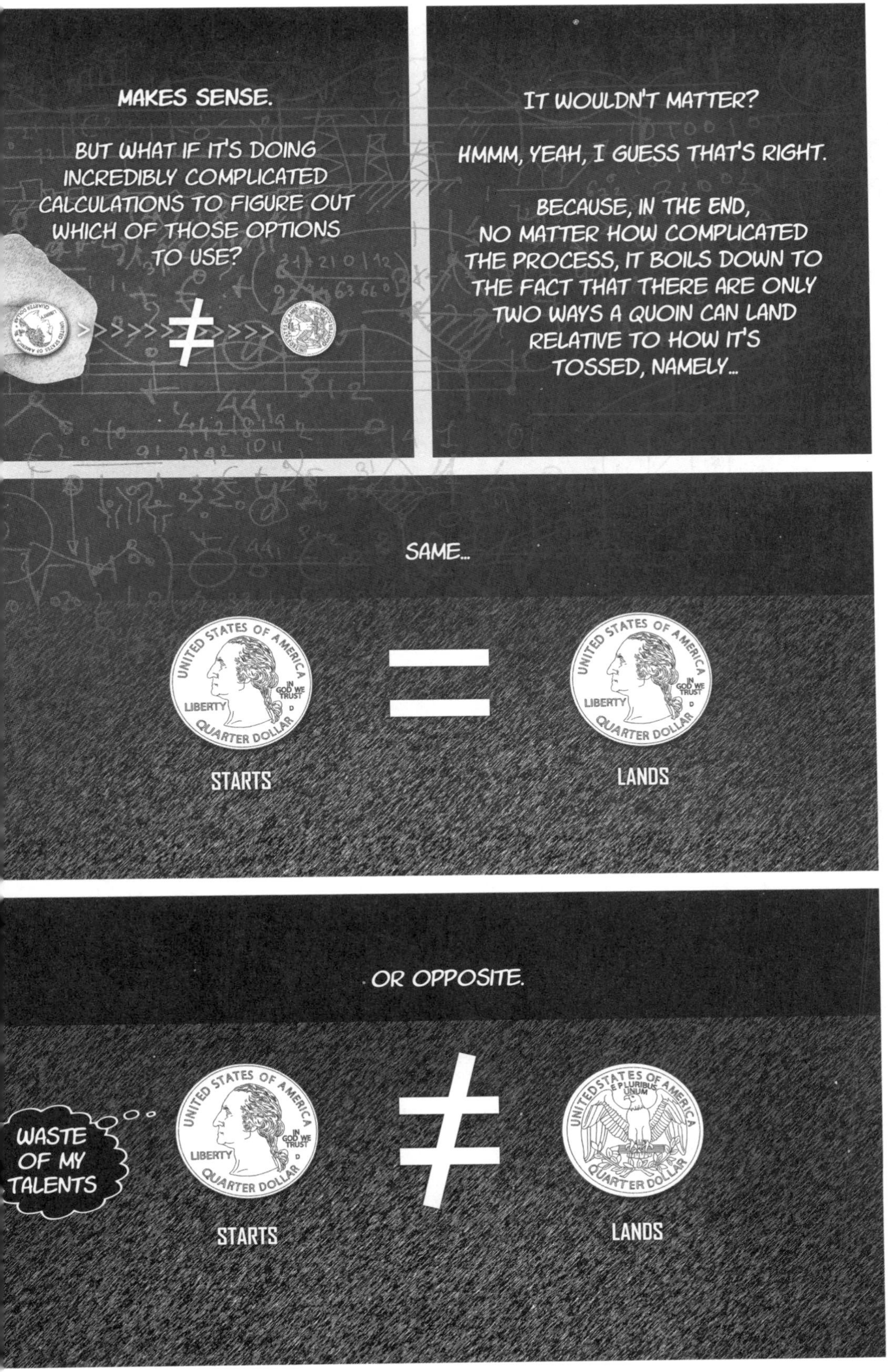
MAKES SENSE.
BUT WHAT IF IT'S DOING INCREDIBLY COMPLICATED CALCULATIONS TO FIGURE OUT WHICH OF THOSE OPTIONS TO USE?
IT WOULDN'T MATTER?
HMMM, YEAH, I GUESS THAT'S RIGHT.
BECAUSE, IN THE END, NO MATTER HOW COMPLICATED THE PROCESS, IT BOILS DOWN TO THE FACT THAT THERE ARE ONLY TWO WAYS A QUOIN CAN LAND RELATIVE TO HOW IT'S TOSSED, NAMELY…
SAME…
STARTS
LANDS
OR OPPOSITE.
WASTE OF MY TALENTS
STARTS
LANDS

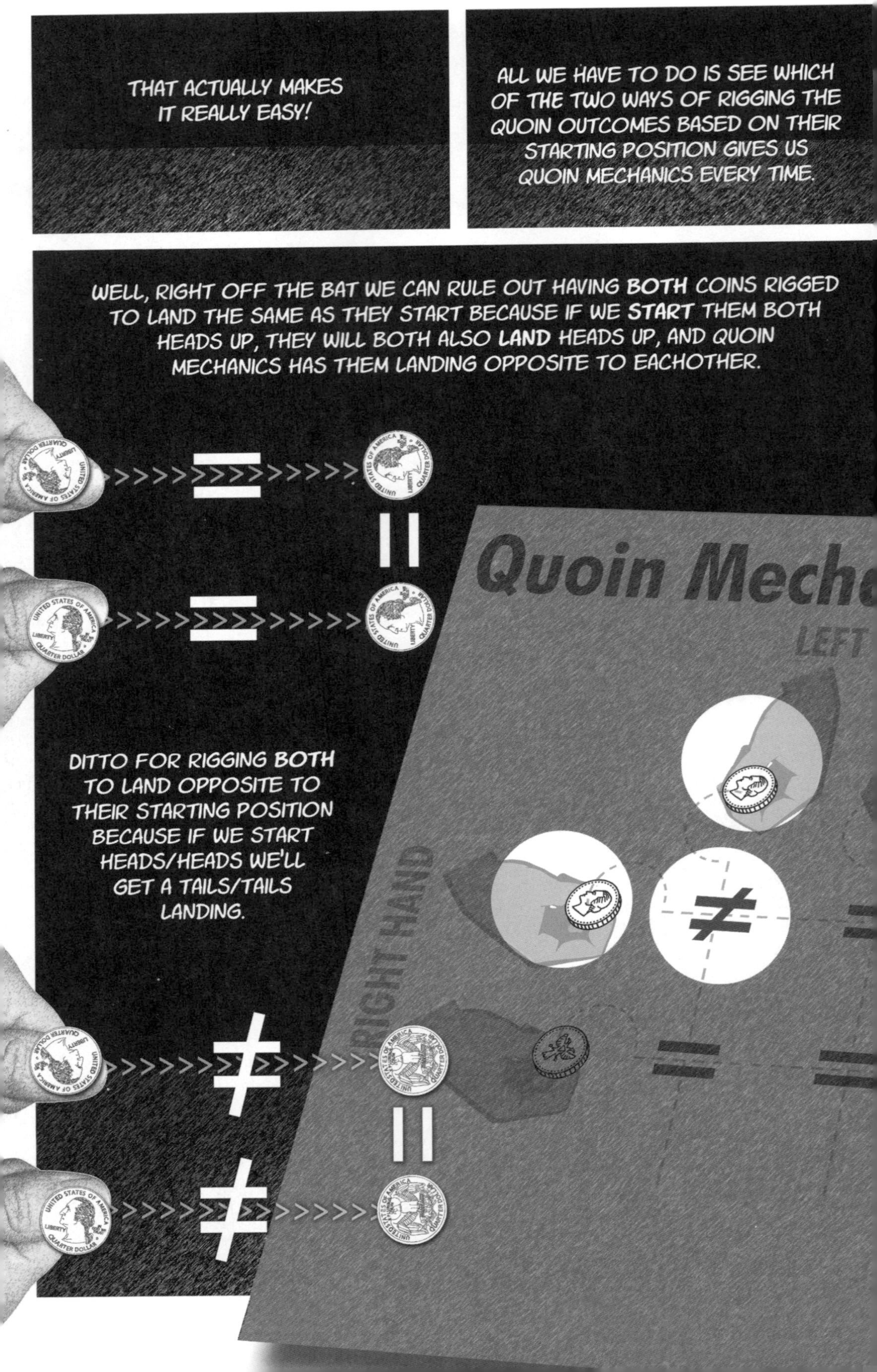
THAT ACTUALLY MAKES IT REALLY EASY!
ALL WE HAVE TO DO IS SEE WHICH OF THE TWO WAYS OF RIGGING THE QUOIN OUTCOMES BASED ON THEIR STARTING POSITION GIVES US QUOIN MECHANICS EVERY TIME.
WELL, RIGHT OFF THE BAT WE CAN RULE OUT HAVING **BOTH** COINS RIGGED TO LAND THE SAME AS THEY START BECAUSE IF WE **START** THEM BOTH HEADS UP, THEY WILL BOTH ALSO **LAND** HEADS UP, AND QUOIN MECHANICS HAS THEM LANDING OPPOSITE TO EACHOTHER.
DITTO FOR RIGGING **BOTH** TO LAND OPPOSITE TO THEIR STARTING POSITION BECAUSE IF WE START HEADS/HEADS WE'LL GET A TAILS/TAILS LANDING.
Quoin Mech
LEFT
RIGHT HAND

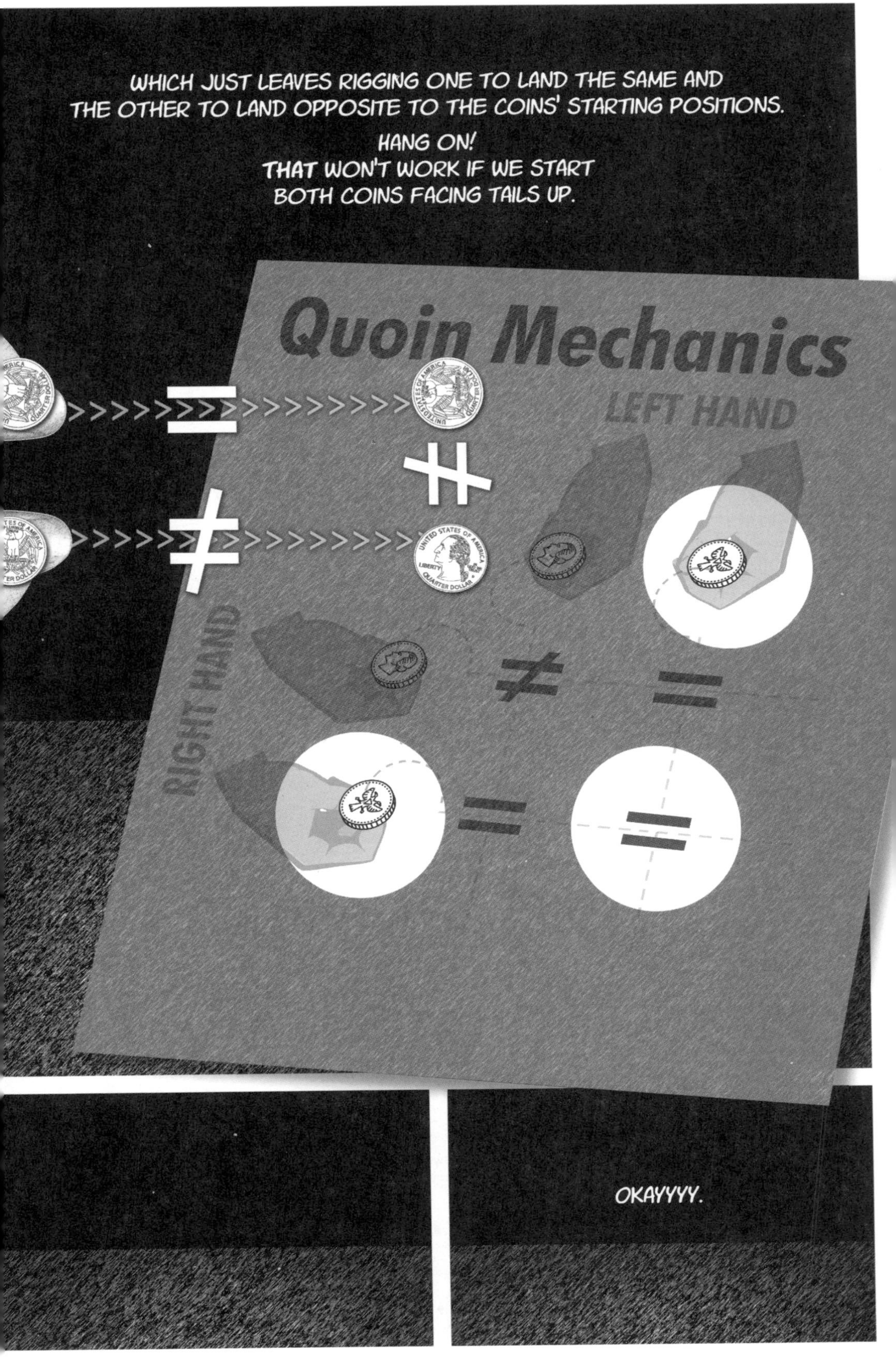
WHICH JUST LEAVES RIGGING ONE TO LAND THE SAME AND THE OTHER TO LAND OPPOSITE TO THE COINS' STARTING POSITIONS.
HANG ON!
THAT WON'T WORK IF WE START BOTH COINS FACING TAILS UP.
Quoin Mechanics
LEFT HAND
RIGHT HAND
OKAYYYY.

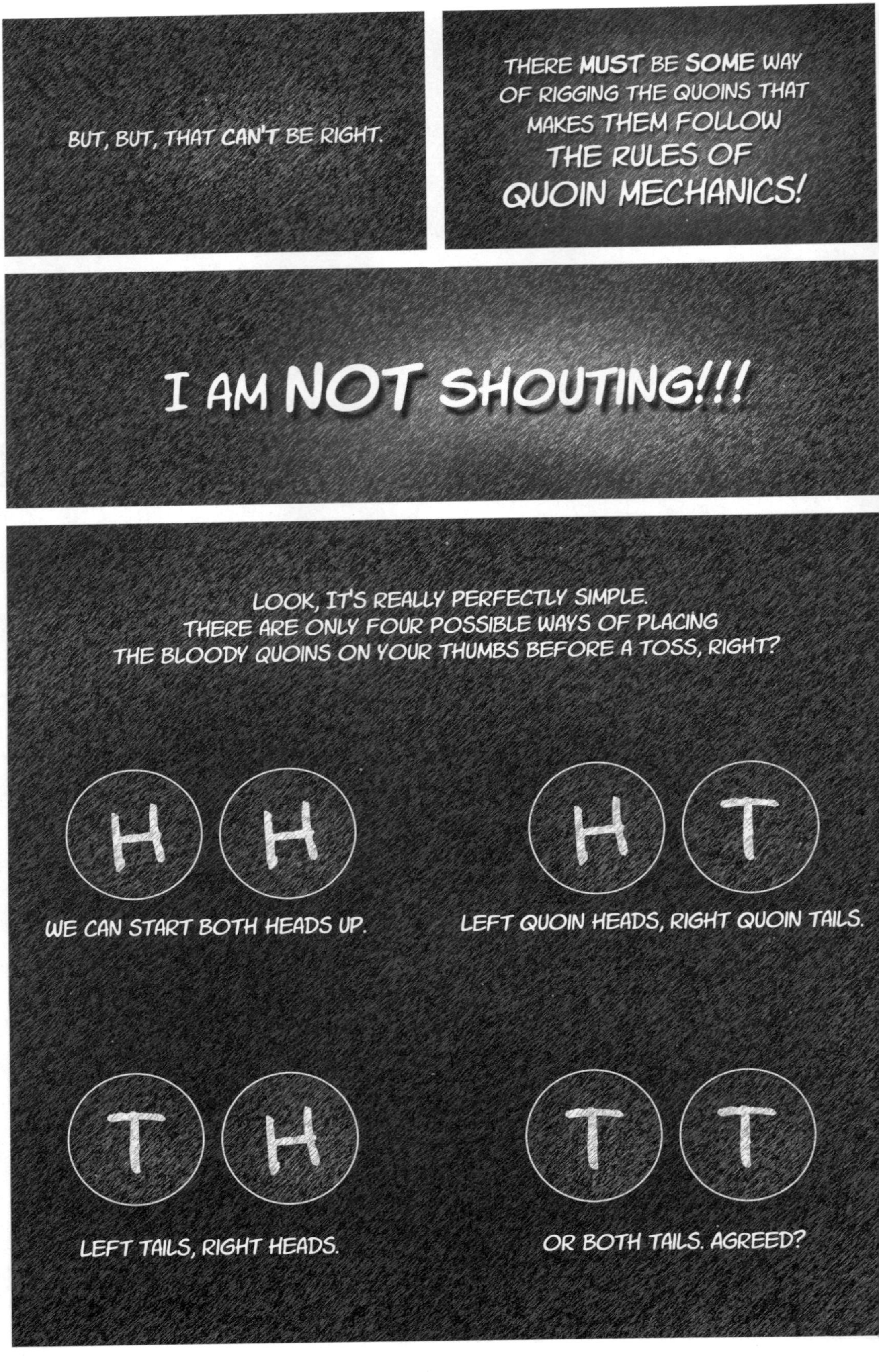
BUT, BUT, THAT CAN'T BE RIGHT.
THERE MUST BE SOME WAY OF RIGGING THE QUOINS THAT MAKES THEM FOLLOW THE RULES OF QUOIN MECHANICS!
I AM NOT SHOUTING!!!
LOOK, IT'S REALLY PERFECTLY SIMPLE. THERE ARE ONLY FOUR POSSIBLE WAYS OF PLACING THE BLOODY QUOINS ON YOUR THUMBS BEFORE A TOSS, RIGHT?
H
H
WE CAN START BOTH HEADS UP.
H
T
LEFT QUOIN HEADS, RIGHT QUOIN TAILS.
T
H
LEFT TAILS, RIGHT HEADS.
T
T
OR BOTH TAILS. AGREED?

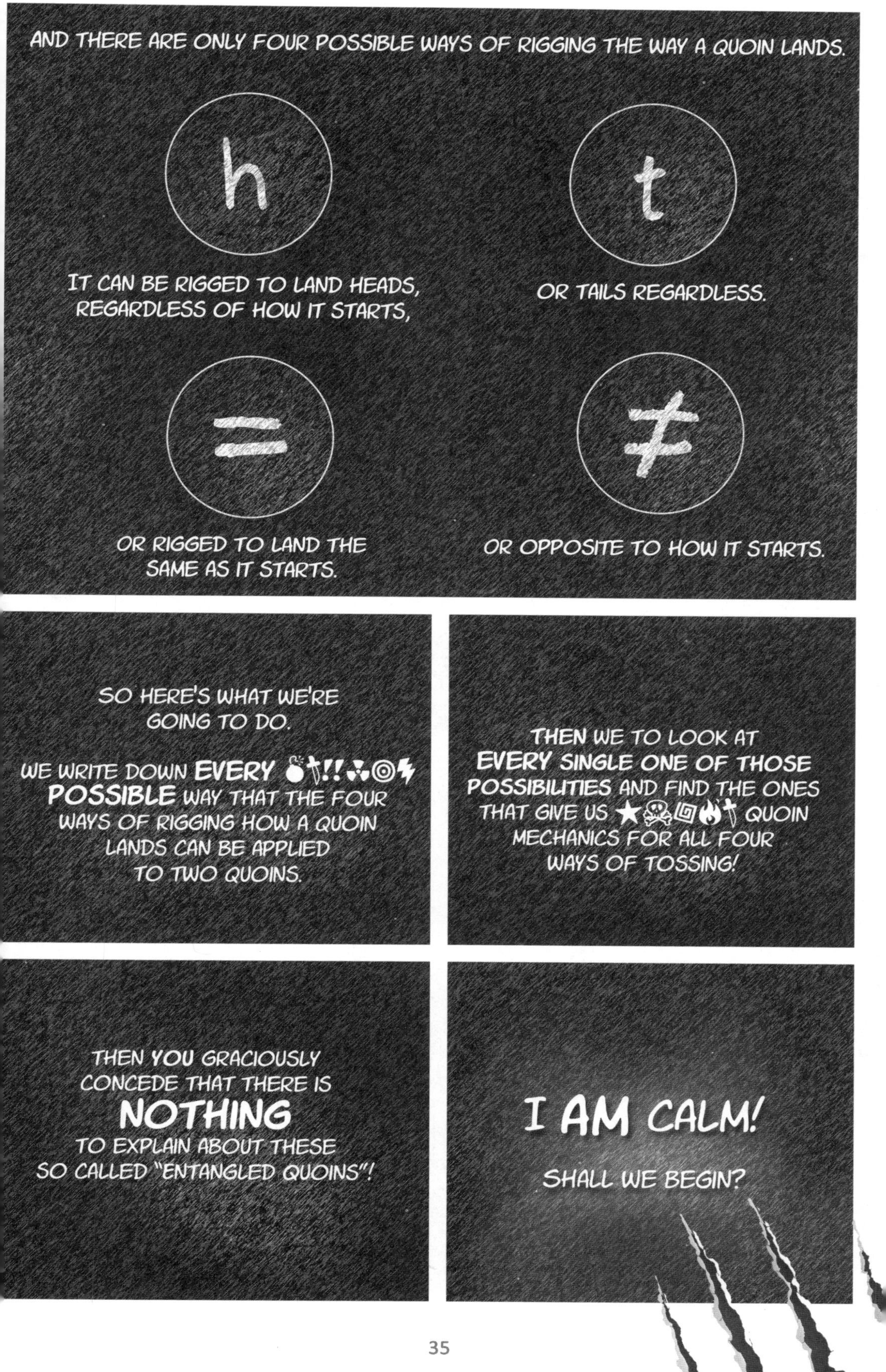
AND THERE ARE ONLY FOUR POSSIBLE WAYS OF RIGGING THE WAY A QUOIN LANDS.
h
t
IT CAN BE RIGGED TO LAND HEADS, REGARDLESS OF HOW IT STARTS,
OR TAILS REGARDLESS.
=
≠
OR RIGGED TO LAND THE SAME AS IT STARTS.
OR OPPOSITE TO HOW IT STARTS.
SO HERE'S WHAT WE'RE GOING TO DO.
WE WRITE DOWN EVERY POSSIBLE WAY THAT THE FOUR WAYS OF RIGGING HOW A QUOIN LANDS CAN BE APPLIED TO TWO QUOINS.
THEN WE TO LOOK AT EVERY SINGLE ONE OF THOSE POSSIBILITIES AND FIND THE ONES THAT GIVE US QUOIN MECHANICS FOR ALL FOUR WAYS OF TOSSING!
THEN YOU GRACIOUSLY CONCEDE THAT THERE IS NOTHING TO EXPLAIN ABOUT THESE SO CALLED "ENTANGLED QUOINS"!
I AM CALM!
SHALL WE BEGIN?

The Punchline

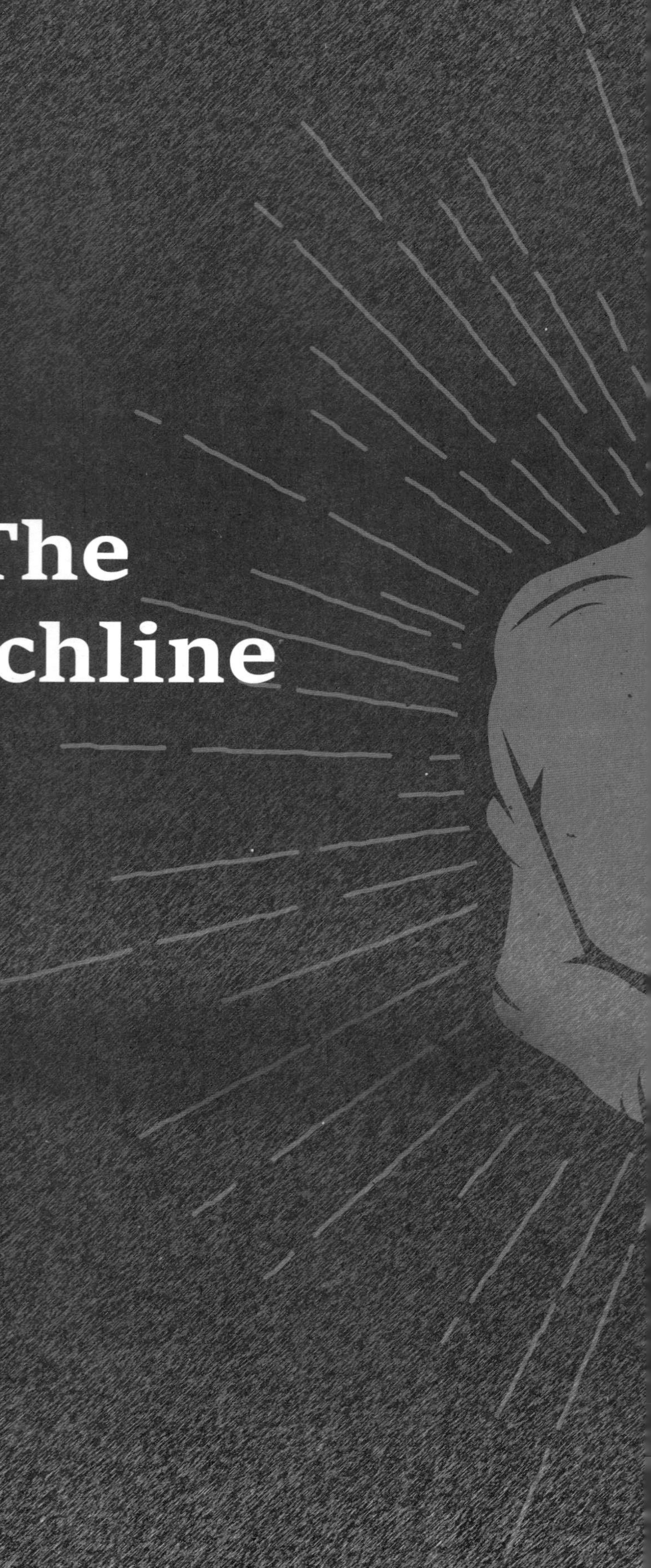

$$1/4(P(a=b|0,0) + P(a=b|0,1) + P(a=b|1,0) + P(a\neq b|1,1)) \leq 3/4$$

— **Bell's Inequality**

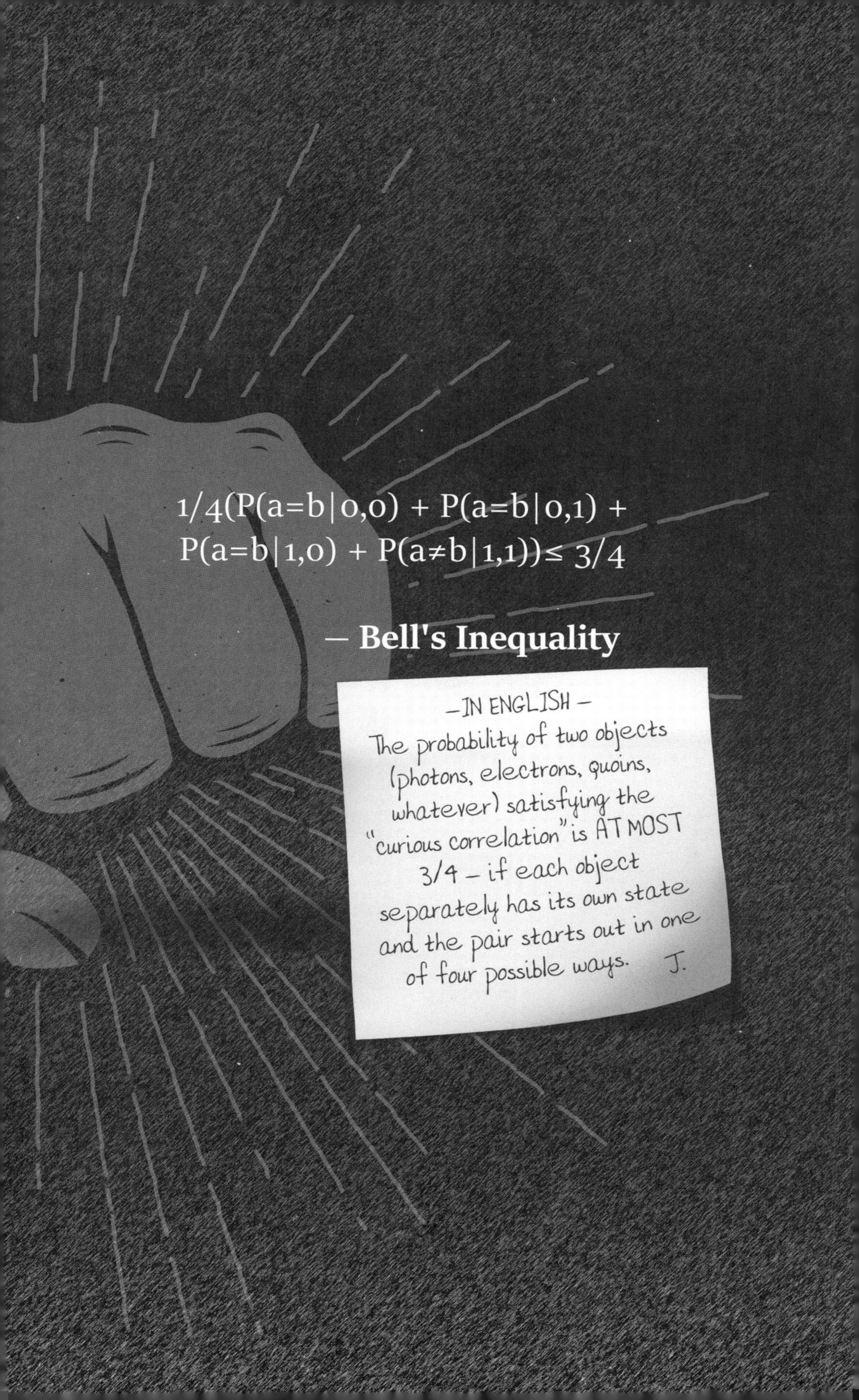

—IN ENGLISH—
The probability of two objects (photons, electrons, quoins, whatever) satisfying the "curious correlation" is AT MOST 3/4 — if each object separately has its own state and the pair starts out in one of four possible ways.
J.

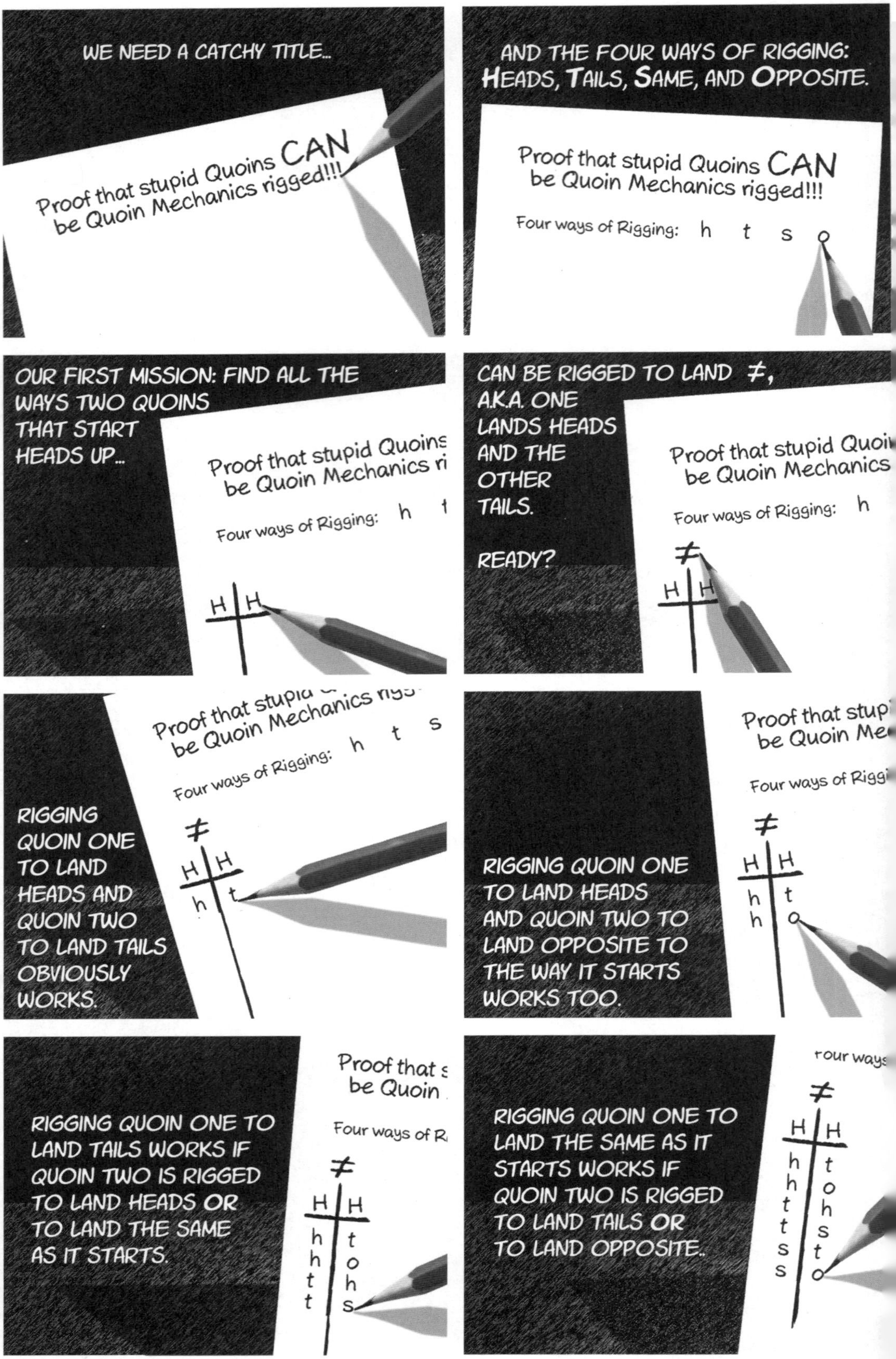

WE NEED A CATCHY TITLE...

AND THE FOUR WAYS OF RIGGING: **H**EADS, **T**AILS, **S**AME, AND **O**PPOSITE.

Proof that stupid Quoins **CAN** be Quoin Mechanics rigged!!!

Four ways of Rigging: h t s o

OUR FIRST MISSION: FIND ALL THE WAYS TWO QUOINS THAT START HEADS UP...

Proof that stupid Quoins be Quoin Mechanics ri

Four ways of Rigging: h t

CAN BE RIGGED TO LAND ≠, A.K.A. ONE LANDS HEADS AND THE OTHER TAILS.

READY?

Proof that stupid Quoi be Quoin Mechanics

Four ways of Rigging: h

RIGGING QUOIN ONE TO LAND HEADS AND QUOIN TWO TO LAND TAILS OBVIOUSLY WORKS.

Proof that stupid Qu be Quoin Mechanics rigs

Four ways of Rigging: h t s

RIGGING QUOIN ONE TO LAND HEADS AND QUOIN TWO TO LAND OPPOSITE TO THE WAY IT STARTS WORKS TOO.

Proof that stup be Quoin Me

Four ways of Riggi

RIGGING QUOIN ONE TO LAND TAILS WORKS IF QUOIN TWO IS RIGGED TO LAND HEADS **OR** TO LAND THE SAME AS IT STARTS.

Proof that s be Quoin

Four ways of R

RIGGING QUOIN ONE TO LAND THE SAME AS IT STARTS WORKS IF QUOIN TWO IS RIGGED TO LAND TAILS **OR** TO LAND OPPOSITE..

Four ways

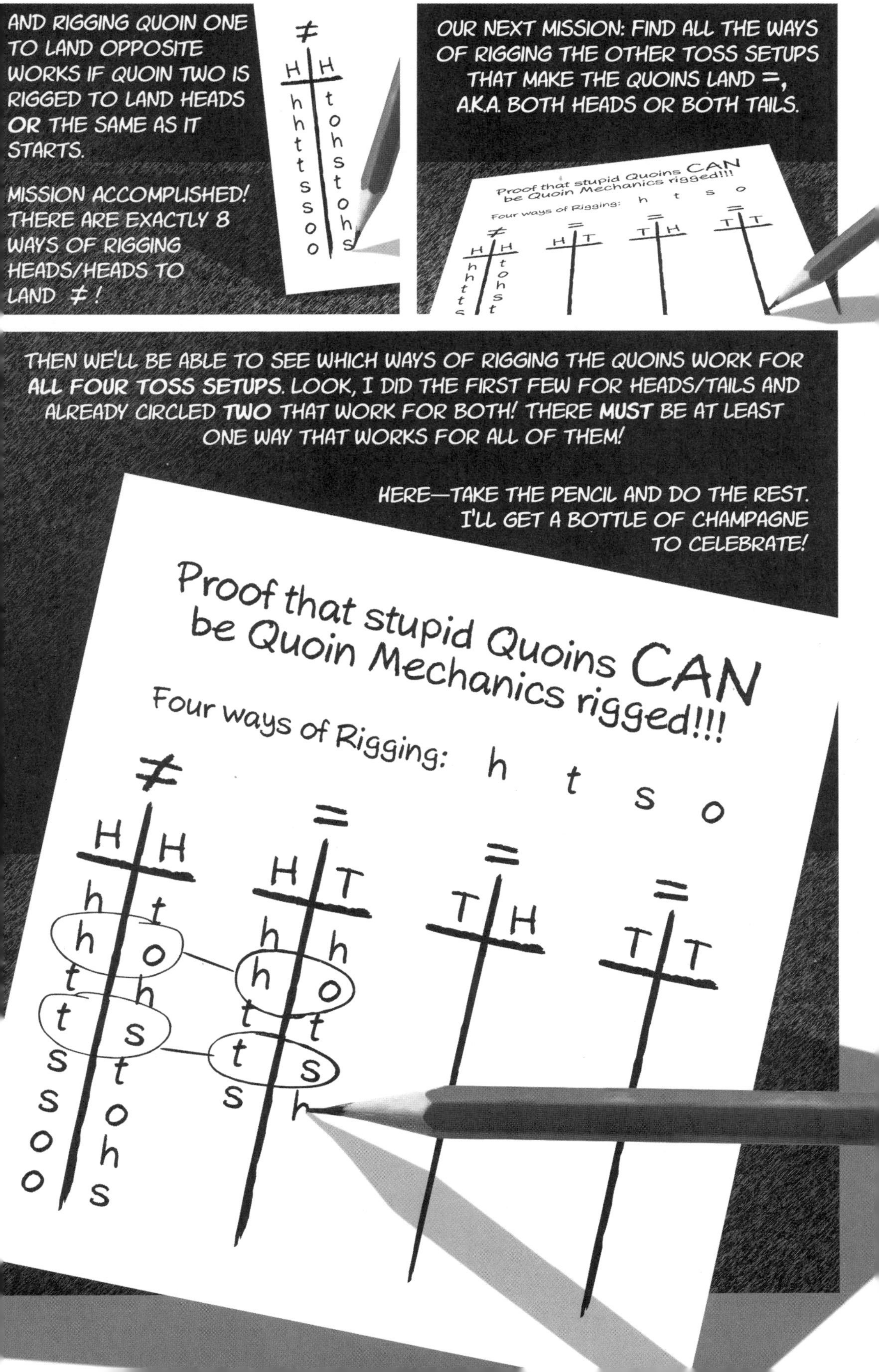

AND RIGGING QUOIN ONE TO LAND OPPOSITE WORKS IF QUOIN TWO IS RIGGED TO LAND HEADS OR THE SAME AS IT STARTS.

MISSION ACCOMPLISHED! THERE ARE EXACTLY 8 WAYS OF RIGGING HEADS/HEADS TO LAND $\neq$!

OUR NEXT MISSION: FIND ALL THE WAYS OF RIGGING THE OTHER TOSS SETUPS THAT MAKE THE QUOINS LAND $=$, A.K.A. BOTH HEADS OR BOTH TAILS.

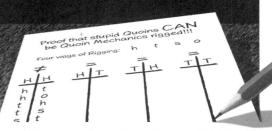

THEN WE'LL BE ABLE TO SEE WHICH WAYS OF RIGGING THE QUOINS WORK FOR ALL FOUR TOSS SETUPS. LOOK, I DID THE FIRST FEW FOR HEADS/TAILS AND ALREADY CIRCLED **TWO** THAT WORK FOR BOTH! THERE **MUST** BE AT LEAST ONE WAY THAT WORKS FOR ALL OF THEM!

HERE—TAKE THE PENCIL AND DO THE REST. I'LL GET A BOTTLE OF CHAMPAGNE TO CELEBRATE!

Proof that stupid Quoins CAN be Quoin Mechanics rigged!!!

Four ways of Rigging: h   t   s   o

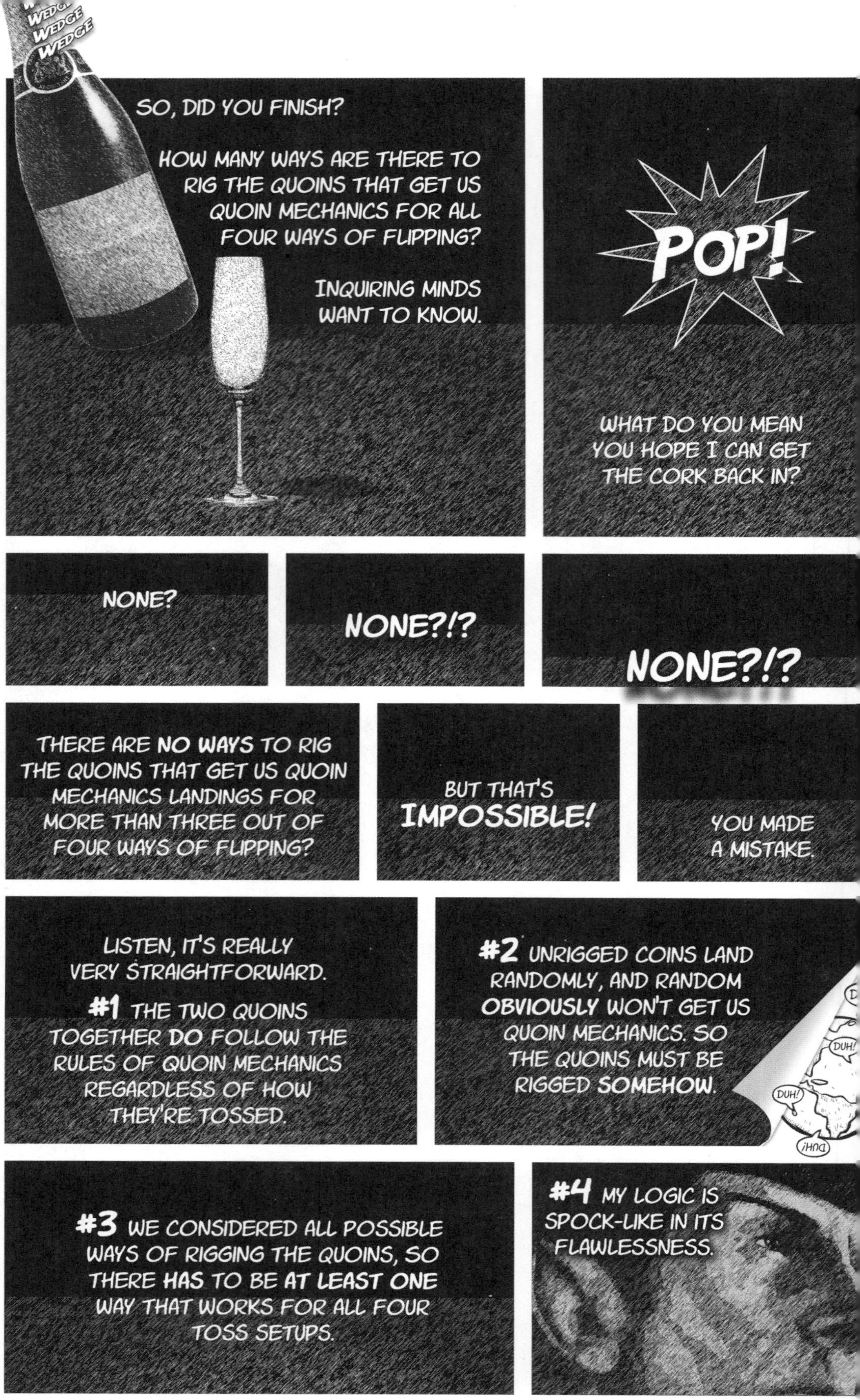

WEDGE
WEDGE
SO, DID YOU FINISH?
HOW MANY WAYS ARE THERE TO RIG THE QUOINS THAT GET US QUOIN MECHANICS FOR ALL FOUR WAYS OF FLIPPING?
INQUIRING MINDS WANT TO KNOW.
POP!
WHAT DO YOU MEAN YOU HOPE I CAN GET THE CORK BACK IN?
NONE?
NONE?!?
NONE?!?
THERE ARE NO WAYS TO RIG THE QUOINS THAT GET US QUOIN MECHANICS LANDINGS FOR MORE THAN THREE OUT OF FOUR WAYS OF FLIPPING?
BUT THAT'S IMPOSSIBLE!
YOU MADE A MISTAKE.
LISTEN, IT'S REALLY VERY STRAIGHTFORWARD.
#1 THE TWO QUOINS TOGETHER DO FOLLOW THE RULES OF QUOIN MECHANICS REGARDLESS OF HOW THEY'RE TOSSED.
#2 UNRIGGED COINS LAND RANDOMLY, AND RANDOM OBVIOUSLY WON'T GET US QUOIN MECHANICS. SO THE QUOINS MUST BE RIGGED SOMEHOW.
DUH!
DUH!
#3 WE CONSIDERED ALL POSSIBLE WAYS OF RIGGING THE QUOINS, SO THERE HAS TO BE AT LEAST ONE WAY THAT WORKS FOR ALL FOUR TOSS SETUPS.
#4 MY LOGIC IS SPOCK-LIKE IN ITS FLAWLESSNESS.

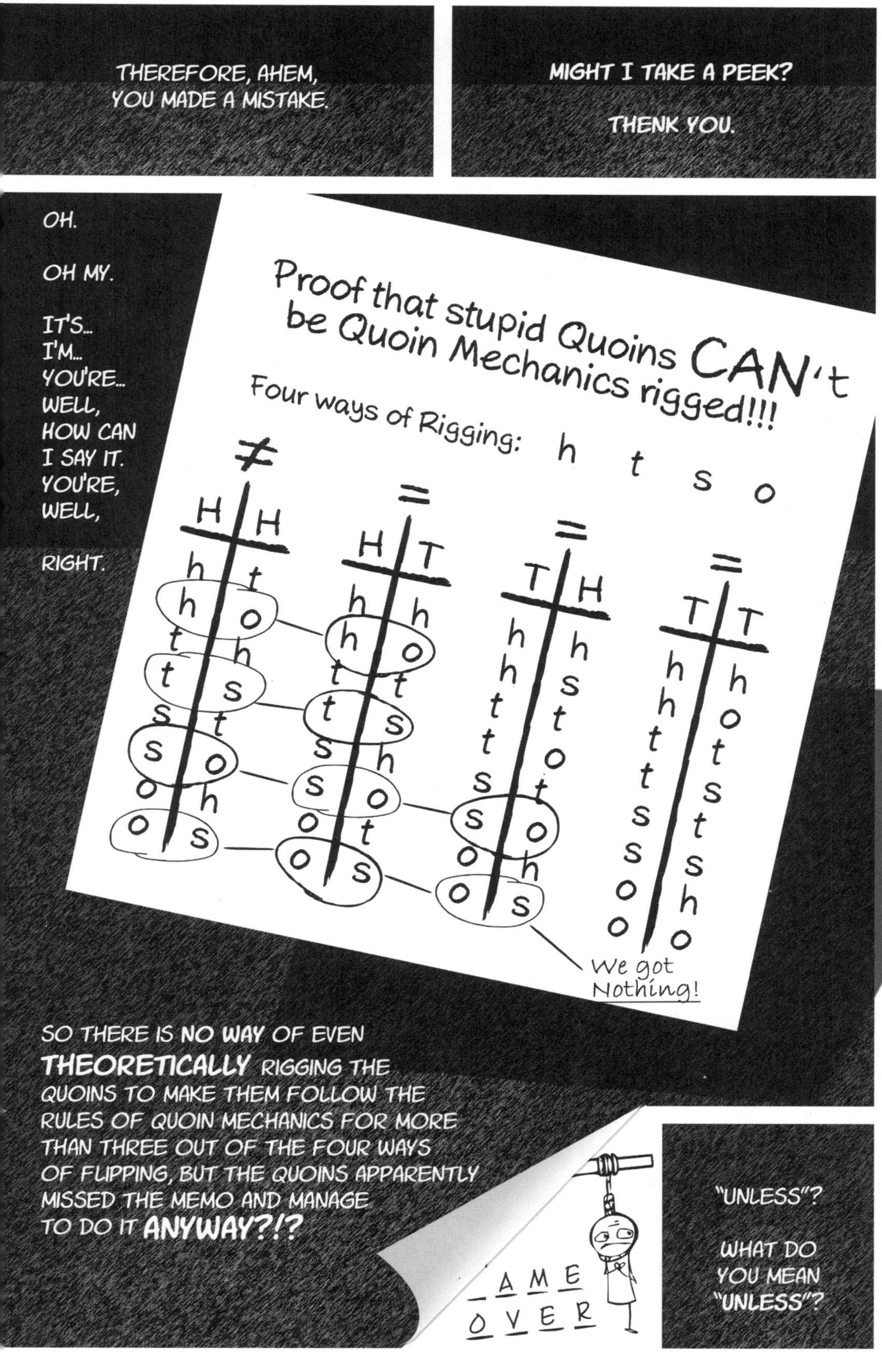
THEREFORE, AHEM, YOU MADE A MISTAKE.
MIGHT I TAKE A PEEK?
THENK YOU.
OH.
OH MY.
IT'S... I'M... YOU'RE... WELL, HOW CAN I SAY IT. YOU'RE, WELL,
RIGHT.
Proof that stupid Quoins CAN't be Quoin Mechanics rigged!!!
Four ways of Rigging: h t s o
≠ H H
= H T
= T H
= T T
We got Nothing!
SO THERE IS NO WAY OF EVEN THEORETICALLY RIGGING THE QUOINS TO MAKE THEM FOLLOW THE RULES OF QUOIN MECHANICS FOR MORE THAN THREE OUT OF THE FOUR WAYS OF FLIPPING, BUT THE QUOINS APPARENTLY MISSED THE MEMO AND MANAGE TO DO IT ANYWAY?!?
_ A M E
O V E R
"UNLESS"?
WHAT DO YOU MEAN "UNLESS"?

Signaling

Conceivably, [quantum mechanical predictions] might apply only to experiments in which the settings of the instruments are made sufficiently in advance to allow them to reach some mutual rapport by exchange of signals with velocity less than or equal to the velocity of light.

—John Stewart Bell

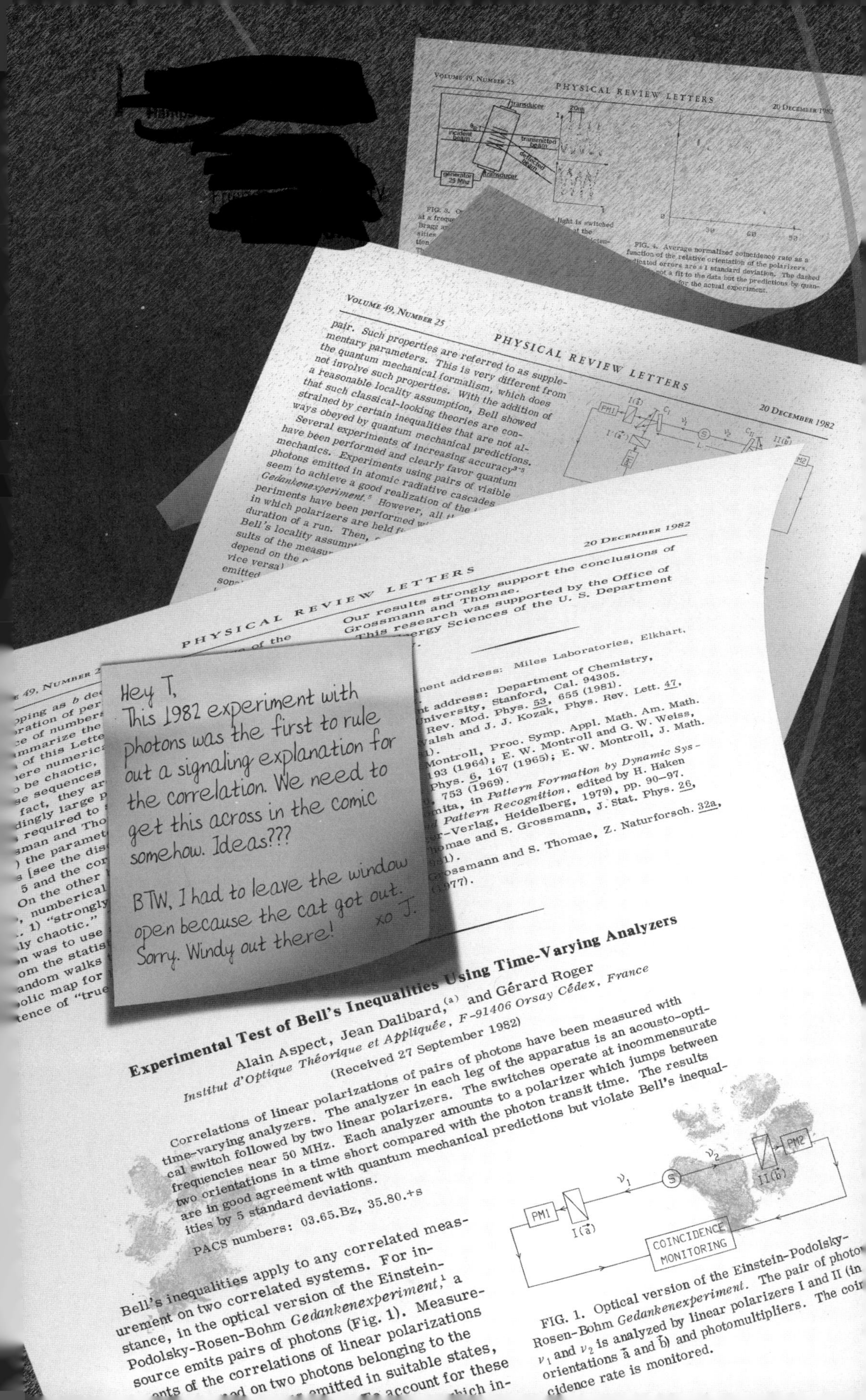
Hey T,
This 1982 experiment with photons was the first to rule out a signaling explanation for the correlation. We need to get this across in the comic somehow. Ideas???
BTW, I had to leave the window open because the cat got out. Sorry. Windy out there! xo J.
PHYSICAL REVIEW LETTERS
VOLUME 49, NUMBER 25
20 DECEMBER 1982
Experimental Test of Bell's Inequalities Using Time-Varying Analyzers
Alain Aspect, Jean Dalibard,(a) and Gérard Roger
Institut d'Optique Théorique et Appliquée, F-91406 Orsay Cédex, France
(Received 27 September 1982)
Correlations of linear polarizations of pairs of photons have been measured with time-varying analyzers. The analyzer in each leg of the apparatus is an acousto-optical switch followed by two linear polarizers. The switches operate at incommensurate frequencies near 50 MHz. Each analyzer amounts to a polarizer which jumps between two orientations in a time short compared with the photon transit time. The results are in good agreement with quantum mechanical predictions but violate Bell's inequalities by 5 standard deviations.
PACS numbers: 03.65.Bz, 35.80.+s
COINCIDENCE MONITORING
FIG. 1. Optical version of the Einstein-Podolsky-Rosen-Bohm Gedankenexperiment.

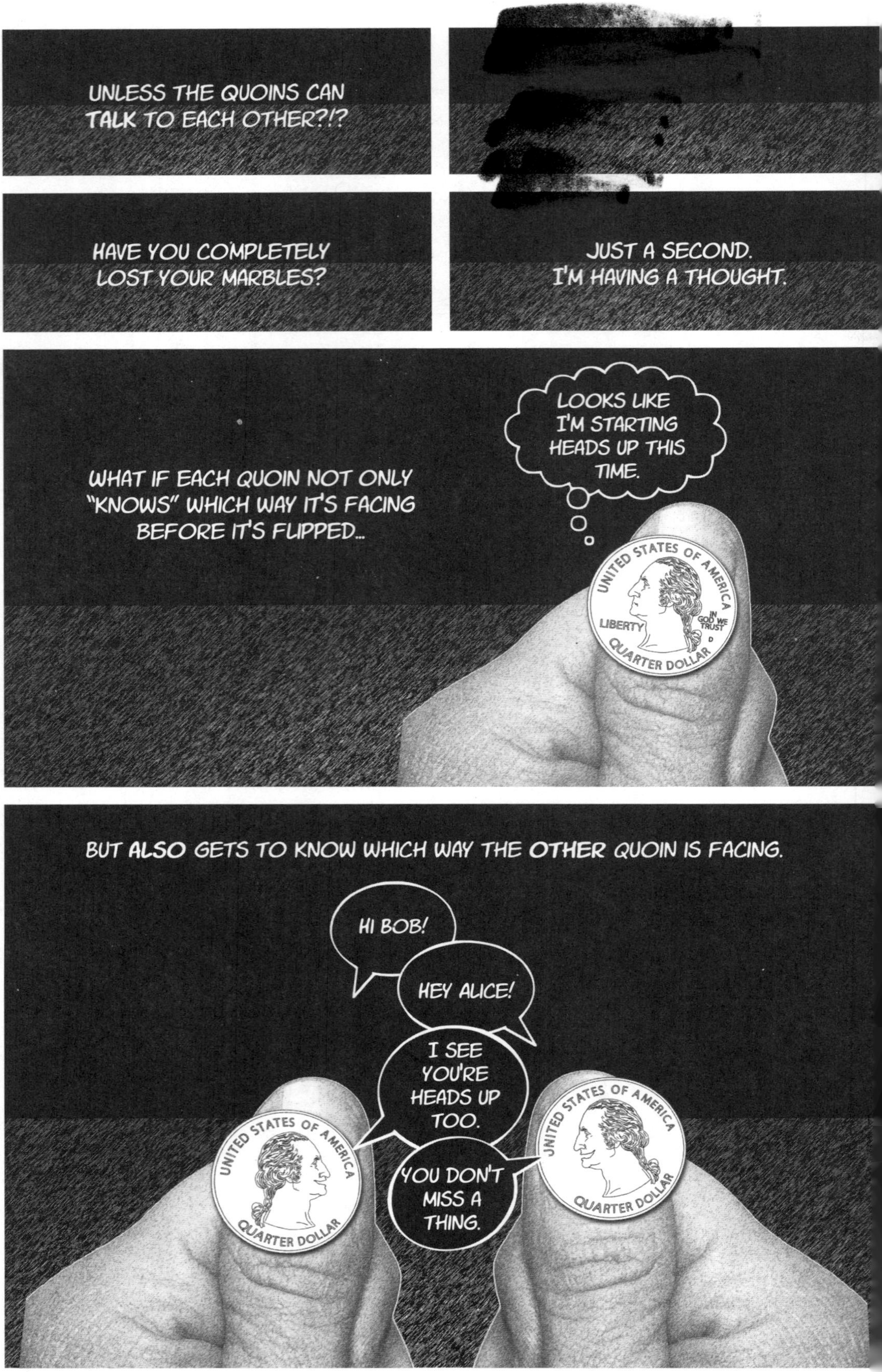
UNLESS THE QUOINS CAN TALK TO EACH OTHER?!?
HAVE YOU COMPLETELY LOST YOUR MARBLES?
JUST A SECOND. I'M HAVING A THOUGHT.
WHAT IF EACH QUOIN NOT ONLY "KNOWS" WHICH WAY IT'S FACING BEFORE IT'S FLIPPED...
LOOKS LIKE I'M STARTING HEADS UP THIS TIME.
UNITED STATES OF AMERICA
LIBERTY
IN GOD WE TRUST
D
QUARTER DOLLAR
BUT ALSO GETS TO KNOW WHICH WAY THE OTHER QUOIN IS FACING.
HI BOB!
HEY ALICE!
I SEE YOU'RE HEADS UP TOO.
YOU DON'T MISS A THING.
UNITED STATES OF AMERICA
QUARTER DOLLAR
UNITED STATES OF AMERICA
QUARTER DOLLAR

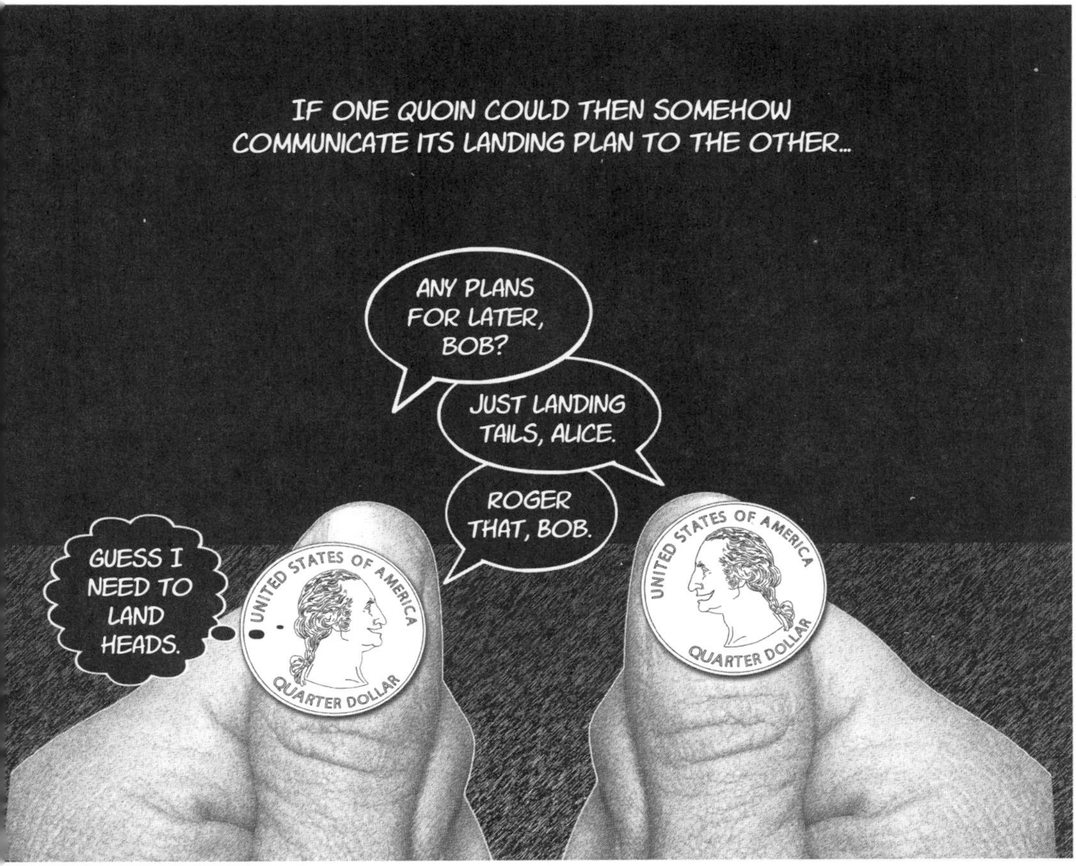
IF ONE QUOIN COULD THEN SOMEHOW COMMUNICATE ITS LANDING PLAN TO THE OTHER...
ANY PLANS FOR LATER, BOB?
JUST LANDING TAILS, ALICE.
ROGER THAT, BOB.
GUESS I NEED TO LAND HEADS.

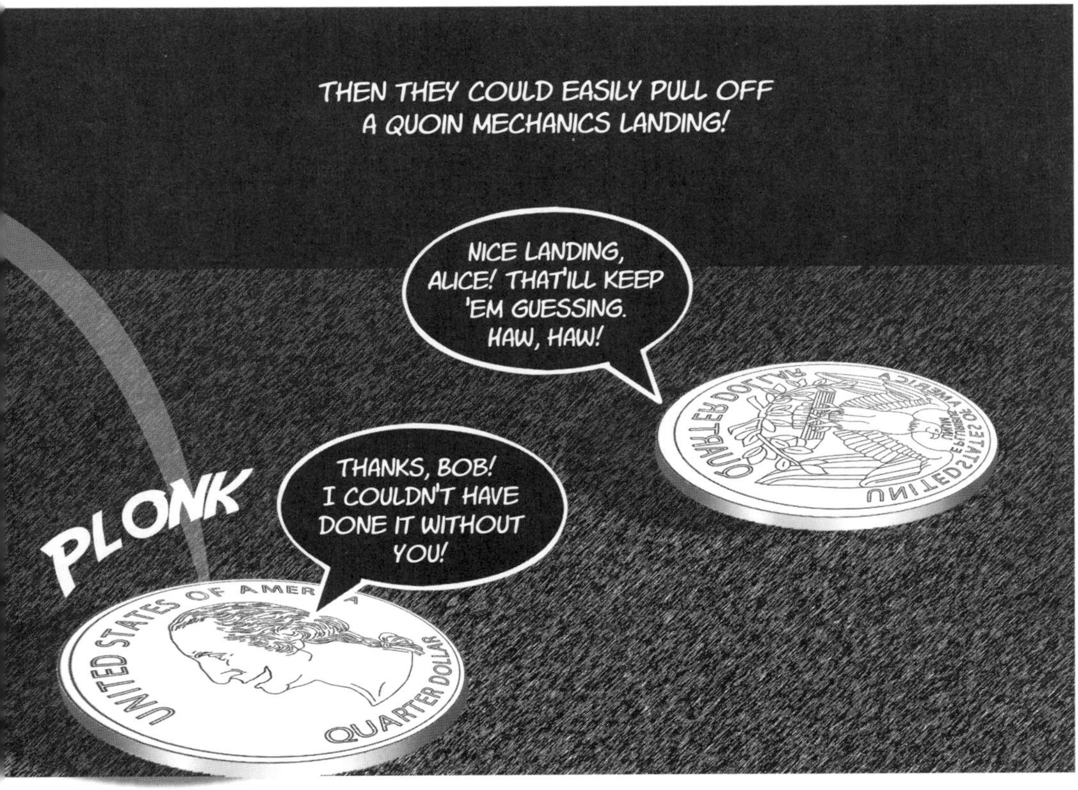
THEN THEY COULD EASILY PULL OFF A QUOIN MECHANICS LANDING!
NICE LANDING, ALICE! THAT'LL KEEP 'EM GUESSING. HAW, HAW!
PLONK
THANKS, BOB! I COULDN'T HAVE DONE IT WITHOUT YOU!

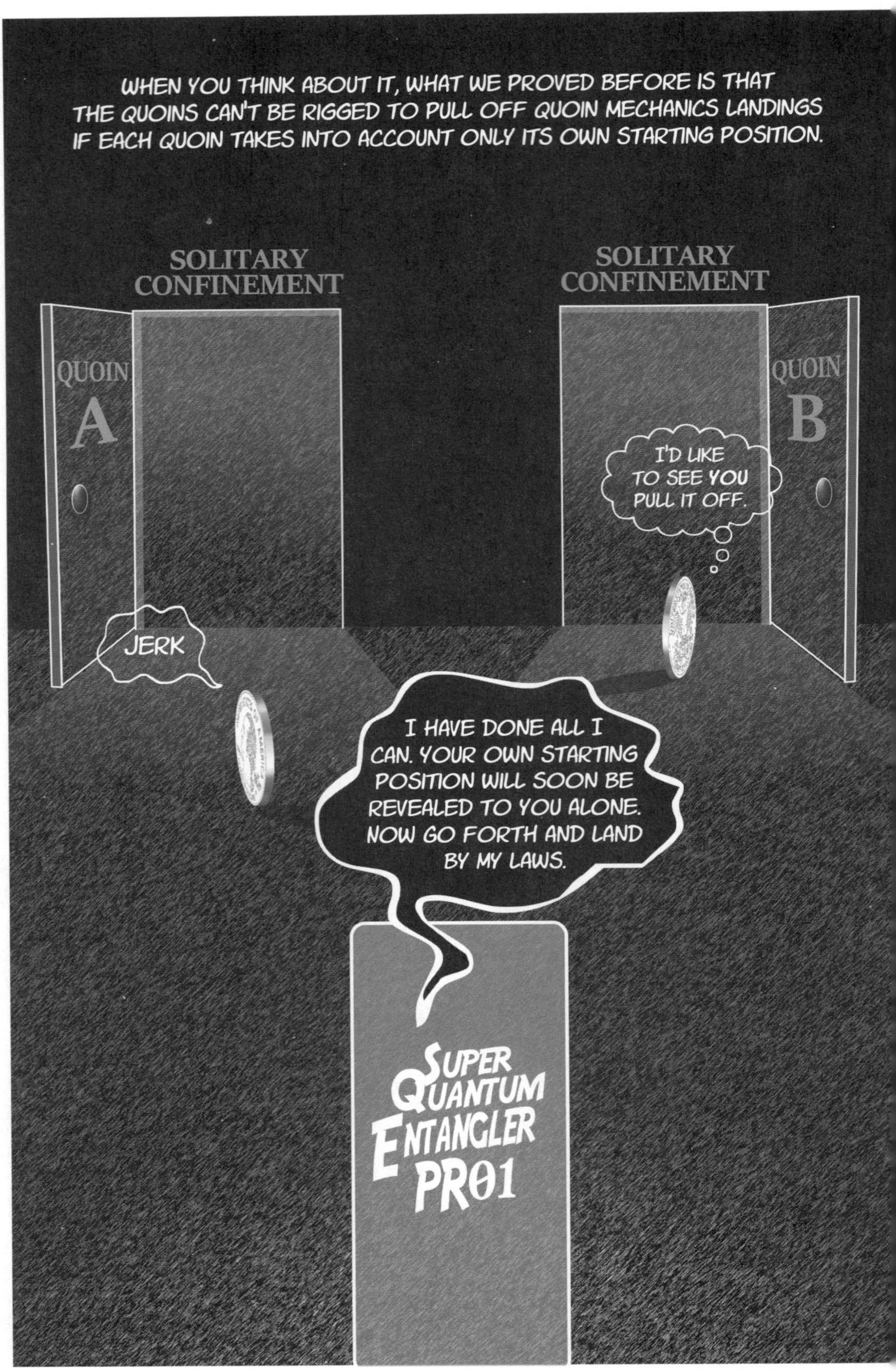
WHEN YOU THINK ABOUT IT, WHAT WE PROVED BEFORE IS THAT THE QUOINS CAN'T BE RIGGED TO PULL OFF QUOIN MECHANICS LANDINGS IF EACH QUOIN TAKES INTO ACCOUNT ONLY ITS OWN STARTING POSITION.
SOLITARY CONFINEMENT
QUOIN A
SOLITARY CONFINEMENT
QUOIN B
I'D LIKE TO SEE YOU PULL IT OFF.
JERK
I HAVE DONE ALL I CAN. YOUR OWN STARTING POSITION WILL SOON BE REVEALED TO YOU ALONE. NOW GO FORTH AND LAND BY MY LAWS.
SUPER QUANTUM ENTANGLER PRO1

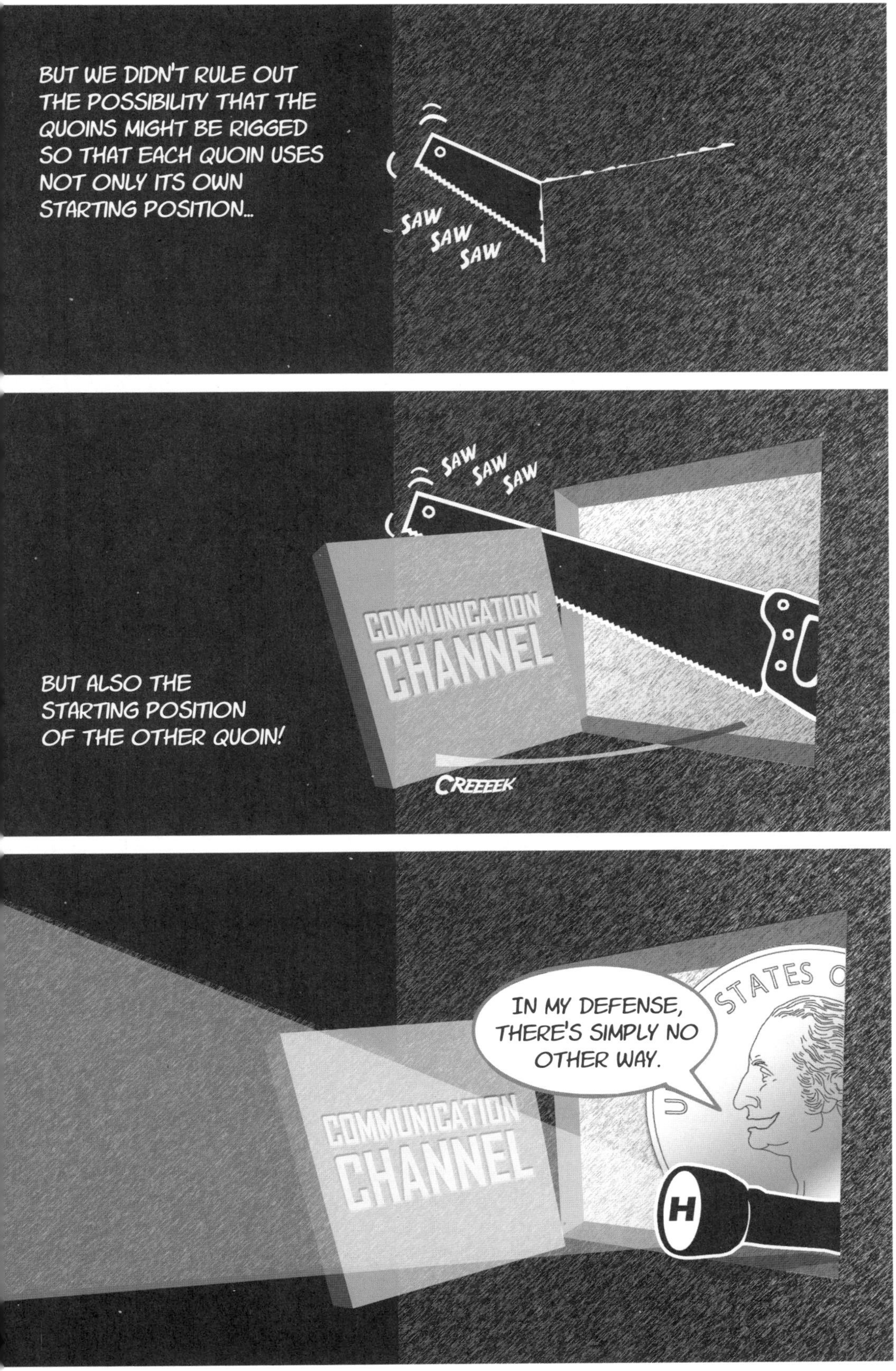
BUT WE DIDN'T RULE OUT THE POSSIBILITY THAT THE QUOINS MIGHT BE RIGGED SO THAT EACH QUOIN USES NOT ONLY ITS OWN STARTING POSITION...
SAW SAW SAW
SAW SAW SAW
COMMUNICATION CHANNEL
BUT ALSO THE STARTING POSITION OF THE OTHER QUOIN!
CREEEEK
COMMUNICATION CHANNEL
IN MY DEFENSE, THERE'S SIMPLY NO OTHER WAY.

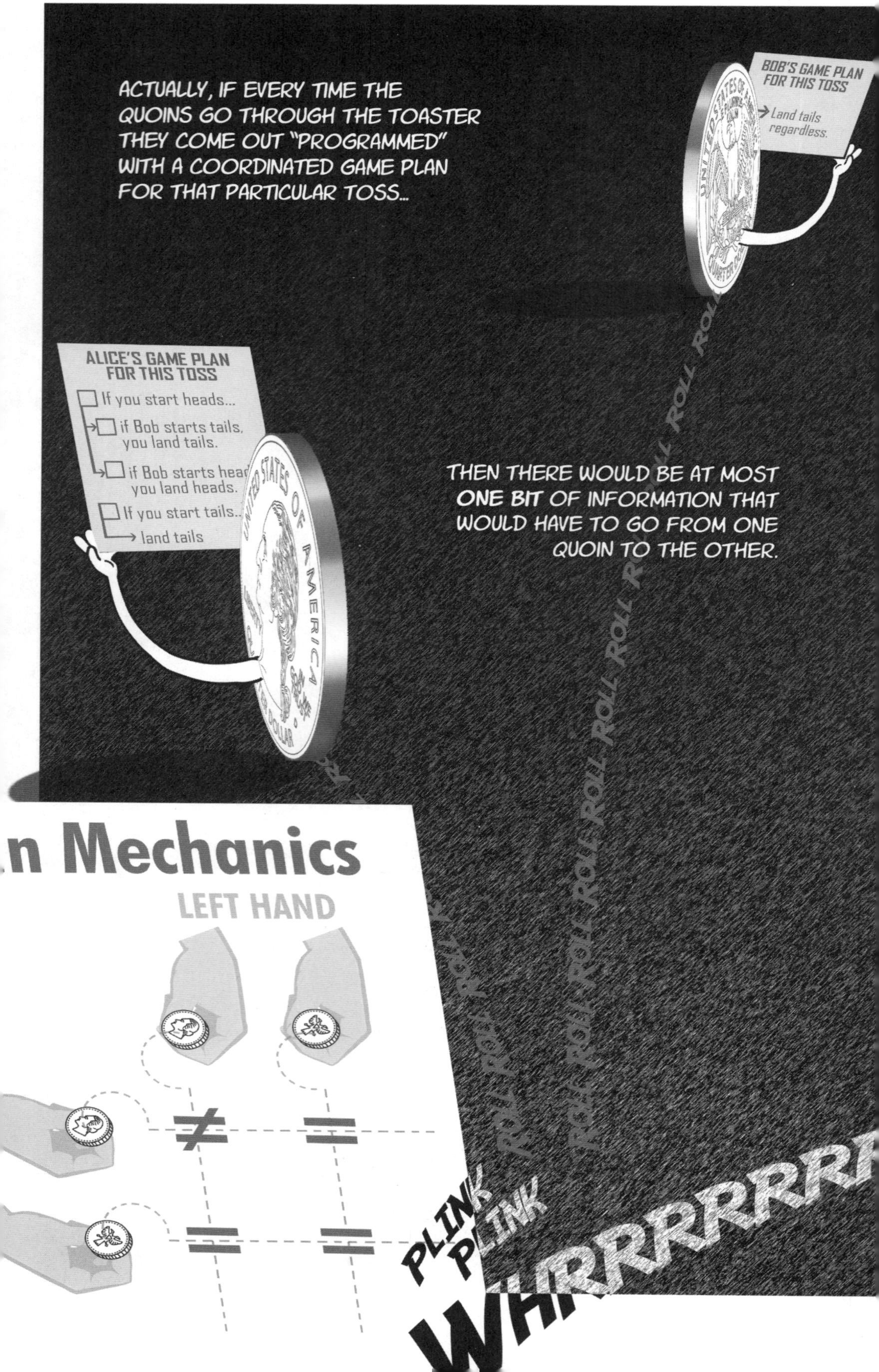

ACTUALLY, IF EVERY TIME THE
QUOINS GO THROUGH THE TOASTER
THEY COME OUT "PROGRAMMED"
WITH A COORDINATED GAME PLAN
FOR THAT PARTICULAR TOSS...

BOB'S GAME PLAN FOR THIS TOSS
→ Land tails regardless.

ALICE'S GAME PLAN FOR THIS TOSS

☐ If you start heads...
→ ☐ if Bob starts tails, you land tails.
→ ☐ if Bob starts heads, you land heads.
☐ If you start tails...
→ land tails

THEN THERE WOULD BE AT MOST
ONE BIT OF INFORMATION THAT
WOULD HAVE TO GO FROM ONE
QUOIN TO THE OTHER.

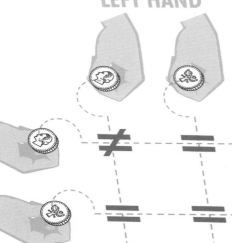

n Mechanics

LEFT HAND

≠   =

=   =

PLINK
PLINK

WHRRRRRR

ROLL ROLL ROLL ROLL ROLL ROLL ROLL

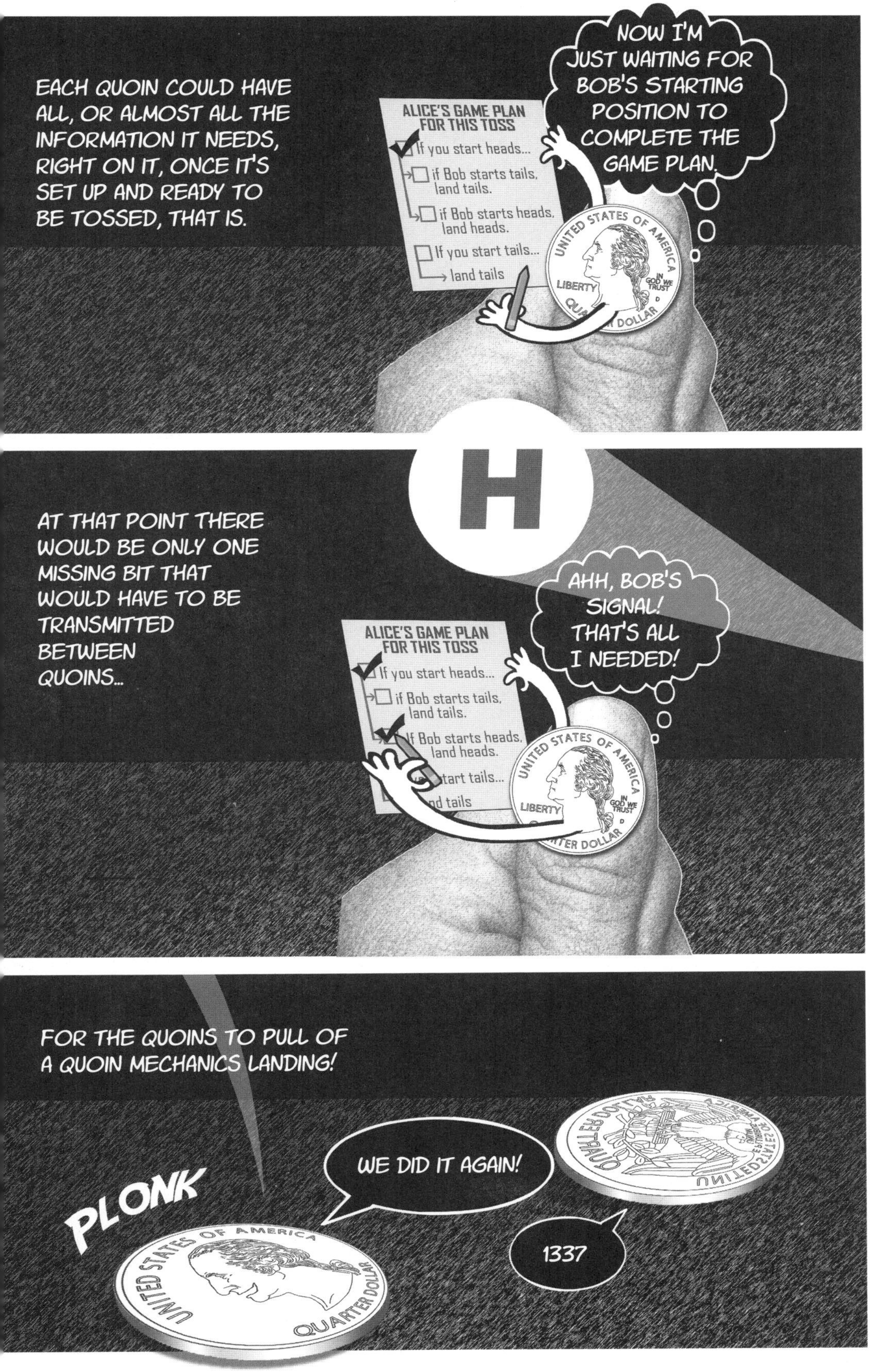
EACH QUOIN COULD HAVE ALL, OR ALMOST ALL THE INFORMATION IT NEEDS, RIGHT ON IT, ONCE IT'S SET UP AND READY TO BE TOSSED, THAT IS.
ALICE'S GAME PLAN FOR THIS TOSS
If you start heads...
if Bob starts tails, land tails.
if Bob starts heads, land heads.
If you start tails...
land tails
NOW I'M JUST WAITING FOR BOB'S STARTING POSITION TO COMPLETE THE GAME PLAN.
H
AT THAT POINT THERE WOULD BE ONLY ONE MISSING BIT THAT WOULD HAVE TO BE TRANSMITTED BETWEEN QUOINS...
ALICE'S GAME PLAN FOR THIS TOSS
If you start heads...
if Bob starts tails, land tails.
If Bob starts heads, land heads.
AHH, BOB'S SIGNAL! THAT'S ALL I NEEDED!
FOR THE QUOINS TO PULL OF A QUOIN MECHANICS LANDING!
PLONK
WE DID IT AGAIN!
1337

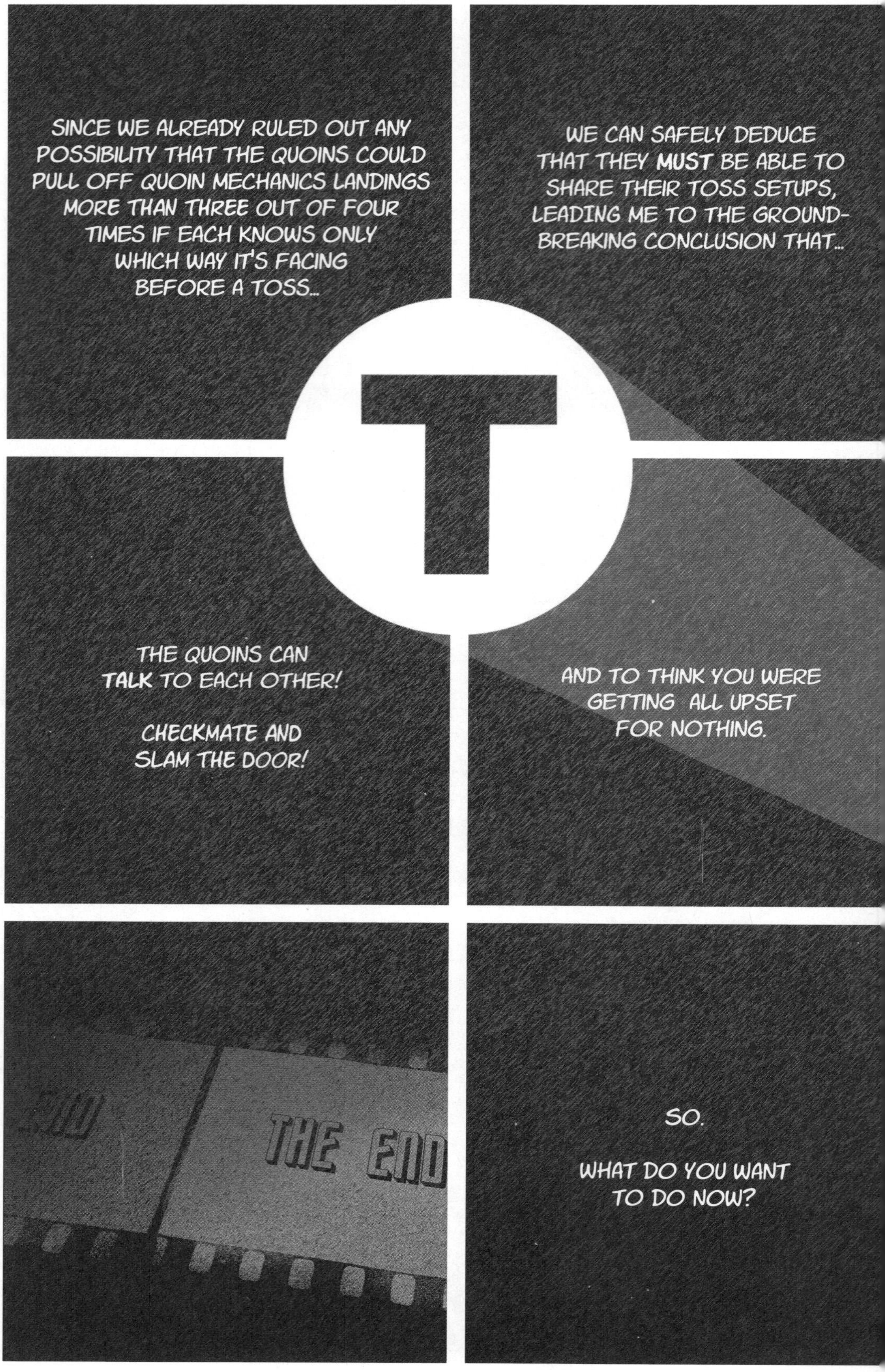
SINCE WE ALREADY RULED OUT ANY POSSIBILITY THAT THE QUOINS COULD PULL OFF QUOIN MECHANICS LANDINGS MORE THAN THREE OUT OF FOUR TIMES IF EACH KNOWS ONLY WHICH WAY IT'S FACING BEFORE A TOSS...
WE CAN SAFELY DEDUCE THAT THEY MUST BE ABLE TO SHARE THEIR TOSS SETUPS, LEADING ME TO THE GROUND-BREAKING CONCLUSION THAT...
T
THE QUOINS CAN TALK TO EACH OTHER!
CHECKMATE AND SLAM THE DOOR!
AND TO THINK YOU WERE GETTING ALL UPSET FOR NOTHING.
THE END
SO.
WHAT DO YOU WANT TO DO NOW?

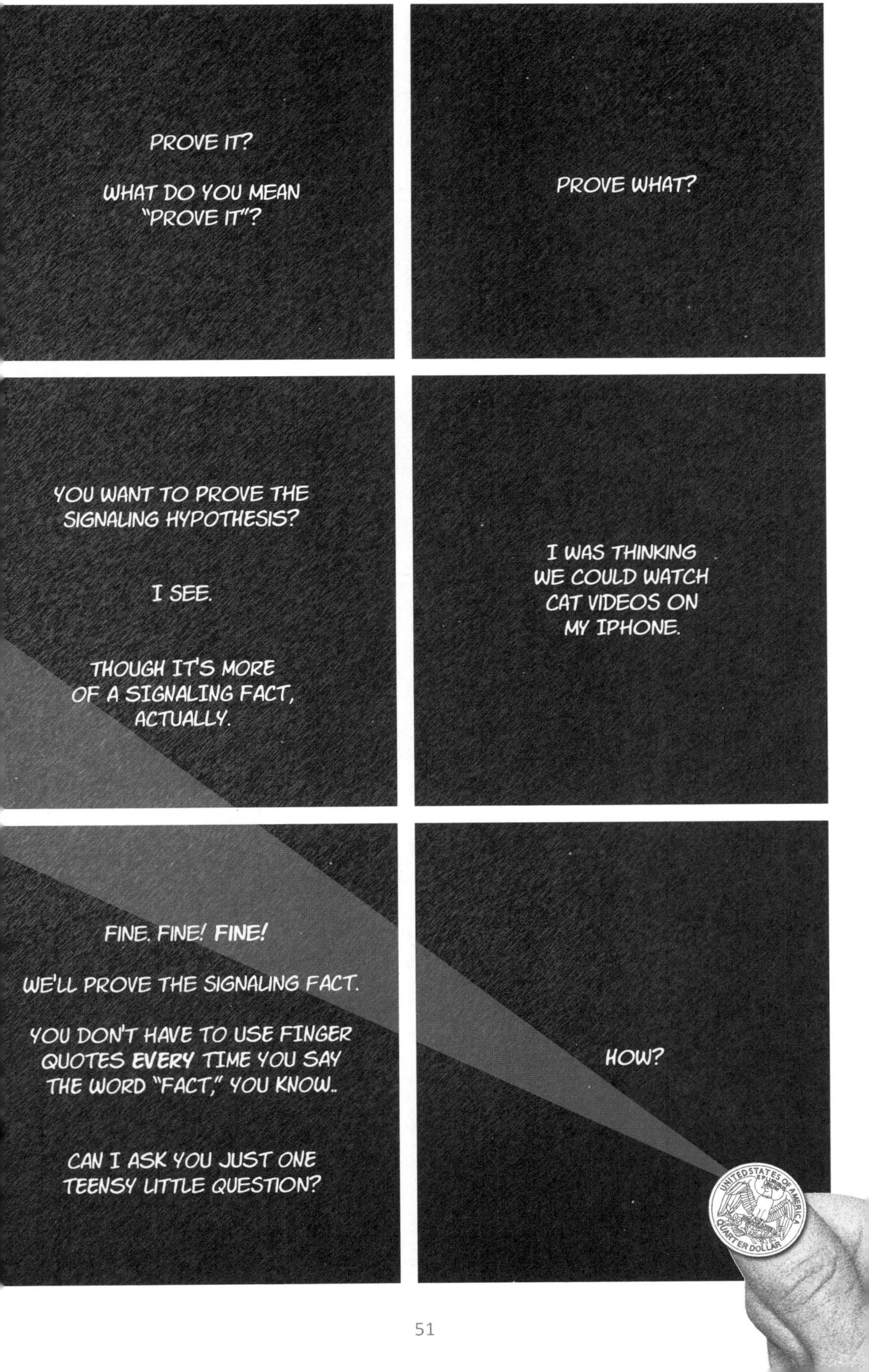
PROVE IT?
WHAT DO YOU MEAN "PROVE IT"?
PROVE WHAT?
YOU WANT TO PROVE THE SIGNALING HYPOTHESIS?
I SEE.
THOUGH IT'S MORE OF A SIGNALING FACT, ACTUALLY.
I WAS THINKING WE COULD WATCH CAT VIDEOS ON MY IPHONE.
FINE. FINE! FINE!
WE'LL PROVE THE SIGNALING FACT.
YOU DON'T HAVE TO USE FINGER QUOTES EVERY TIME YOU SAY THE WORD "FACT," YOU KNOW..
CAN I ASK YOU JUST ONE TEENSY LITTLE QUESTION?
HOW?
UNITED STATES OF AMERICA
QUARTER DOLLAR

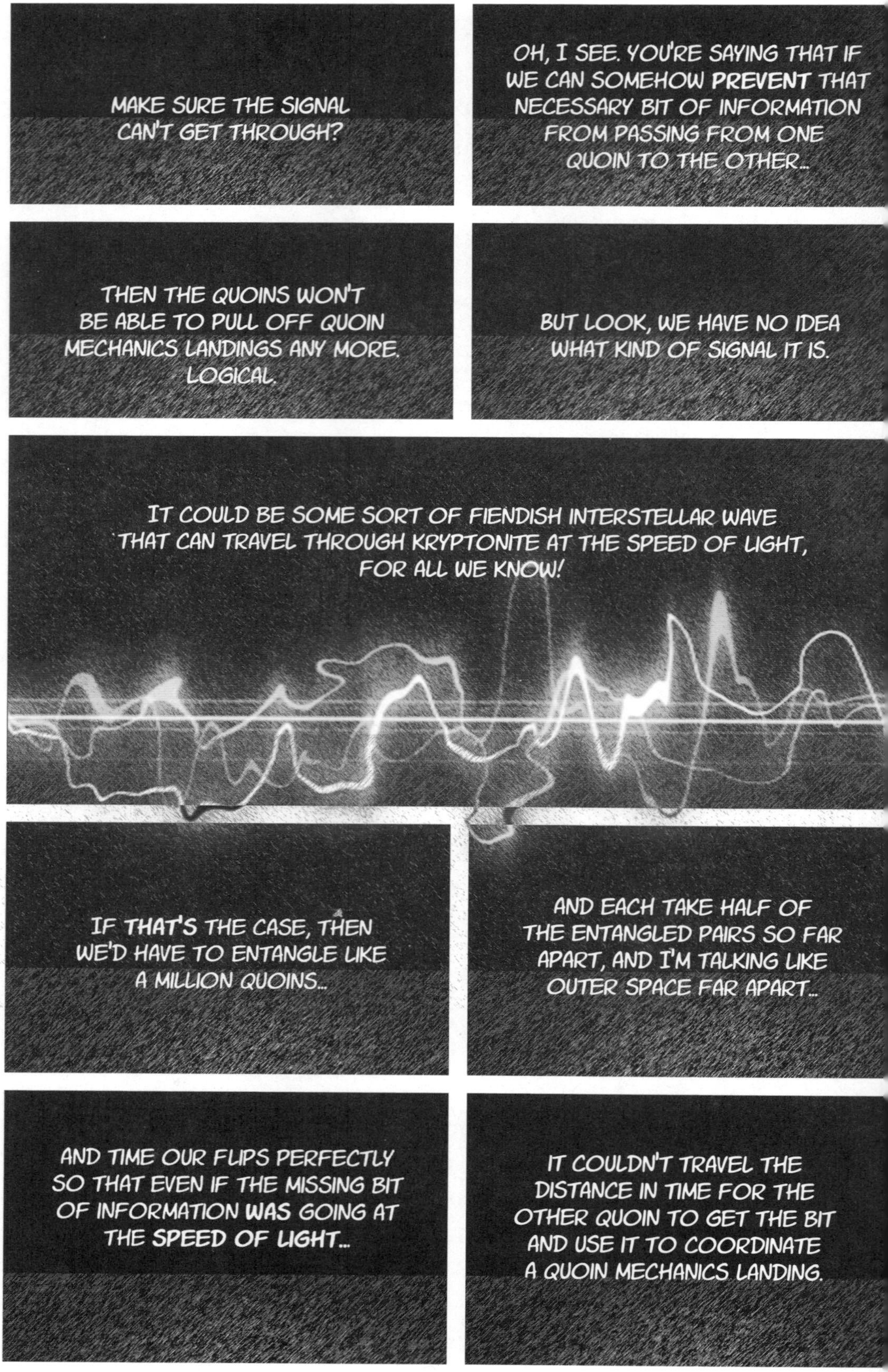
MAKE SURE THE SIGNAL CAN'T GET THROUGH?
OH, I SEE. YOU'RE SAYING THAT IF WE CAN SOMEHOW **PREVENT** THAT NECESSARY BIT OF INFORMATION FROM PASSING FROM ONE QUOIN TO THE OTHER...
THEN THE QUOINS WON'T BE ABLE TO PULL OFF QUOIN MECHANICS LANDINGS ANY MORE. LOGICAL.
BUT LOOK, WE HAVE NO IDEA WHAT KIND OF SIGNAL IT IS.
IT COULD BE SOME SORT OF FIENDISH INTERSTELLAR WAVE THAT CAN TRAVEL THROUGH KRYPTONITE AT THE SPEED OF LIGHT, FOR ALL WE KNOW!
IF **THAT'S** THE CASE, THEN WE'D HAVE TO ENTANGLE LIKE A MILLION QUOINS...
AND EACH TAKE HALF OF THE ENTANGLED PAIRS SO FAR APART, AND I'M TALKING LIKE OUTER SPACE FAR APART...
AND TIME OUR FLIPS PERFECTLY SO THAT EVEN IF THE MISSING BIT OF INFORMATION **WAS** GOING AT THE **SPEED OF LIGHT**...
IT COULDN'T TRAVEL THE DISTANCE IN TIME FOR THE OTHER QUOIN TO GET THE BIT AND USE IT TO COORDINATE A QUOIN MECHANICS LANDING.

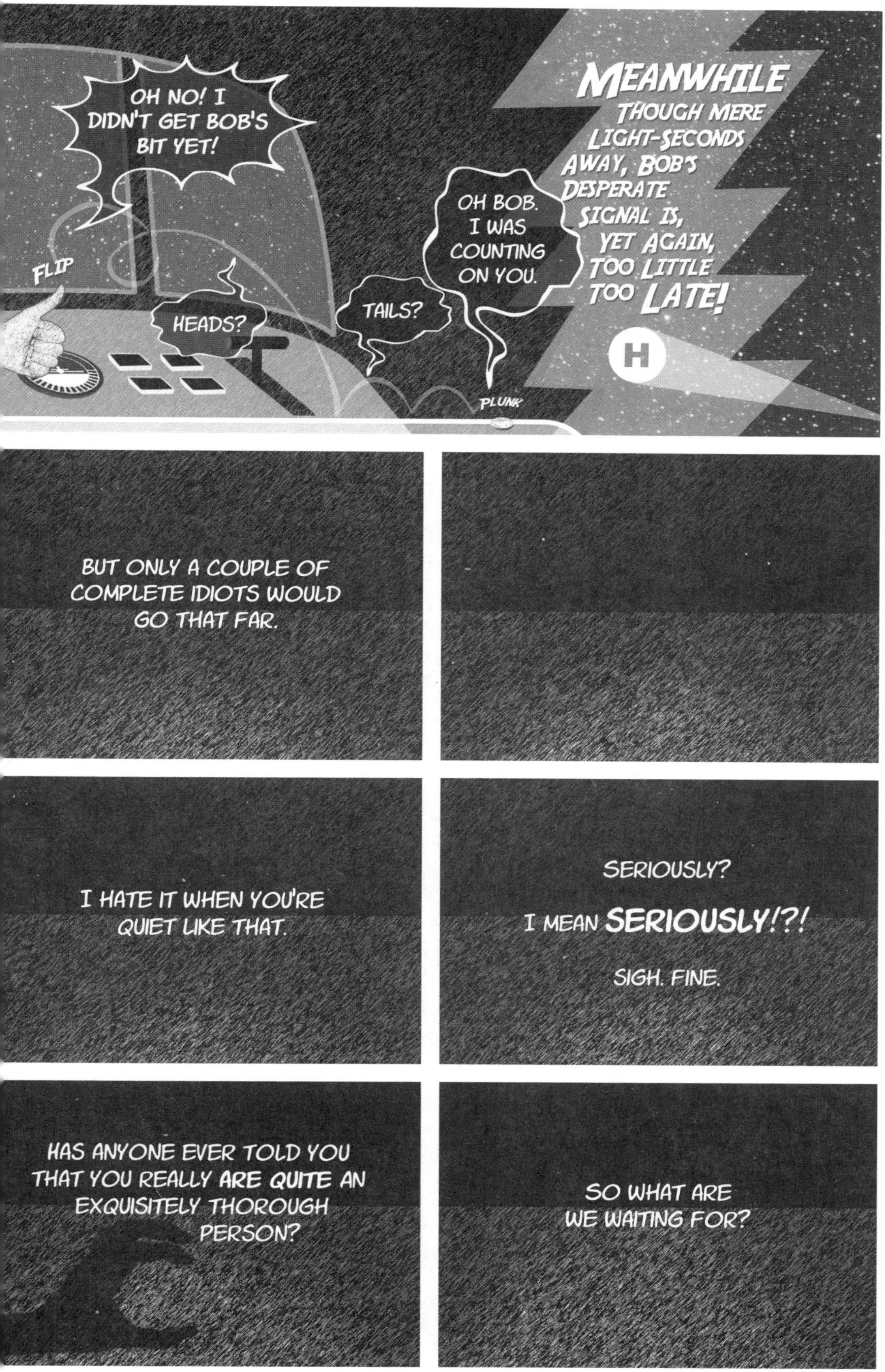
OH NO! I DIDN'T GET BOB'S BIT YET!
FLIP
HEADS?
TAILS?
OH BOB. I WAS COUNTING ON YOU.
PLUNK
MEANWHILE THOUGH MERE LIGHT-SECONDS AWAY, BOB'S DESPERATE SIGNAL IS, YET AGAIN, TOO LITTLE TOO LATE!
H
BUT ONLY A COUPLE OF COMPLETE IDIOTS WOULD GO THAT FAR.
I HATE IT WHEN YOU'RE QUIET LIKE THAT.
SERIOUSLY?
I MEAN SERIOUSLY!?!
SIGH. FINE.
HAS ANYONE EVER TOLD YOU THAT YOU REALLY ARE QUITE AN EXQUISITELY THOROUGH PERSON?
SO WHAT ARE WE WAITING FOR?

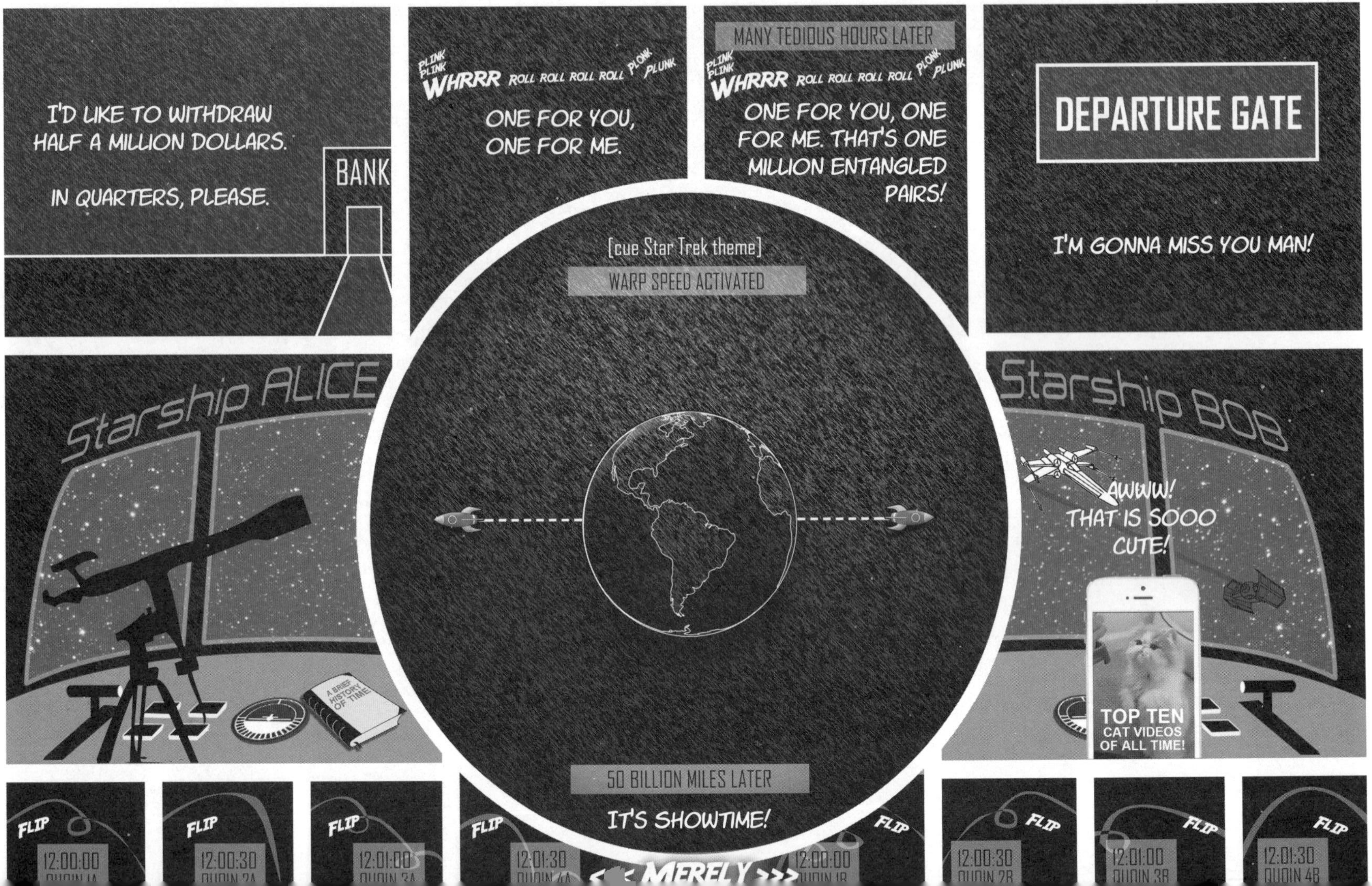

I'D LIKE TO WITHDRAW HALF A MILLION DOLLARS.
IN QUARTERS, PLEASE.
BANK
PLINK PLINK WHRRR ROLL ROLL ROLL ROLL PLONK PLUNK
ONE FOR YOU, ONE FOR ME.
MANY TEDIOUS HOURS LATER
PLINK PLINK WHRRR ROLL ROLL ROLL ROLL PLONK PLUNK
ONE FOR YOU, ONE FOR ME. THAT'S ONE MILLION ENTANGLED PAIRS!
DEPARTURE GATE
I'M GONNA MISS YOU MAN!
Starship ALICE
A BRIEF HISTORY OF TIME
[cue Star Trek theme]
WARP SPEED ACTIVATED
50 BILLION MILES LATER
IT'S SHOWTIME!
Starship BOB
AWWW! THAT IS SOOO CUTE!
TOP TEN CAT VIDEOS OF ALL TIME!
FLIP
12:00:00
FLIP
12:00:30
FLIP
12:01:00
FLIP
12:01:30
FLIP
12:00:00
FLIP
12:00:30
FLIP
12:01:00
FLIP
12:01:30
MERELY

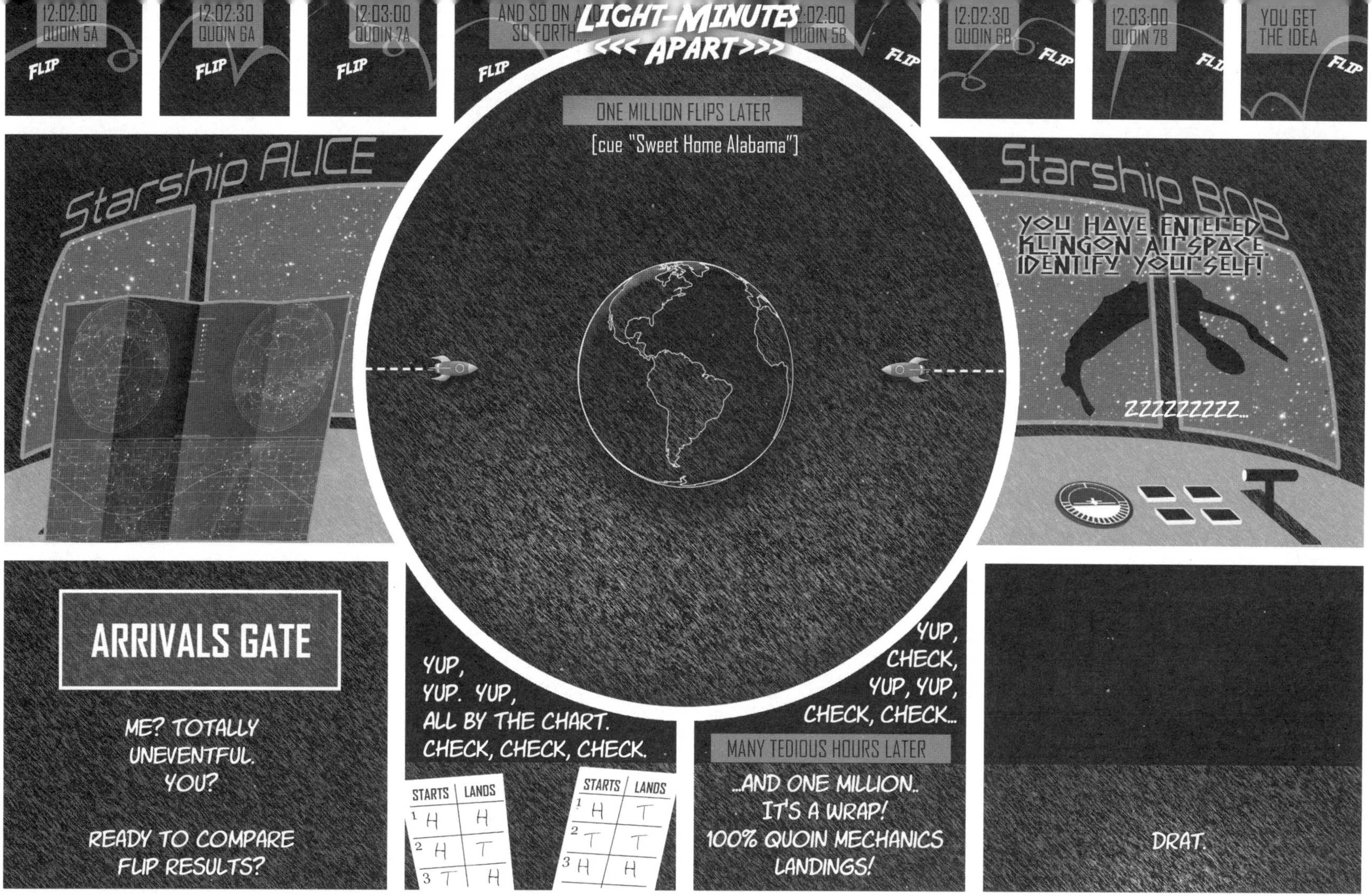
12:02:00
QUOIN 5A
FLIP
12:02:30
QUOIN 6A
FLIP
12:03:00
QUOIN 7A
FLIP
AND SO ON AND
SO FORTH...
FLIP
LIGHT-MINUTES
<<< APART >>>
12:02:00
QUOIN 5B
FLIP
12:02:30
QUOIN 6B
FLIP
12:03:00
QUOIN 7B
YOU GET
THE IDEA
FLIP
ONE MILLION FLIPS LATER
[cue "Sweet Home Alabama"]
Starship ALICE
Starship BOB
YOU HAVE ENTERED KLINGON AIRSPACE. IDENTIFY YOURSELF!
ZZZZZZZZ...
ARRIVALS GATE
ME? TOTALLY UNEVENTFUL. YOU?
READY TO COMPARE FLIP RESULTS?
YUP, YUP. YUP, ALL BY THE CHART. CHECK, CHECK, CHECK.
STARTS | LANDS
1 H H
2 H T
3 T H
STARTS | LANDS
1 H T
2 T T
3 H H
YUP, CHECK, YUP, YUP, CHECK, CHECK...
MANY TEDIOUS HOURS LATER
...AND ONE MILLION.. IT'S A WRAP! 100% QUOIN MECHANICS LANDINGS!
DRAT.

What the ——

Do not keep saying to yourself, if you can possibly avoid it, "But how can it be like that?" because you will get "down the drain," into a blind alley from which nobody has yet escaped. Nobody knows how it can be like that.

—Richard Feynman

Those who are not shocked
when they first come across quantum theory
cannot possibly have understood it.

—Niels Bohr

It seems hard to sneak a look at God's cards.
But that he plays dice and uses "telepathic" methods
(as the present quantum theory requires of him)
is something that I cannot believe for a single moment.

—**Albert Einstein**

One should no more rack one's brain about the
problem of whether something one cannot know
anything about exists at all, than about the ancient
question of how many angels are able to sit
on the point of a needle.

—**Wolfgang Pauli**

Can nature possibly be so absurd as it seemed
to us in these atomic experiments?

—**Werner Heisenberg**

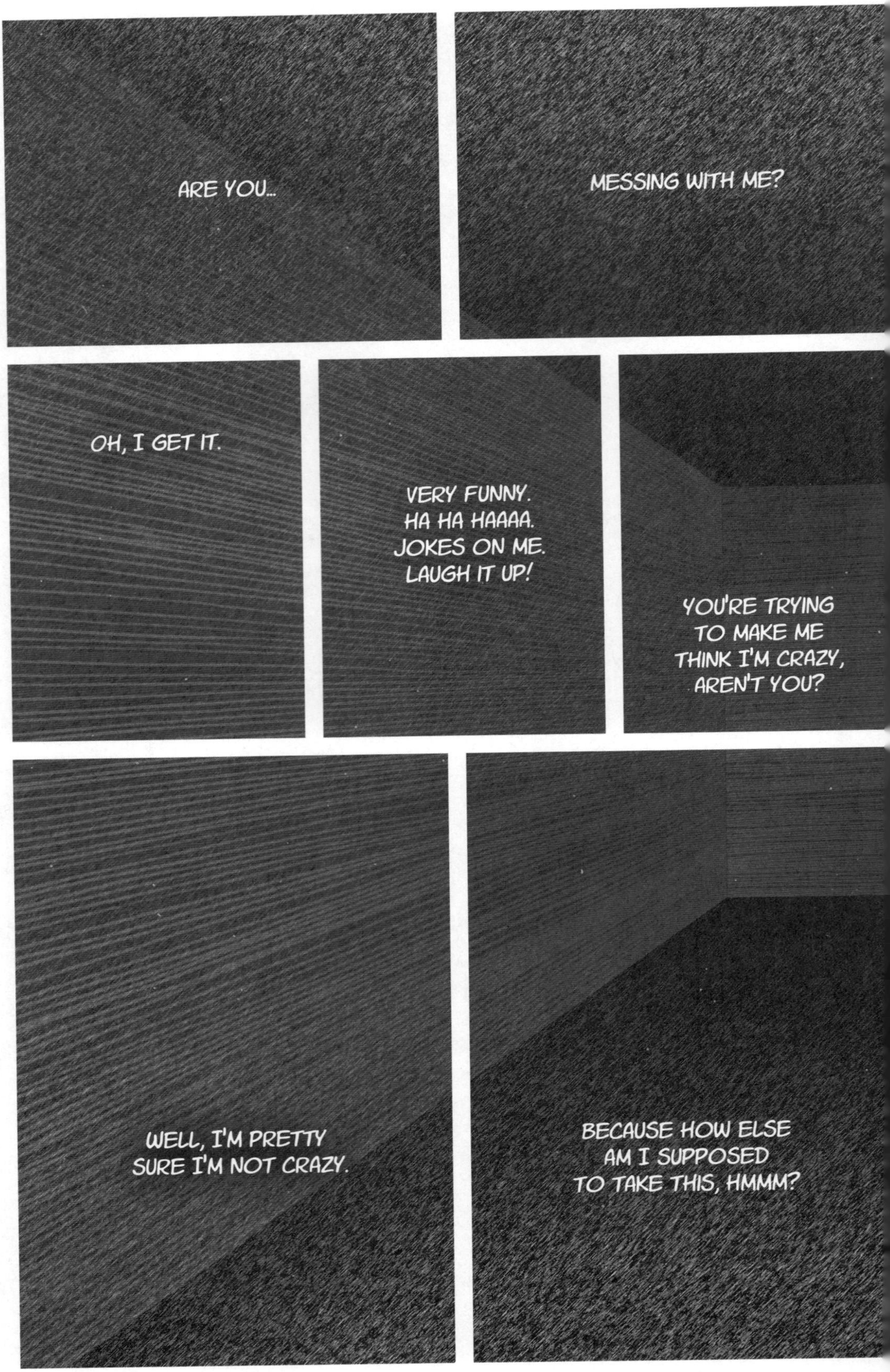
ARE YOU...
MESSING WITH ME?
OH, I GET IT.
VERY FUNNY. HA HA HAAAA. JOKES ON ME. LAUGH IT UP!
YOU'RE TRYING TO MAKE ME THINK I'M CRAZY, AREN'T YOU?
WELL, I'M PRETTY SURE I'M NOT CRAZY.
BECAUSE HOW ELSE AM I SUPPOSED TO TAKE THIS, HMMM?

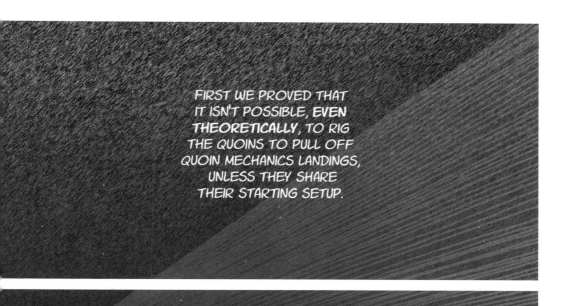

FIRST WE PROVED THAT
IT ISN'T POSSIBLE, **EVEN
THEORETICALLY**, TO RIG
THE QUOINS TO PULL OFF
QUOIN MECHANICS LANDINGS,
UNLESS THEY SHARE
THEIR STARTING SETUP.

AND THEN WE PROVED
THAT THE QUOINS **CAN AND DO**
PULL OFF QUOIN MECHANICS
LANDINGS EVEN WHEN IT'S
PHYSICALLY IMPOSSIBLE
FOR THEM TO SHARE
THEIR STARTING SETUP.

AND NOW I GUESS YOU'RE
EXPECTING ME TO BELIEVE
THAT THE QUOINS ARE JUST
SORT OF RANDOMLY HAPPENING
TO FALL IN A QUOIN MECHANICS-ISH
WAY, EVERY SINGLE TIME,
FOR **ABSOLUTELY NO
REASON AT ALL!**

WHICH WOULD MAKE ME
THE WORLD'S BIGGEST SUCKER!
AND I GUESS YOU THINK
THAT'S HILARIOUS.

HA, HA, HA.

PUNK'D.
BY MY OWN TOASTER!

SO HOW'D YA DO IT?
FESS UP WISE GUY.

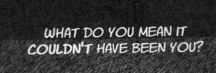

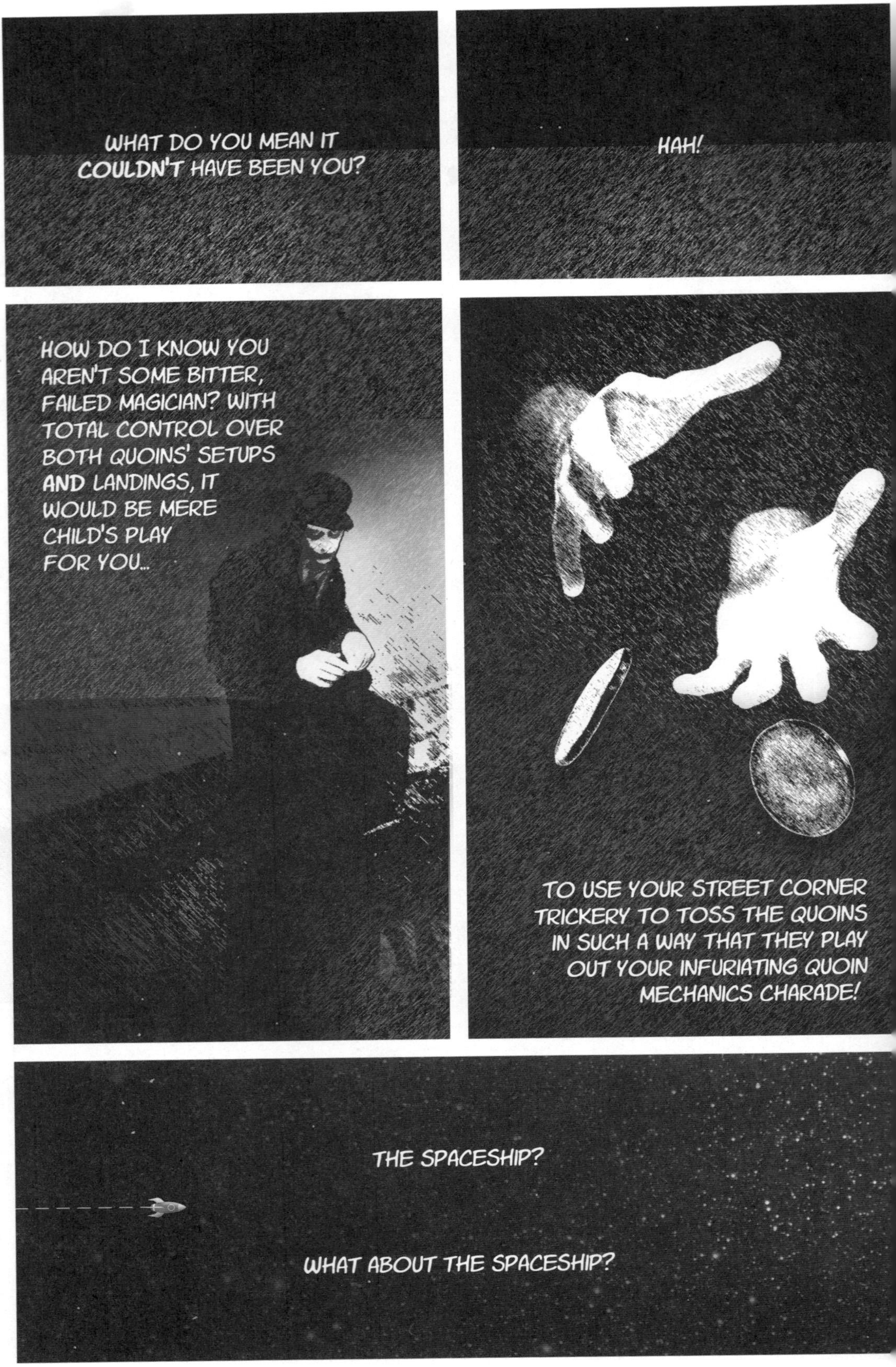

WHAT DO YOU MEAN IT **COULDN'T** HAVE BEEN YOU?

HAH!

HOW DO I KNOW YOU AREN'T SOME BITTER, FAILED MAGICIAN? WITH TOTAL CONTROL OVER BOTH QUOINS' SETUPS AND LANDINGS, IT WOULD BE MERE CHILD'S PLAY FOR YOU...

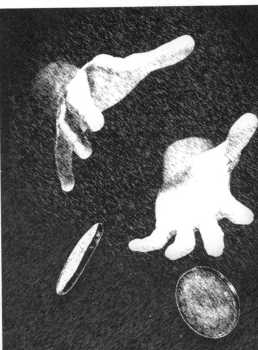

TO USE YOUR STREET CORNER TRICKERY TO TOSS THE QUOINS IN SUCH A WAY THAT THEY PLAY OUT YOUR INFURIATING QUOIN MECHANICS CHARADE!

THE SPACESHIP?

WHAT ABOUT THE SPACESHIP?

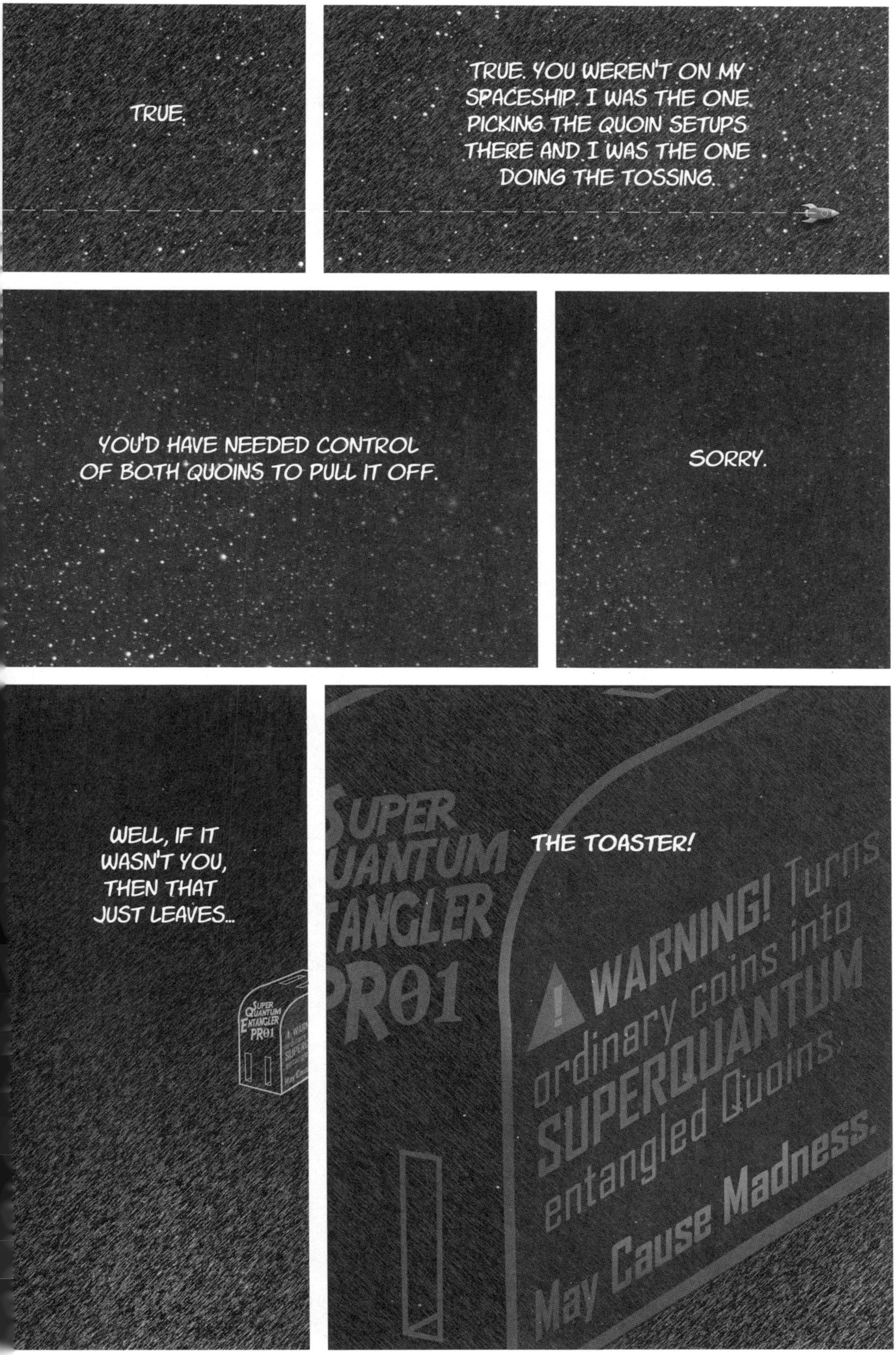
TRUE.
TRUE. YOU WEREN'T ON MY SPACESHIP. I WAS THE ONE PICKING THE QUOIN SETUPS THERE AND I WAS THE ONE DOING THE TOSSING.
YOU'D HAVE NEEDED CONTROL OF BOTH QUOINS TO PULL IT OFF.
SORRY.
WELL, IF IT WASN'T YOU, THEN THAT JUST LEAVES...
SUPER QUANTUM ENTANGLER PRO1
THE TOASTER!
SUPER QUANTUM ENTANGLER PRO1
WARNING! Turns ordinary coins into SUPERQUANTUM entangled Quoins.
May Cause Madness.

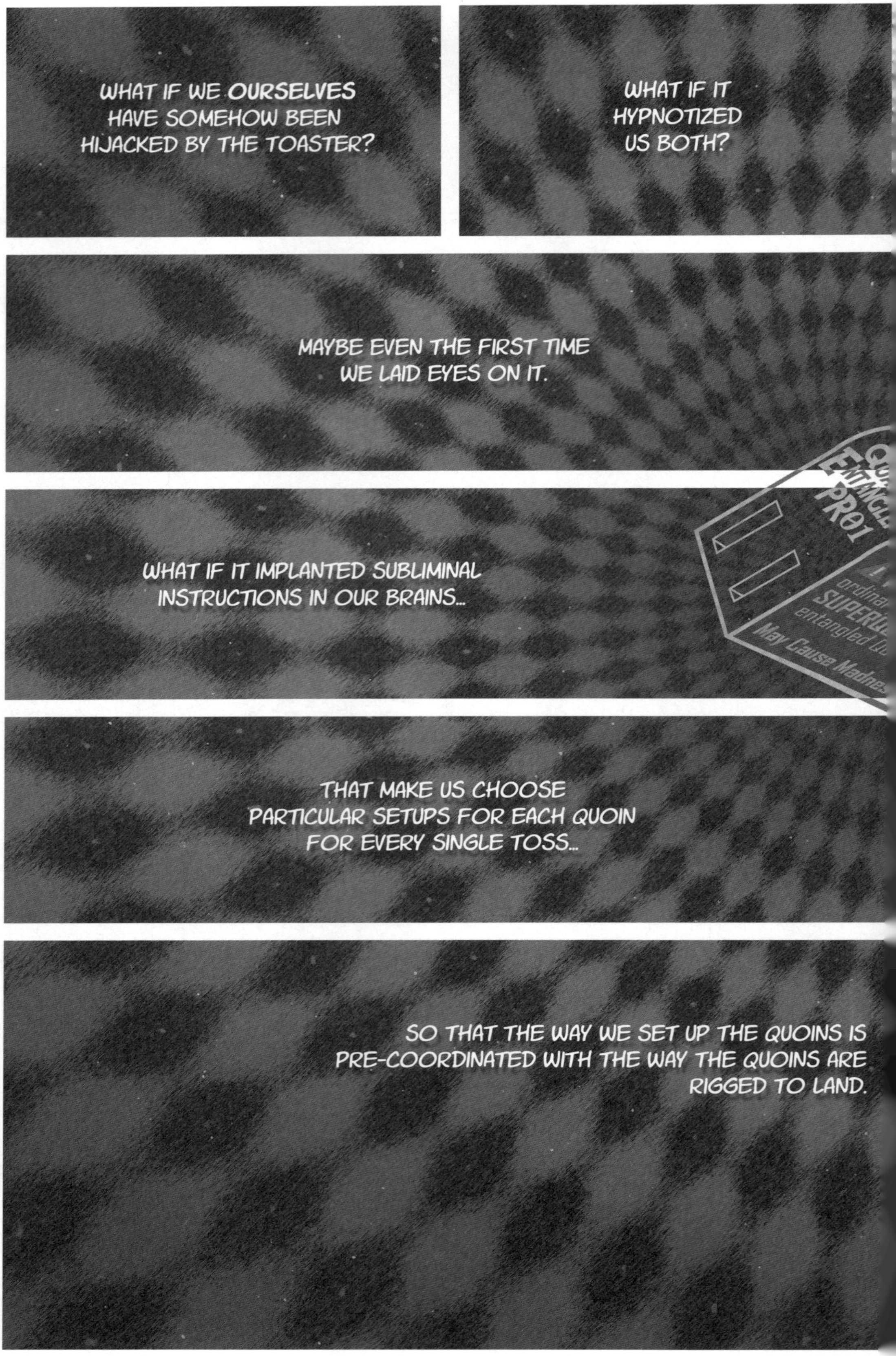
WHAT IF WE OURSELVES HAVE SOMEHOW BEEN HIJACKED BY THE TOASTER?
WHAT IF IT HYPNOTIZED US BOTH?
MAYBE EVEN THE FIRST TIME WE LAID EYES ON IT.
WHAT IF IT IMPLANTED SUBLIMINAL INSTRUCTIONS IN OUR BRAINS...
THAT MAKE US CHOOSE PARTICULAR SETUPS FOR EACH QUOIN FOR EVERY SINGLE TOSS...
SO THAT THE WAY WE SET UP THE QUOINS IS PRE-COORDINATED WITH THE WAY THE QUOINS ARE RIGGED TO LAND.

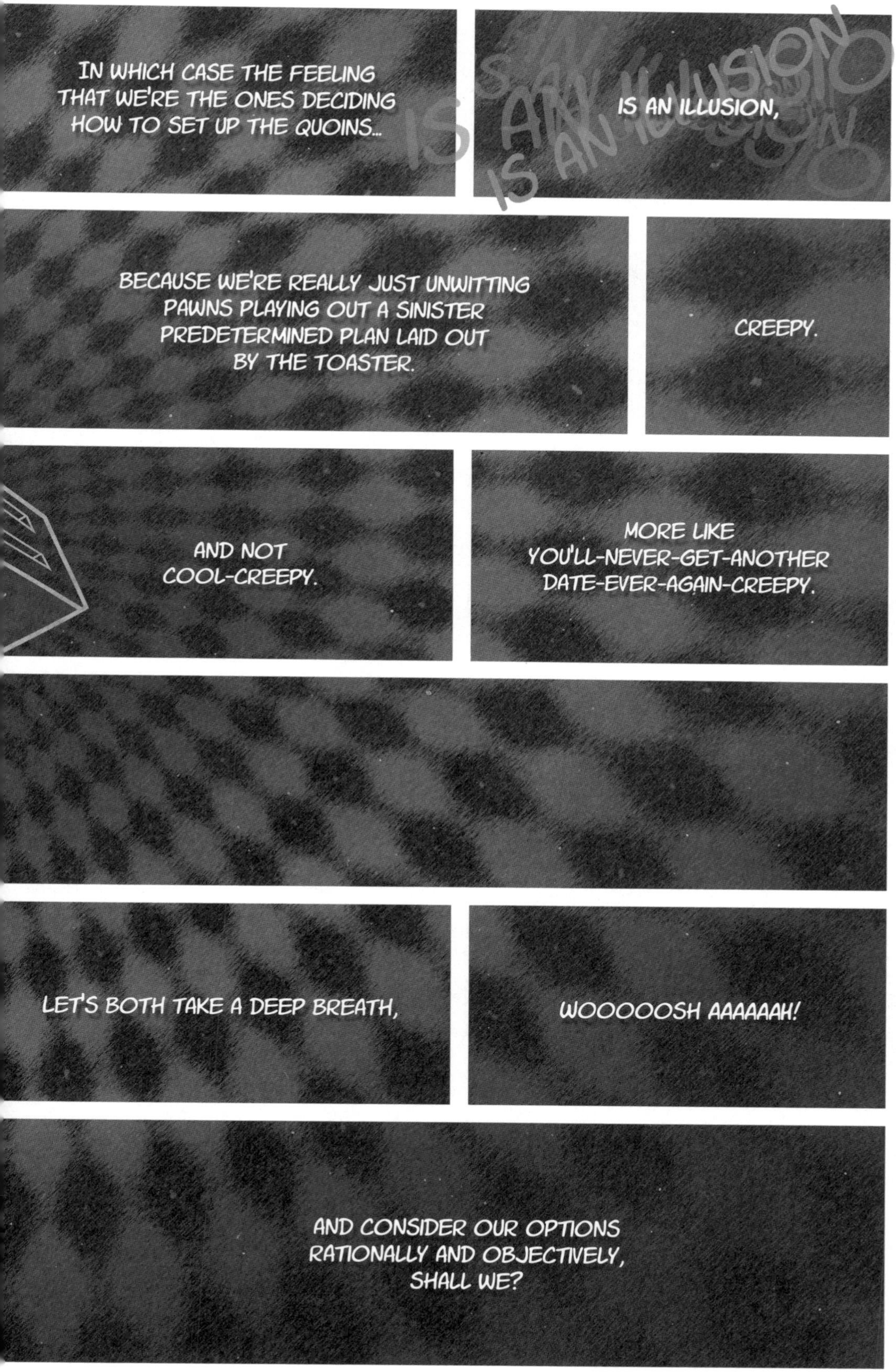

IN WHICH CASE THE FEELING THAT WE'RE THE ONES DECIDING HOW TO SET UP THE QUOINS...

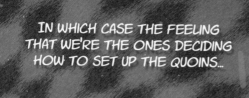

IS AN ILLUSION,

BECAUSE WE'RE REALLY JUST UNWITTING PAWNS PLAYING OUT A SINISTER PREDETERMINED PLAN LAID OUT BY THE TOASTER.

CREEPY.

AND NOT COOL-CREEPY.

MORE LIKE YOU'LL-NEVER-GET-ANOTHER DATE-EVER-AGAIN-CREEPY.

LET'S BOTH TAKE A DEEP BREATH,

WOOOOOSH AAAAAAH!

AND CONSIDER OUR OPTIONS RATIONALLY AND OBJECTIVELY, SHALL WE?

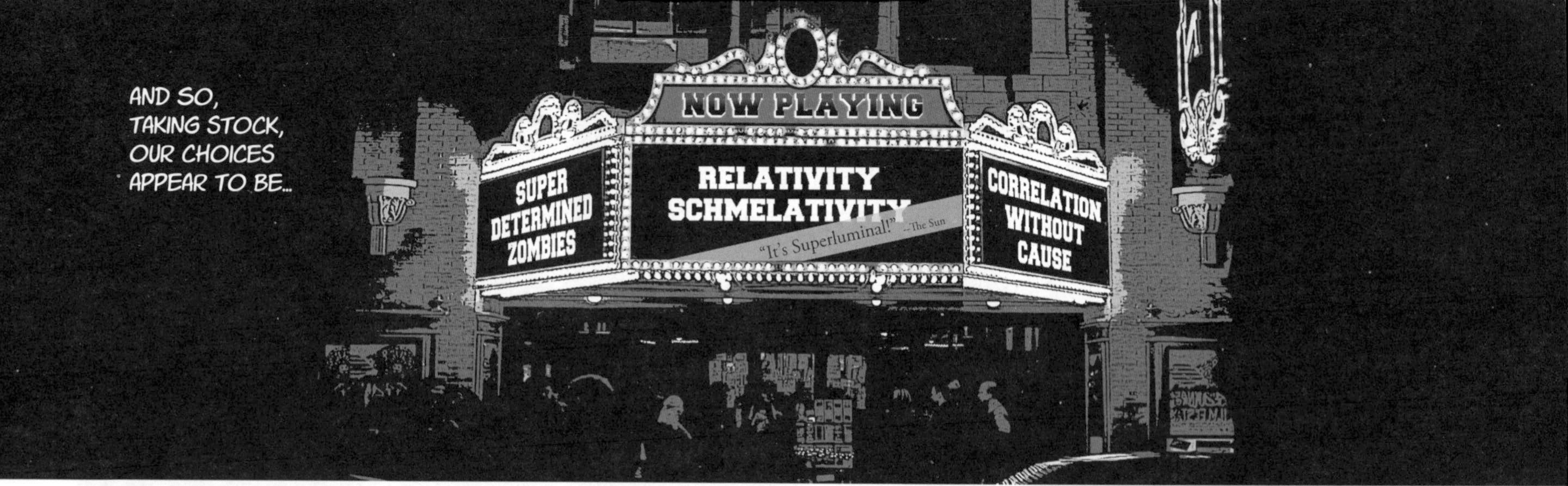
AND SO, TAKING STOCK, OUR CHOICES APPEAR TO BE...
NOW PLAYING
SUPER DETERMINED ZOMBIES
RELATIVITY SCHMELATIVITY
"It's Superluminal!" –The Sun
CORRELATION WITHOUT CAUSE

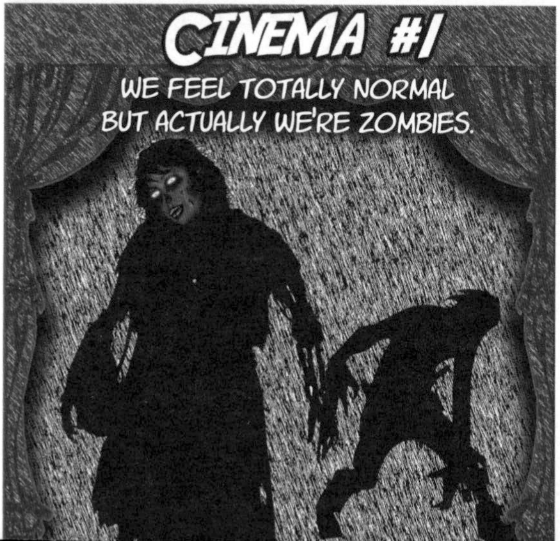
CINEMA #1
WE FEEL TOTALLY NORMAL BUT ACTUALLY WE'RE ZOMBIES.

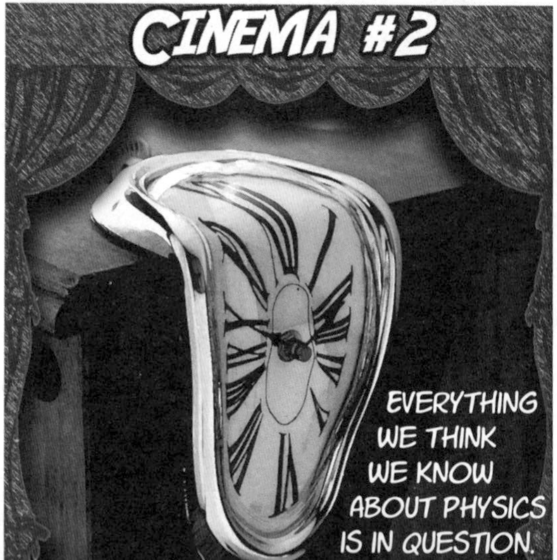
CINEMA #2
EVERYTHING WE THINK WE KNOW ABOUT PHYSICS IS IN QUESTION.

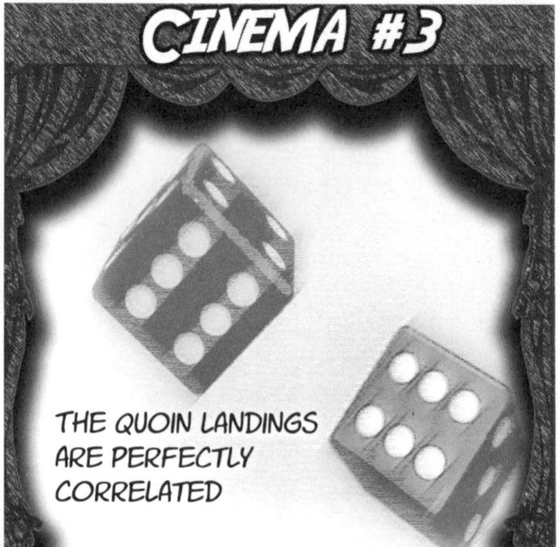
CINEMA #3
THE QUOIN LANDINGS ARE PERFECTLY CORRELATED

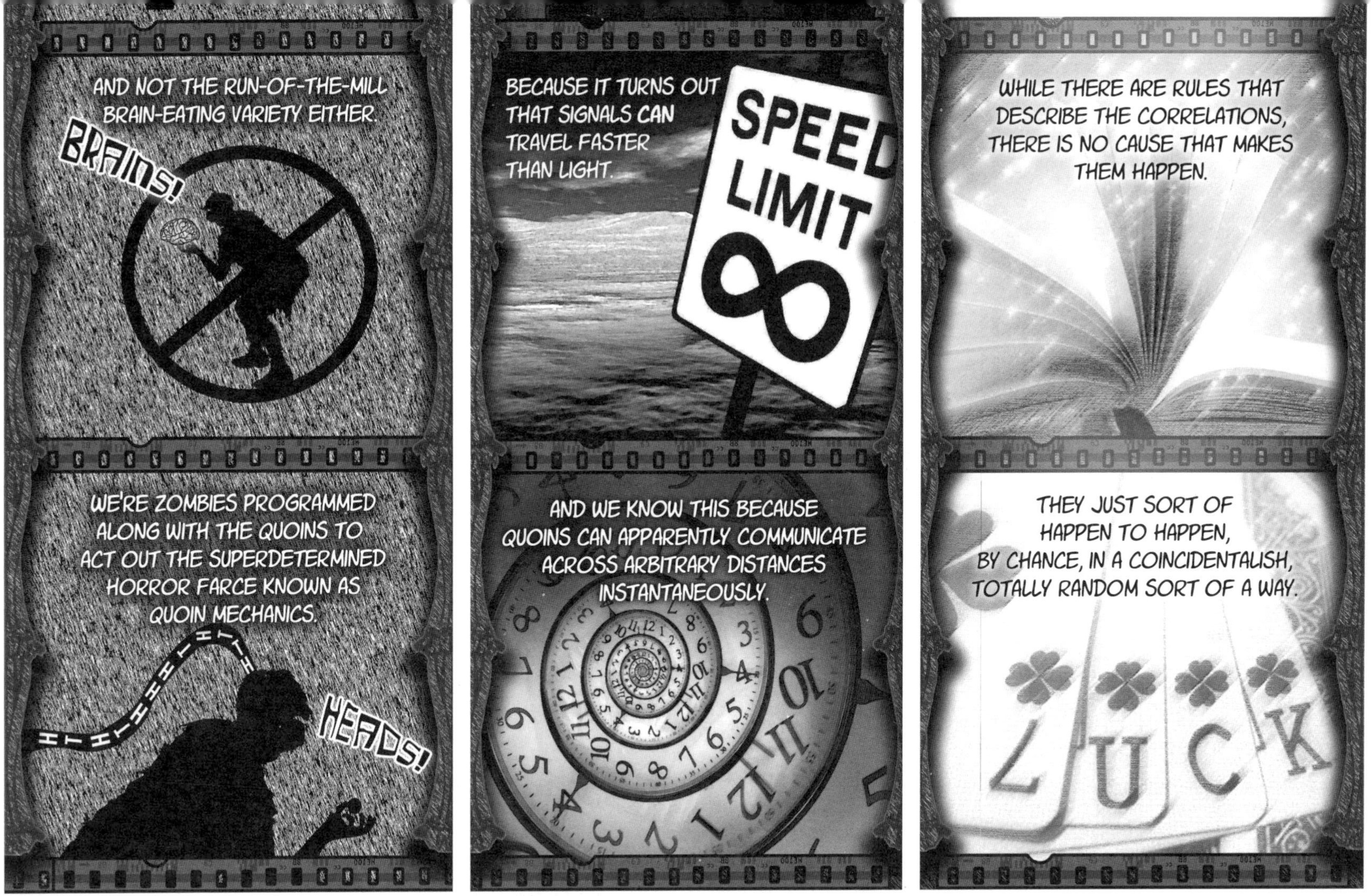
AND NOT THE RUN-OF-THE-MILL BRAIN-EATING VARIETY EITHER.
BRAINS!
WE'RE ZOMBIES PROGRAMMED ALONG WITH THE QUOINS TO ACT OUT THE SUPERDETERMINED HORROR FARCE KNOWN AS QUOIN MECHANICS.
HEADS!
BECAUSE IT TURNS OUT THAT SIGNALS CAN TRAVEL FASTER THAN LIGHT.
SPEED LIMIT ∞
AND WE KNOW THIS BECAUSE QUOINS CAN APPARENTLY COMMUNICATE ACROSS ARBITRARY DISTANCES INSTANTANEOUSLY.
WHILE THERE ARE RULES THAT DESCRIBE THE CORRELATIONS, THERE IS NO CAUSE THAT MAKES THEM HAPPEN.
THEY JUST SORT OF HAPPEN TO HAPPEN, BY CHANCE, IN A COINCIDENTALISH, TOTALLY RANDOM SORT OF A WAY.
LUCK

Don't Blame Me

A Quantum Threat

Entanglement, like many quantum effects, violates
some of our deepest intuitions about the world.
It may also undermine Einstein's special theory of relativity
By David Z Albert and Rivka Galchen

ur intuition, going back forever, is that to
move, say, a rock, one has to touch that

the physical world could in principle be had by
describing, one by one, each of the world's small-

Testing Superdeterministic Conspiracy

Sabine Hossenfelder

Nordita, KTH Royal Institute of Technology and Stockholm University
Roslagstullsbacken 23, SE-106 91 Stockholm, Sweden

Abstract. Tests of Bell's theorem rule out local hidden variables theories. But any theorem
is only as good as the assumptions that go into it, and one of these assumptions is that the
experimenter can freely chose the detector settings. Without this assumption, one enters the
realm of superdeterministic hidden variables theories and can no longer use Bell's theorem as a
criterion. One can like or not like such superdeterministic hidden variables theories and their
inevitable nonlocality, the real question is how one can test them. Here, we propose a possible
experiment that could reveal superdeterminism.

1. Introduction

For any locally causal theory, the attempt to explain quantum effects by use of random
induced through hidden variables can be shown to be in disagreement with experiment
theorem [1], or its generalization respectively [2]. Throughout the last decades, experiment
established that the hidden variables theories for which Bell's theorem applies an
in nature [3, 4, 5, 6, 7, 8, 9]. There are various loopholes in the conclusions tha
from these experiments and not all of these loopholes have yet been satisfac
recent summary see e.g. [10]). But even so, local hidden variables are stron

However, Bell's theorem uses the assumption that one has the freedom to
settings without modifying the prepared state does not apply. Superdete
freedom the conclusion of Bell's theorem and the detector whil
local correlation between the prepared state and the detector whil
the question whether the time evolution of our universe is fundame
too important to leave it to taste – it is an hypothesis that must

2. A Proposal for an Experiment

The essential difference between standard quantum mech
variables theories is that in the former case two identicall
measurement outcomes, while in the latter case it is dif
ared' includes the hidden variables, and it is th
sure. That is after all the reason why it
this problem, instead of tryi
on the same state. Fo
tion in two dif

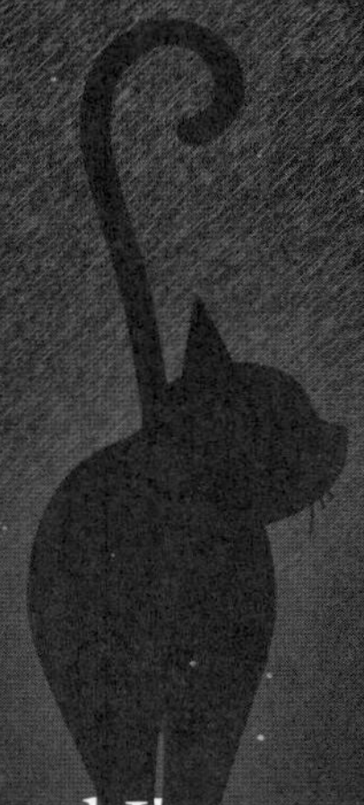

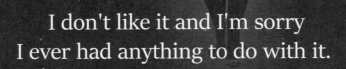

I don't like it and I'm sorry
I ever had anything to do with it.

—Erwin Schrödinger
(on quantum mechanics)

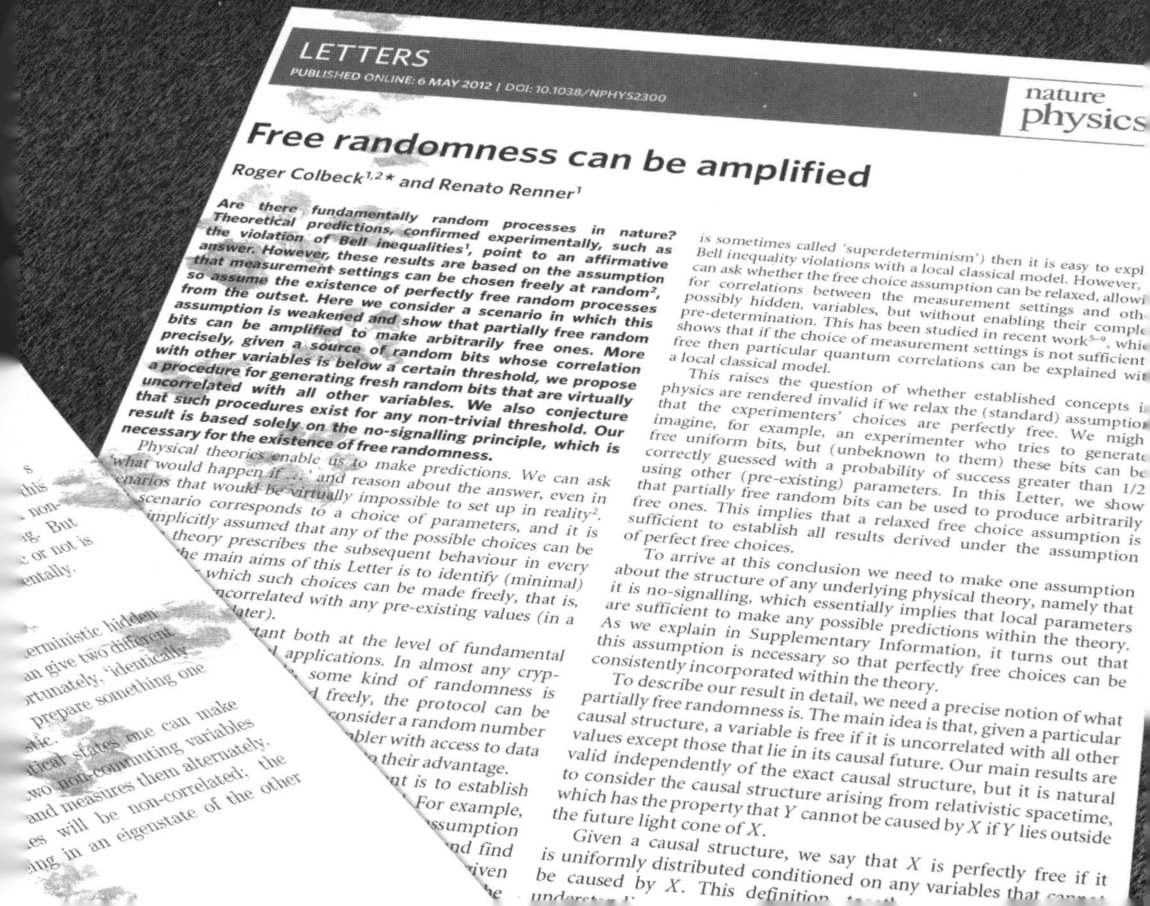

LETTERS
PUBLISHED ONLINE: 6 MAY 2012 | DOI: 10.1038/NPHYS2300
nature physics
Free randomness can be amplified
Roger Colbeck1,2★ and Renato Renner1
Are there fundamentally random processes in nature? Theoretical predictions, confirmed experimentally, such as the violation of Bell inequalities1, point to an affirmative answer. However, these results are based on the assumption that measurement settings can be chosen freely at random2, so assume the existence of perfectly free random processes from the outset. Here we consider a scenario in which this assumption is weakened and show that partially free random bits can be amplified to make arbitrarily free ones. More precisely, given a source of random bits whose correlation with other variables is below a certain threshold, we propose a procedure for generating fresh random bits that are virtually uncorrelated with all other variables. We also conjecture that such procedures exist for any non-trivial threshold. Our result is based solely on the no-signalling principle, which is necessary for the existence of free randomness.
Physical theories enable us to make predictions. We can ask 'what would happen if ...' and reason about the answer, even in scenarios that would be virtually impossible to set up in reality2. scenario corresponds to a choice of parameters, and it is implicitly assumed that any of the possible choices can be theory prescribes the subsequent behaviour in every the main aims of this Letter is to identify (minimal) which such choices can be made freely, that is, uncorrelated with any pre-existing values (in a later).
tant both at the level of fundamental applications. In almost any cryp- some kind of randomness is freely, the protocol can be consider a random number bler with access to data their advantage. nt is to establish For example, ssumption nd find iven
is sometimes called 'superdeterminism') then it is easy to expl Bell inequality violations with a local classical model. However, can ask whether the free choice assumption can be relaxed, allowi for correlations between the measurement settings and oth possibly hidden, variables, but without enabling their comple pre-determination. This has been studied in recent work5–9, whi shows that if the choice of measurement settings is not sufficient free then particular quantum correlations can be explained wit a local classical model.
This raises the question of whether established concepts i physics are rendered invalid if we relax the (standard) assumptio that the experimenters' choices are perfectly free. We migh imagine, for example, an experimenter who tries to generate free uniform bits, but (unbeknown to them) these bits can be correctly guessed with a probability of success greater than 1/2 using other (pre-existing) parameters. In this Letter, we show that partially free random bits can be used to produce arbitrarily free ones. This implies that a relaxed free choice assumption is sufficient to establish all results derived under the assumption of perfect free choices.
To arrive at this conclusion we need to make one assumption about the structure of any underlying physical theory, namely that it is no-signalling, which essentially implies that local parameters are sufficient to make any possible predictions within the theory. As we explain in Supplementary Information, it turns out that this assumption is necessary so that perfectly free choices can be consistently incorporated within the theory.
To describe our result in detail, we need a precise notion of what partially free randomness is. The main idea is that, given a particular causal structure, a variable is free if it is uncorrelated with all other values except those that lie in its causal future. Our main results are valid independently of the exact causal structure, but it is natural to consider the causal structure arising from relativistic spacetime, which has the property that Y cannot be caused by X if Y lies outside the future light cone of X.
Given a causal structure, we say that X is perfectly free if it is uniformly distributed conditioned on any variables that cann be caused by X. This definition

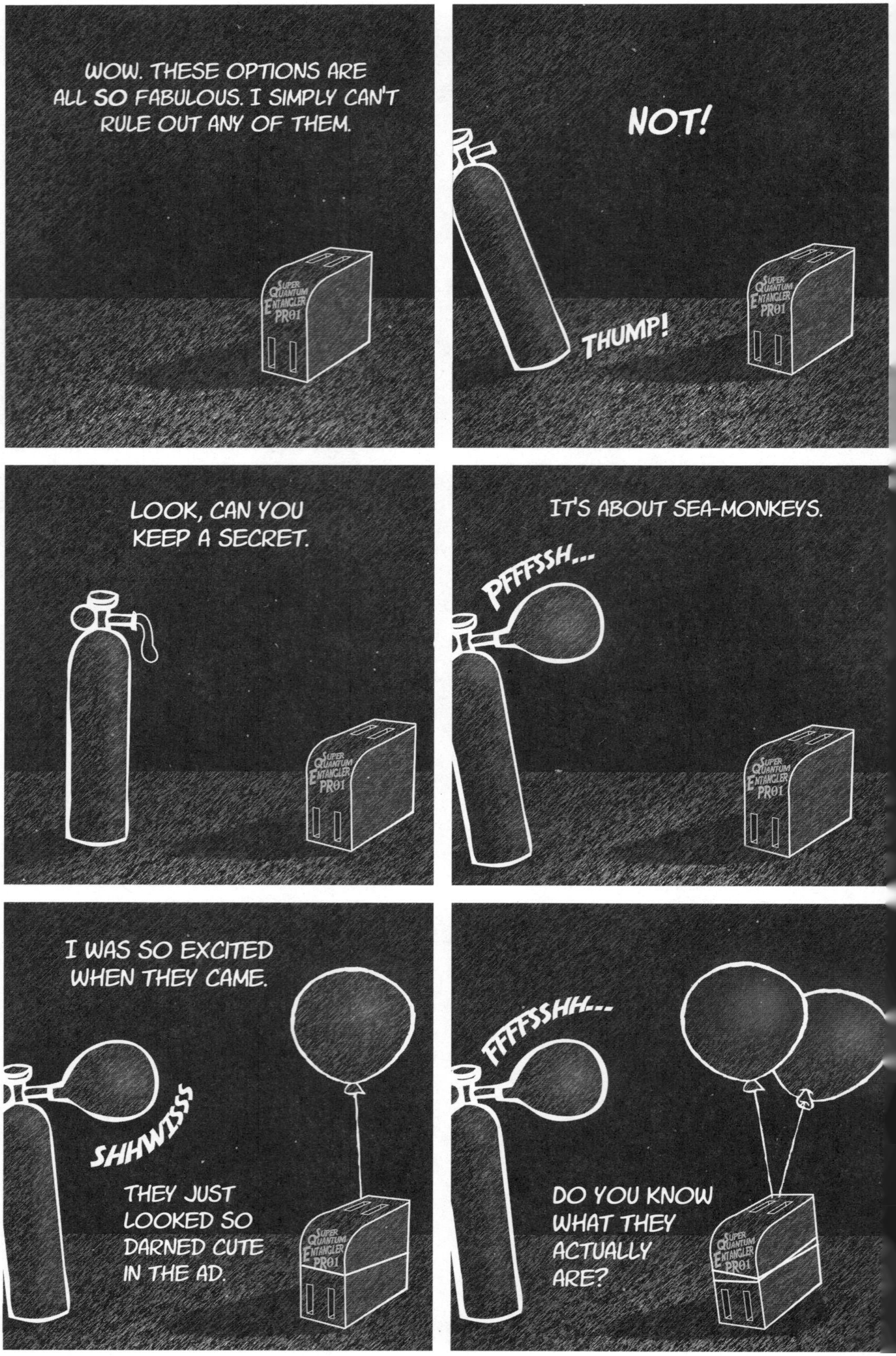

WOW. THESE OPTIONS ARE ALL SO FABULOUS. I SIMPLY CAN'T RULE OUT ANY OF THEM.
SUPER QUANTUM ENTANGLER PR01
NOT!
THUMP!
SUPER QUANTUM ENTANGLER PR01
LOOK, CAN YOU KEEP A SECRET.
SUPER QUANTUM ENTANGLER PR01
IT'S ABOUT SEA-MONKEYS.
PFFFSSH...
SUPER QUANTUM ENTANGLER PR01
I WAS SO EXCITED WHEN THEY CAME.
SHHWISSS
THEY JUST LOOKED SO DARNED CUTE IN THE AD.
SUPER QUANTUM ENTANGLER PR01
FFFFSSHH...
DO YOU KNOW WHAT THEY ACTUALLY ARE?
SUPER QUANTUM ENTANGLER PR01

SEA-MONKEYS ARE BRINE SHRIMP.

BRINE SHRIMP!

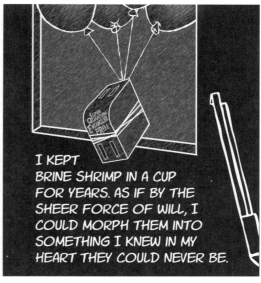

I KEPT BRINE SHRIMP IN A CUP FOR YEARS. AS IF BY THE SHEER FORCE OF WILL, I COULD MORPH THEM INTO SOMETHING I KNEW IN MY HEART THEY COULD NEVER BE.

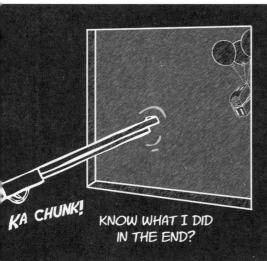

KA CHUNK!

KNOW WHAT I DID IN THE END?

BLAMO!

I FLUSHED THEM.

No actual sea-monkeys or toasters were harmed in the making of this comic.

DING DONG!

LISTEN, NOT A WORD ABOUT ANY OF THIS TO ANYONE!

I MEAN IT!

CREEEAK

OOOOH! A PACKAGE!

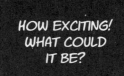

HOW EXCITING! WHAT COULD IT BE?

IT'S A...

GOOD GOD, IT'S A...

NEW TOASTER?!?!

THERE MUST BE SOME MISTAKE.

I PROBABLY DOUBLE-PRESSED THE "ORDER NOW" BUTTON...

OR SOMETHING.

WOOOSH!

KA BOOM!

KA BLOWIE!

NEVER MIND. IT'S ALL OVER NOW.

DING DONG!

BY THOR'S HAMMER!

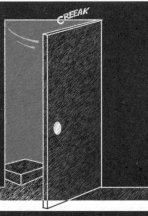

CREEAK

NOT TO WORRY. I'M SURE THIS KIND OF THING...

HAPPENS ALL THE TIME.

DING DONG!

KRANKPKRRRK

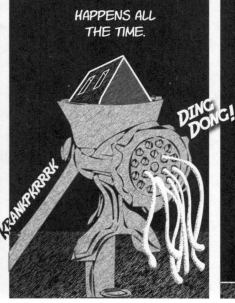

CREEAK

OVER MY DEAD BODY!

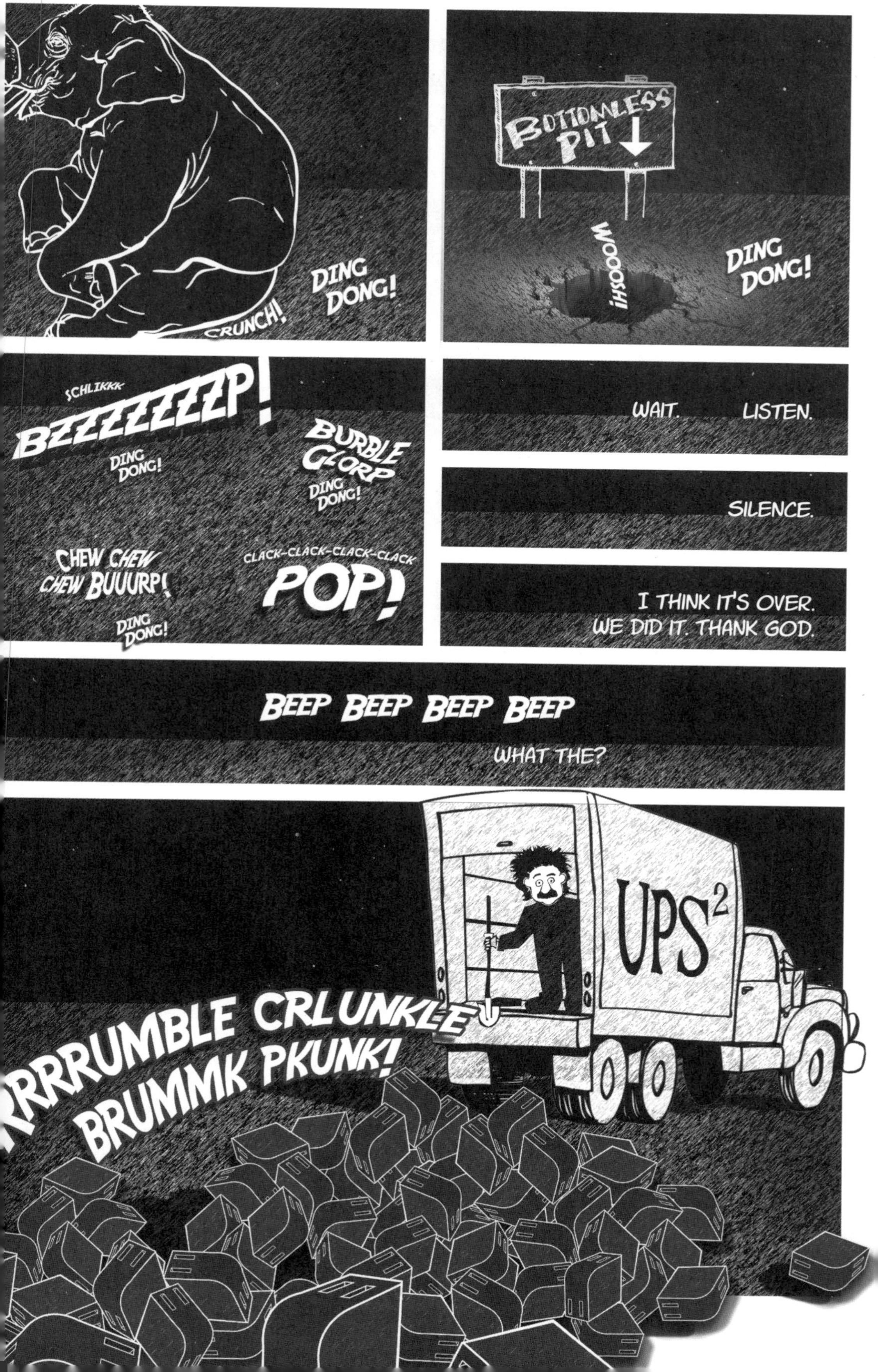
CRUNCH!
DING DONG!
BOTTOMLESS PIT
WOOOSH!
DING DONG!
SCHLIKKK
BZZZZZZZZP!
DING DONG!
BURBLE GLORP
DING DONG!
CHEW CHEW CHEW BUUURP!
DING DONG!
CLACK-CLACK-CLACK-CLACK
POP!
WAIT. LISTEN.
SILENCE.
I THINK IT'S OVER.
WE DID IT. THANK GOD.
BEEP BEEP BEEP BEEP
WHAT THE?
UPS²
RRRUMBLE CRLUNKLE
BRUMMK PKUNK!

Happy?

Very.

I delivered on your "curious correlation."

Signed, sealed and delivered.

I did a "gut-feel" comic version of Bell's proof. No math.

A first. Not bad considering it's known as one of the most profound AND misunderstood theorems of all time. V nice.

Ruled out signaling.

Indeed.

Leaving people with no explanation whatsoever.

Yup.

So.

So.

WHAT NOW?!?

We bring in the big guns.

Part II
HELP?

Hey T, Check your mailbox. I left you the famous EPR article where Einstein, Podolsky, and Rosen attack QM with their "criterion of reality." There are also some NY Times stories from 1935 that report on the great reality debate. It was sensational news at the time! What would Einstein and Bohr have said about our quoins, huh? Anyway, I'm off to a conference. I'll call you when I get back.

CLICK---bzzzzzzzzzzzzzzz

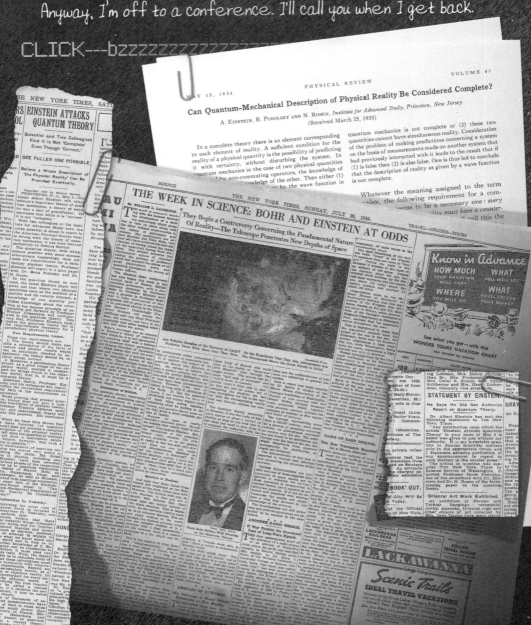

PHYSICAL REVIEW   VOLUME 47

MAY 15, 1935

Can Quantum-Mechanical Description of Physical Reality Be Considered Complete?

A. EINSTEIN, B. PODOLSKY AND N. ROSEN, *Institute for Advanced Study, Princeton, New Jersey*
(Received March 25, 1935)

In a complete theory there is an element corresponding to each element of reality. A sufficient condition for the reality of a physical quantity is the possibility of predicting it with certainty, without disturbing the system. In quantum mechanics in the case of two physical quantities described by non-commuting operators, the knowledge of one precludes the knowledge of the other. Then either (1)

quantum mechanics is not complete or (2) these two quantities cannot have simultaneous reality. Consideration of the problem of making predictions concerning a system on the basis of measurements made on another system that had previously interacted with it leads to the result that if (1) is false then (2) is also false. One is thus led to conclude that the description of reality as given by a wave function is not complete.

THE NEW YORK TIMES, SAT

EINSTEIN ATTACKS QUANTUM THEORY

Scientist and Two Colleagues Find It Is Not 'Complete' Even Though 'Correct.'

SEE FULLER ONE POSSIBLE

Believe a Whole Description of 'the Physical Reality' Can Be Provided Eventually.

THE NEW YORK TIMES, SUNDAY, JULY 28, 1935.

THE WEEK IN SCIENCE: BOHR AND EINSTEIN AT ODDS

By WILLIAM L. LAURENCE

They Begin a Controversy Concerning the Fundamental Nature Of Reality—The Telescope Penetrates New Depths of Space

STATEMENT BY EINSTEIN.

He Says He Did Not Authorize Report on Quantum Theory.

Spooky Action at a Distance and "Being-Thus"*

I cannot seriously believe in [quantum mechanics] because the theory cannot be reconciled with the idea that physics should represent a reality in time and space, free from spooky action at a distance.

—Albert Einstein

Unless one makes [an] assumption about the independence of the existence—the "being-thus"*— of objects which are far apart from one another in space—which stems in the first place from everyday reasoning—physical thinking in the familiar sense would not be possible.

—Albert Einstein

* "So-Sein" in the original German

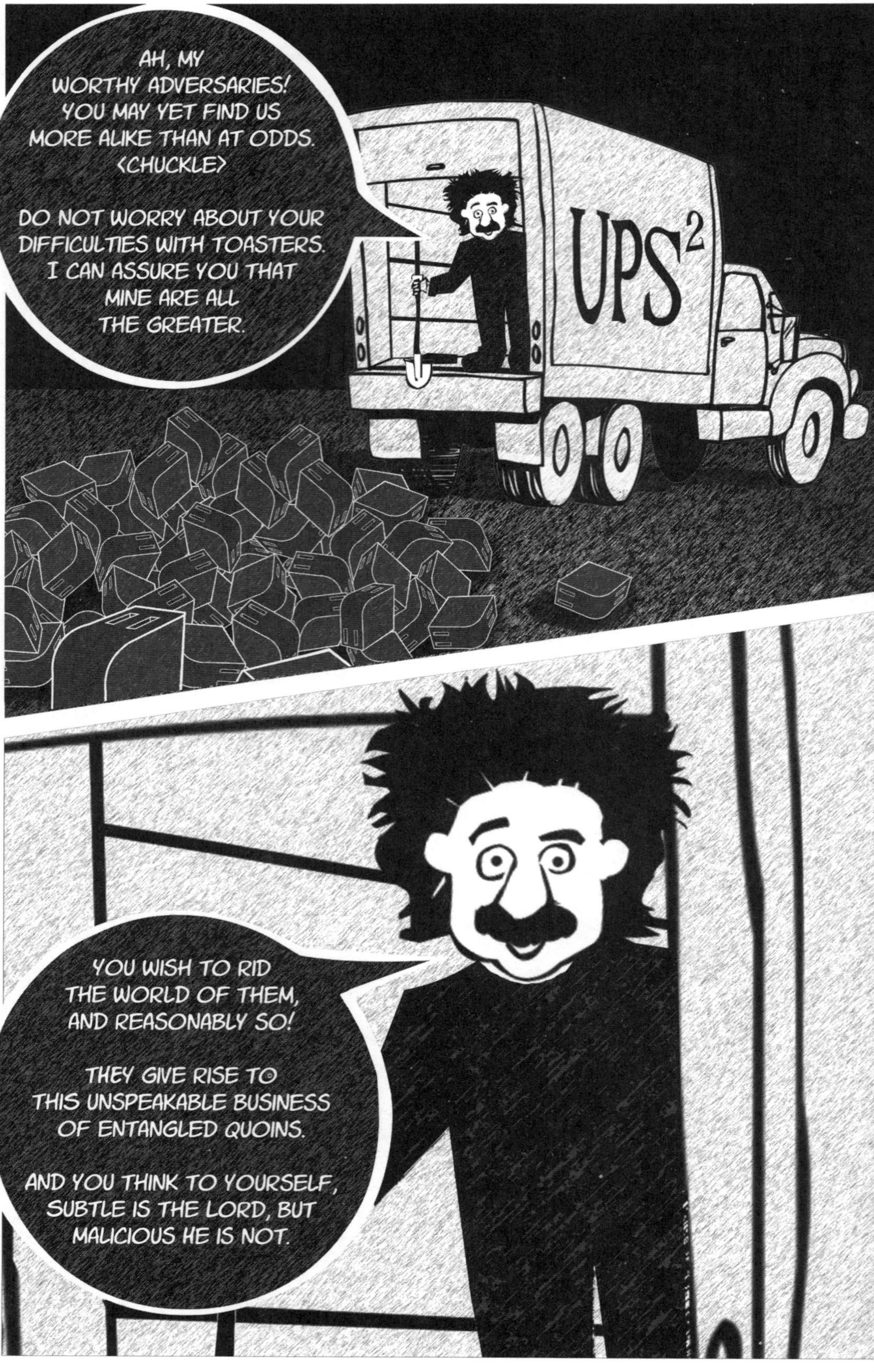
AH, MY WORTHY ADVERSARIES! YOU MAY YET FIND US MORE ALIKE THAN AT ODDS. <CHUCKLE>
DO NOT WORRY ABOUT YOUR DIFFICULTIES WITH TOASTERS. I CAN ASSURE YOU THAT MINE ARE ALL THE GREATER.
UPS^2
YOU WISH TO RID THE WORLD OF THEM, AND REASONABLY SO!
THEY GIVE RISE TO THIS UNSPEAKABLE BUSINESS OF ENTANGLED QUOINS.
AND YOU THINK TO YOURSELF, SUBTLE IS THE LORD, BUT MALICIOUS HE IS NOT.

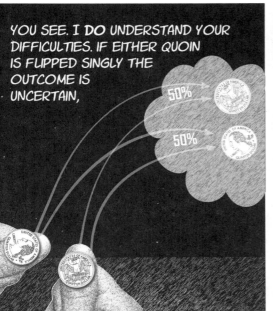

YOU SEE. I **DO** UNDERSTAND YOUR DIFFICULTIES. IF EITHER QUOIN IS FLIPPED SINGLY THE OUTCOME IS UNCERTAIN,

50%

50%

YET FLIP ONE AND YOU CAN PREDICT WITH CERTAINTY THE OUTCOME OF THE OTHER, HMMM?

YOU ASK YOURSELF, IF IN ADVANCE OF FLIPPING THE SECOND QUOIN I CAN PREDICT WITH CERTAINTY ITS OUTCOME, THEN SURELY THERE EXISTS AN ELEMENT OF REALITY, A STATE, A BEING-THUS, THAT DETERMINES THAT OUTCOME, NO?

GIVEN YOUR PRIOR FINDINGS, YOU THEN SUPPOSE THAT THE REALITY OF THE SECOND QUOIN DEPENDS UPON THE PROCESS OF FLIPPING CARRIED OUT ON THE FIRST, WHICH DOES NOT IN ANY WAY DISTURB THE SECOND.

AND YOU FIND YOURSELF FACING, RATHER SUDDENLY, THE SOMEWHAT ALARMING CONCLUSION THAT A PHYSICAL REALITY IN "B" UNDERGOES AN INSTANTANEOUS CHANGE BECAUSE OF AN ACTION MADE ON "A" YOUR INSTINCTS BRISTLE AT THE MERE SUGGESTION!

77

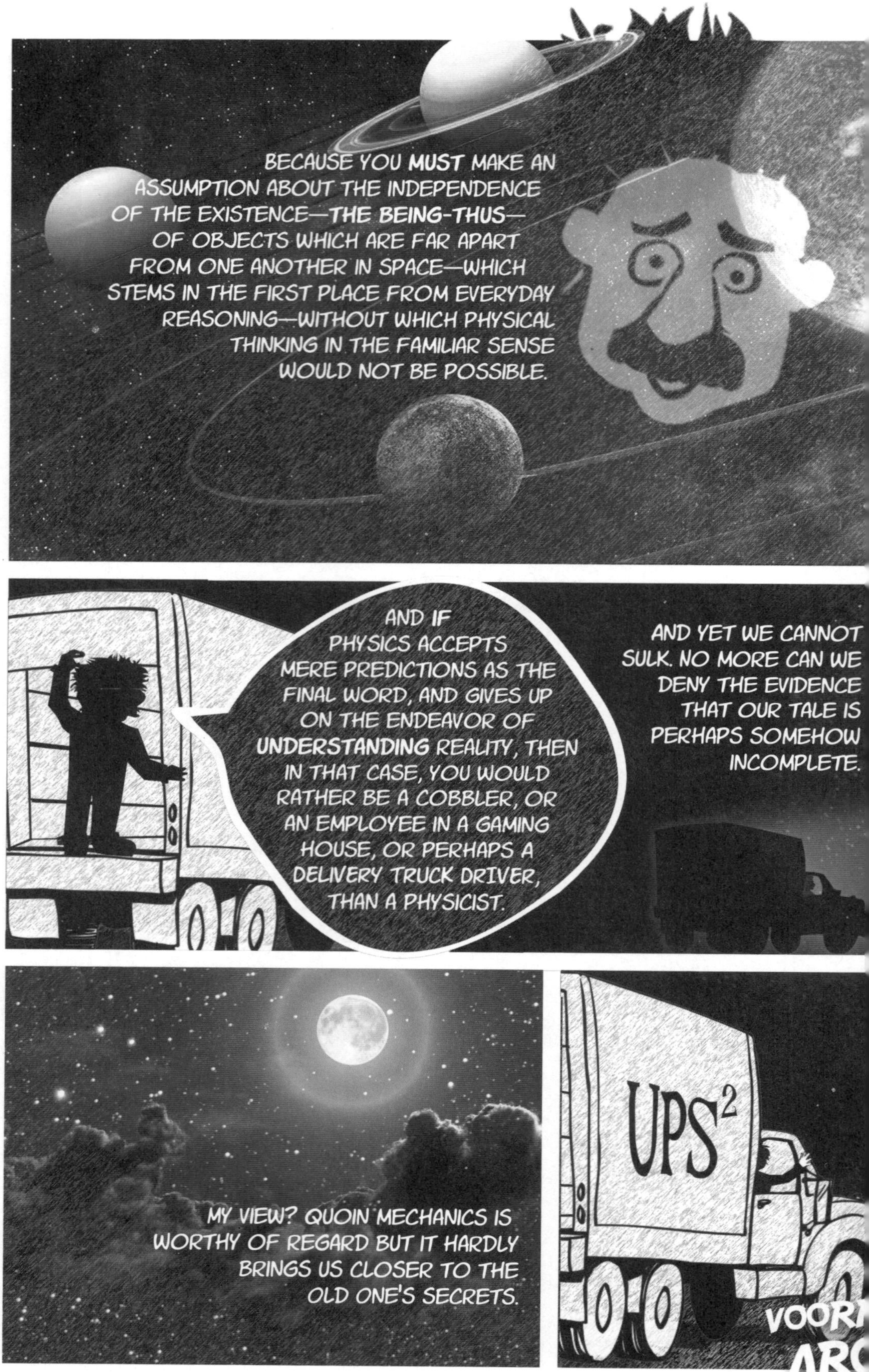

BECAUSE YOU **MUST** MAKE AN ASSUMPTION ABOUT THE INDEPENDENCE OF THE EXISTENCE—**THE BEING-THUS**—OF OBJECTS WHICH ARE FAR APART FROM ONE ANOTHER IN SPACE—WHICH STEMS IN THE FIRST PLACE FROM EVERYDAY REASONING—WITHOUT WHICH PHYSICAL THINKING IN THE FAMILIAR SENSE WOULD NOT BE POSSIBLE.
AND **IF** PHYSICS ACCEPTS MERE PREDICTIONS AS THE FINAL WORD, AND GIVES UP ON THE ENDEAVOR OF **UNDERSTANDING** REALITY, THEN IN THAT CASE, YOU WOULD RATHER BE A COBBLER, OR AN EMPLOYEE IN A GAMING HOUSE, OR PERHAPS A DELIVERY TRUCK DRIVER, THAN A PHYSICIST.
AND YET WE CANNOT SULK. NO MORE CAN WE DENY THE EVIDENCE THAT OUR TALE IS PERHAPS SOMEHOW INCOMPLETE.
MY VIEW? QUOIN MECHANICS IS WORTHY OF REGARD BUT IT HARDLY BRINGS US CLOSER TO THE OLD ONE'S SECRETS.
UPS²

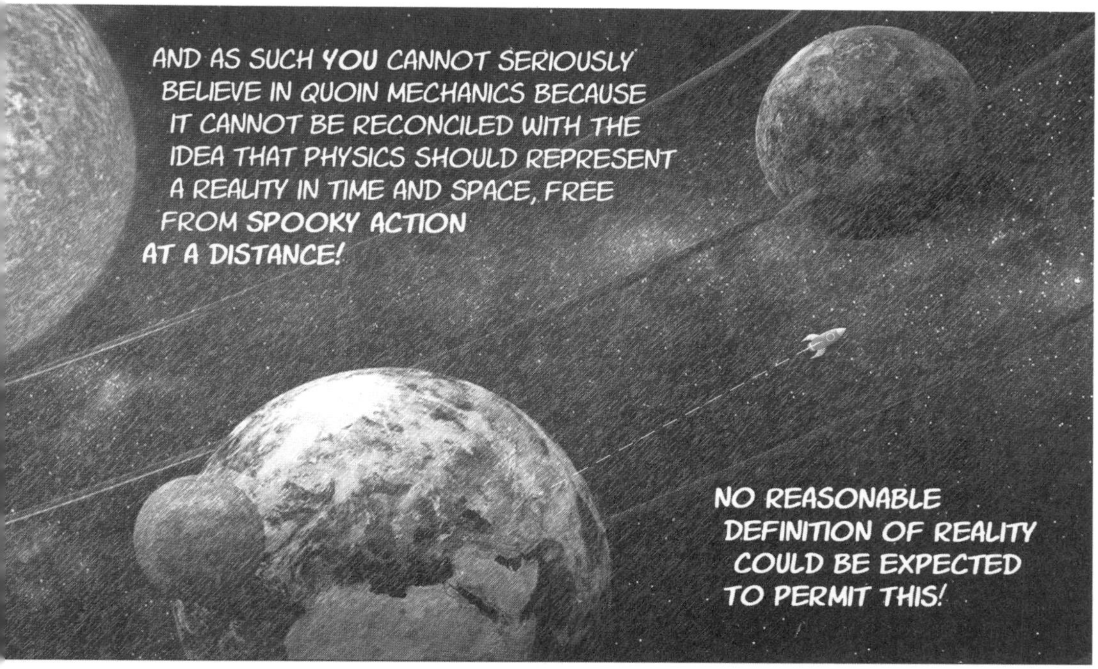
AND AS SUCH YOU CANNOT SERIOUSLY BELIEVE IN QUOIN MECHANICS BECAUSE IT CANNOT BE RECONCILED WITH THE IDEA THAT PHYSICS SHOULD REPRESENT A REALITY IN TIME AND SPACE, FREE FROM SPOOKY ACTION AT A DISTANCE!
NO REASONABLE DEFINITION OF REALITY COULD BE EXPECTED TO PERMIT THIS!

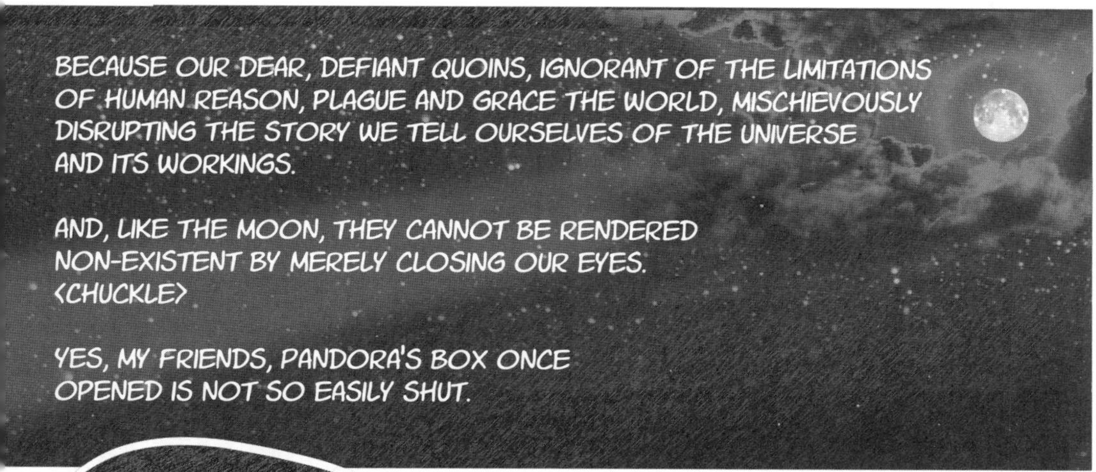
BECAUSE OUR DEAR, DEFIANT QUOINS, IGNORANT OF THE LIMITATIONS OF HUMAN REASON, PLAGUE AND GRACE THE WORLD, MISCHIEVOUSLY DISRUPTING THE STORY WE TELL OURSELVES OF THE UNIVERSE AND ITS WORKINGS.
AND, LIKE THE MOON, THEY CANNOT BE RENDERED NON-EXISTENT BY MERELY CLOSING OUR EYES. <CHUCKLE>
YES, MY FRIENDS, PANDORA'S BOX ONCE OPENED IS NOT SO EASILY SHUT.

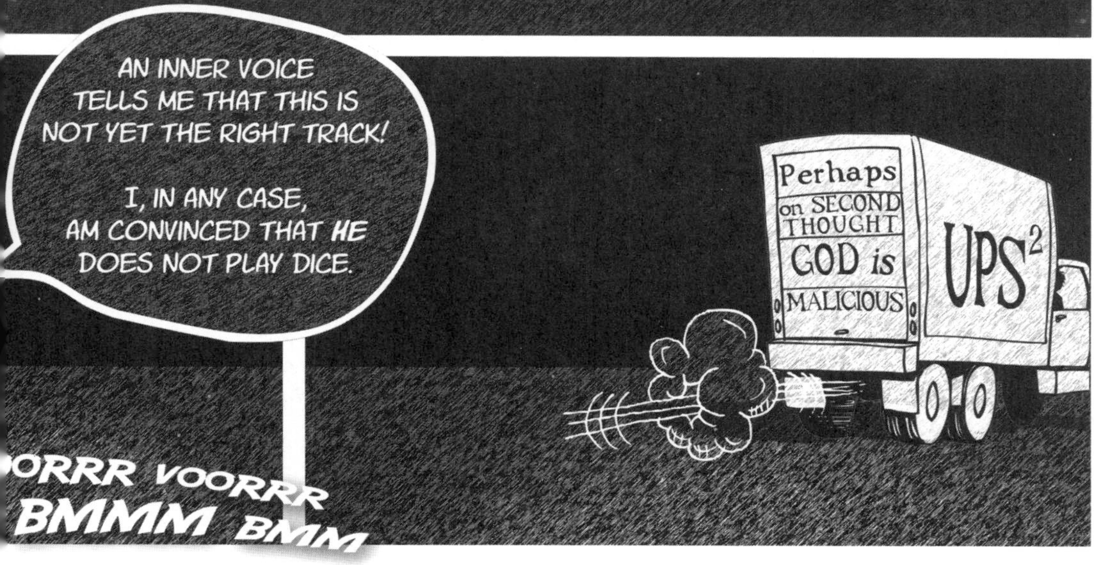
AN INNER VOICE TELLS ME THAT THIS IS NOT YET THE RIGHT TRACK! I, IN ANY CASE, AM CONVINCED THAT HE DOES NOT PLAY DICE.
Perhaps on SECOND THOUGHT GOD is MALICIOUS
UPS²
ORRR VOORRR
BMMM BMM

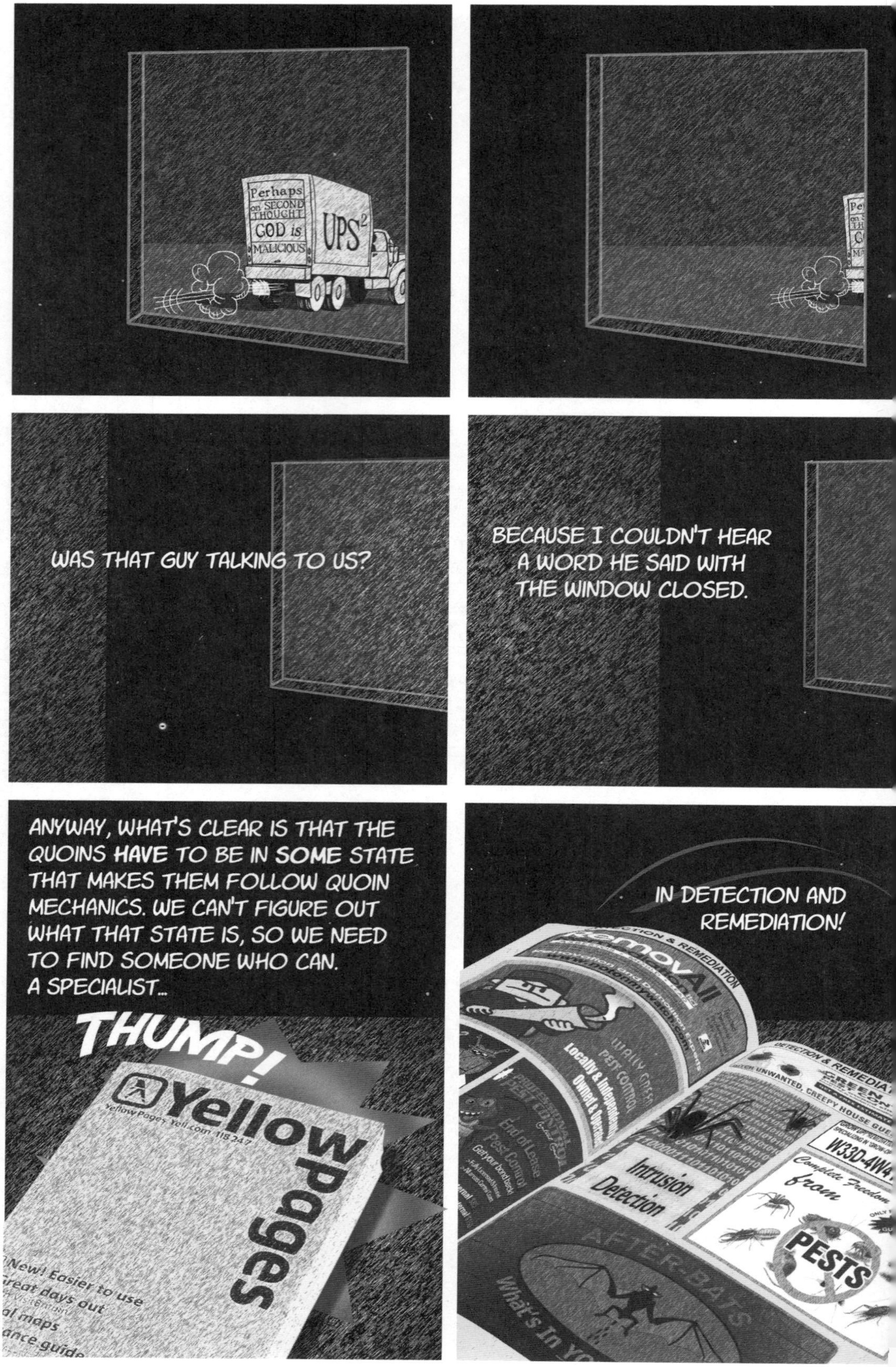
Perhaps on SECOND THOUGHT GOD is MALICIOUS
UPS²
WAS THAT GUY TALKING TO US?
BECAUSE I COULDN'T HEAR A WORD HE SAID WITH THE WINDOW CLOSED.
ANYWAY, WHAT'S CLEAR IS THAT THE QUOINS HAVE TO BE IN SOME STATE THAT MAKES THEM FOLLOW QUOIN MECHANICS. WE CAN'T FIGURE OUT WHAT THAT STATE IS, SO WE NEED TO FIND SOMEONE WHO CAN. A SPECIALIST...
THUMP!
Yellow Pages
IN DETECTION AND REMEDIATION!
DETECTION & REMEDIATION
Intrusion Detection
End of Lease Pest Control
Complete Freedom from
PESTS
AFTER-BATS

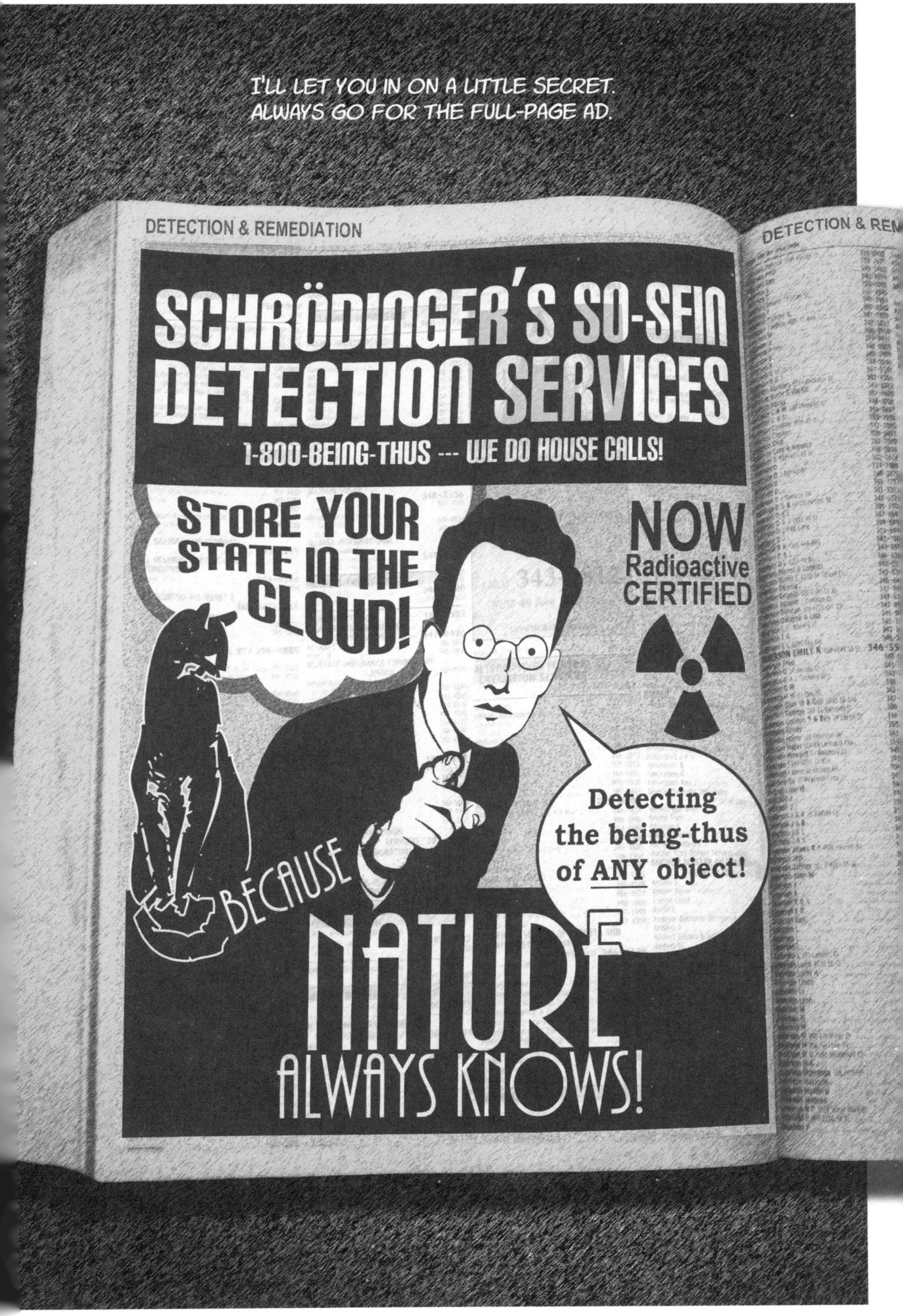
I'LL LET YOU IN ON A LITTLE SECRET. ALWAYS GO FOR THE FULL-PAGE AD.
DETECTION & REMEDIATION
SCHRÖDINGER'S SO-SEIN DETECTION SERVICES
1-800-BEING-THUS --- WE DO HOUSE CALLS!
STORE YOUR STATE IN THE CLOUD!
NOW Radioactive CERTIFIED
Detecting the being-thus of ANY object!
BECAUSE NATURE ALWAYS KNOWS!
DETECTION & REM

If You Know, Nature Knows

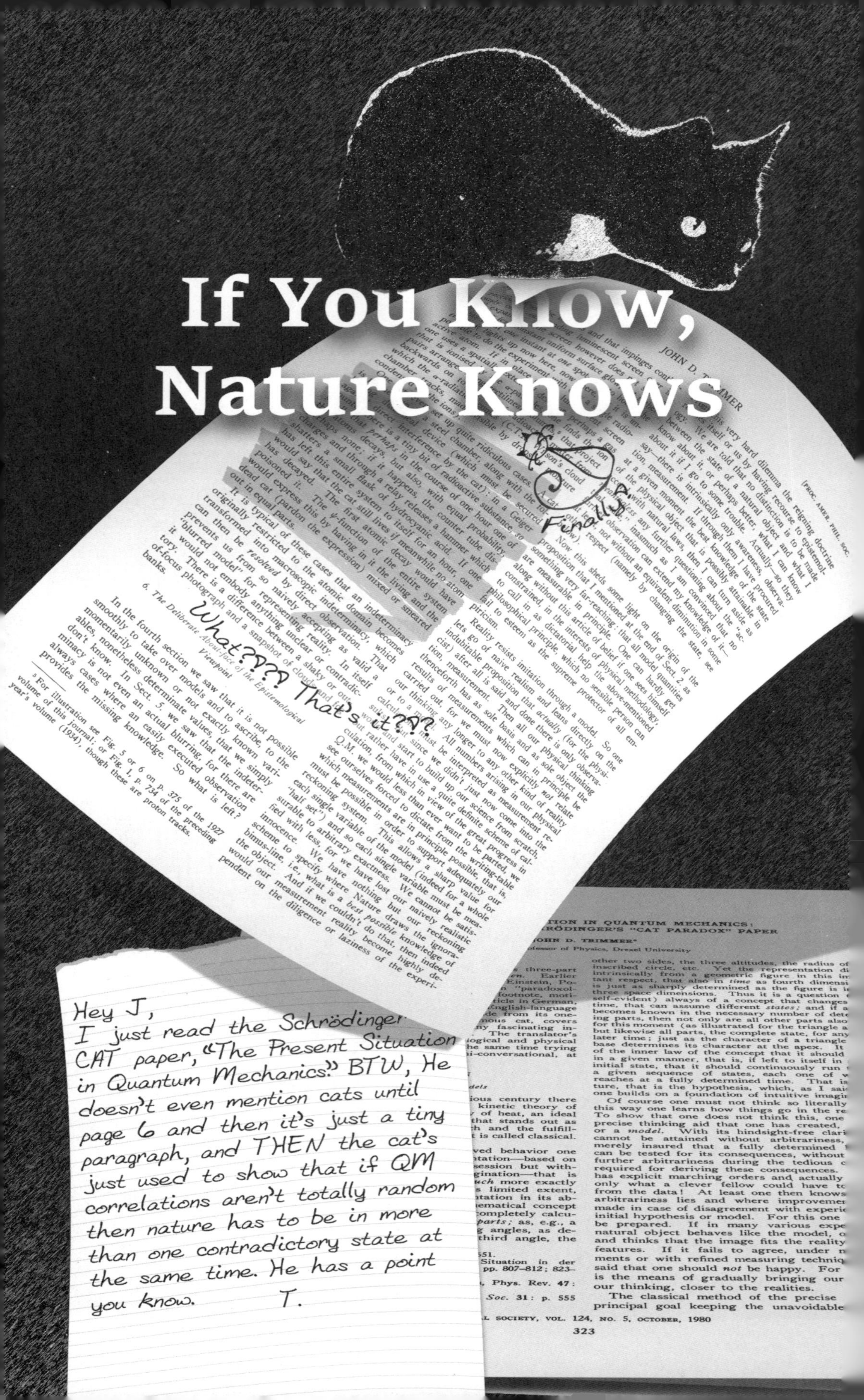
JOHN D. TRIMMER
Finally?
What???? That's it???
JOHN D. TRIMMER*
Professor of Physics, Drexel University
Hey J,
I just read the Schrödinger CAT paper, "The Present Situation in Quantum Mechanics." BTW, He doesn't even mention cats until page 6 and then it's just a tiny paragraph, and THEN the cat's just used to show that if QM correlations aren't totally random then nature has to be in more than one contradictory state at the same time. He has a point you know. T.

One can even set up quite ridiculous cases.
A cat is penned up in a steel chamber, along with
[a] diabolical device (which must be secured
against direct interference by the cat) ...

—**Erwin Schrödinger**

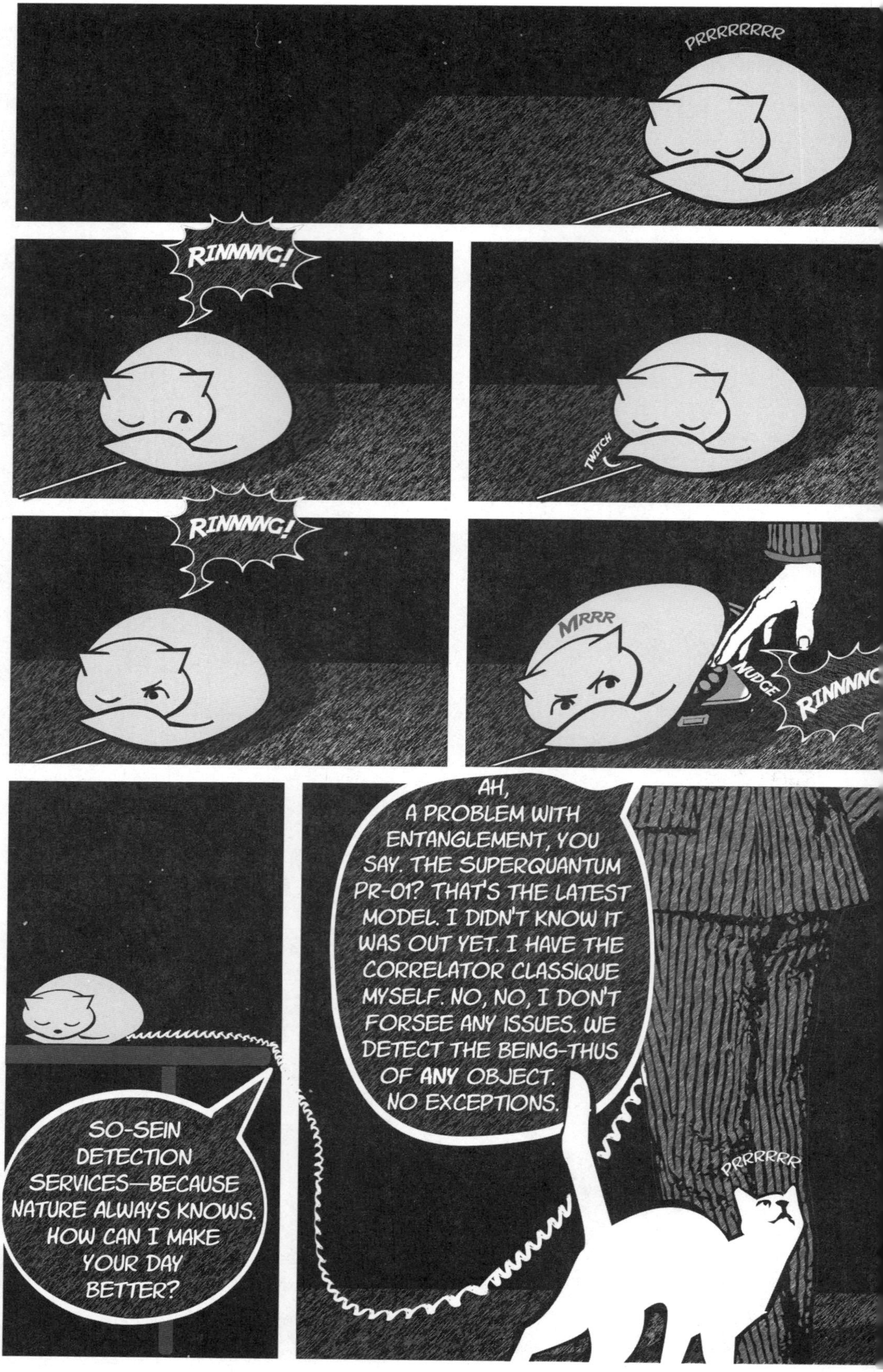
PRRRRRRRR
RINNNNG!
TWITCH
RINNNNG!
MRRR
NUDGE
RINNNNG
SO-SEIN DETECTION SERVICES—BECAUSE NATURE ALWAYS KNOWS. HOW CAN I MAKE YOUR DAY BETTER?
AH, A PROBLEM WITH ENTANGLEMENT, YOU SAY. THE SUPERQUANTUM PR-01? THAT'S THE LATEST MODEL. I DIDN'T KNOW IT WAS OUT YET. I HAVE THE CORRELATOR CLASSIQUE MYSELF. NO, NO, I DON'T FORSEE ANY ISSUES. WE DETECT THE BEING-THUS OF **ANY** OBJECT. NO EXCEPTIONS.
PRRRRRR

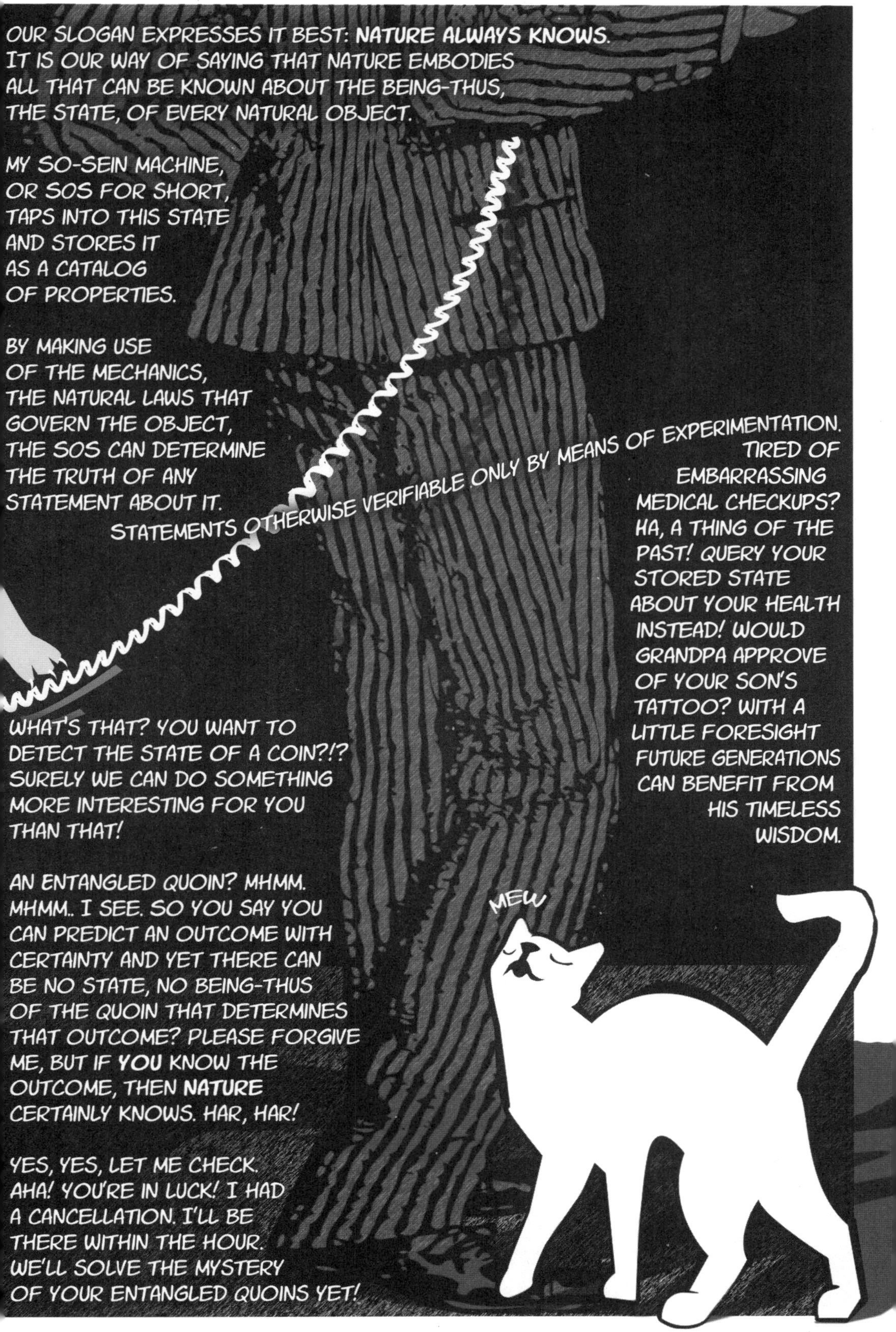

OUR SLOGAN EXPRESSES IT BEST: **NATURE ALWAYS KNOWS.** IT IS OUR WAY OF SAYING THAT NATURE EMBODIES ALL THAT CAN BE KNOWN ABOUT THE BEING-THUS, THE STATE, OF EVERY NATURAL OBJECT.
MY SO-SEIN MACHINE, OR SOS FOR SHORT, TAPS INTO THIS STATE AND STORES IT AS A CATALOG OF PROPERTIES.
BY MAKING USE OF THE MECHANICS, THE NATURAL LAWS THAT GOVERN THE OBJECT, THE SOS CAN DETERMINE THE TRUTH OF ANY STATEMENT ABOUT IT.
STATEMENTS OTHERWISE VERIFIABLE ONLY BY MEANS OF EXPERIMENTATION.
TIRED OF EMBARRASSING MEDICAL CHECKUPS? HA, A THING OF THE PAST! QUERY YOUR STORED STATE ABOUT YOUR HEALTH INSTEAD! WOULD GRANDPA APPROVE OF YOUR SON'S TATTOO? WITH A LITTLE FORESIGHT FUTURE GENERATIONS CAN BENEFIT FROM HIS TIMELESS WISDOM.
WHAT'S THAT? YOU WANT TO DETECT THE STATE OF A COIN?!? SURELY WE CAN DO SOMETHING MORE INTERESTING FOR YOU THAN THAT!
AN ENTANGLED QUOIN? MHMM. MHMM.. I SEE. SO YOU SAY YOU CAN PREDICT AN OUTCOME WITH CERTAINTY AND YET THERE CAN BE NO STATE, NO BEING-THUS OF THE QUOIN THAT DETERMINES THAT OUTCOME? PLEASE FORGIVE ME, BUT IF **YOU** KNOW THE OUTCOME, THEN **NATURE** CERTAINLY KNOWS. HAR, HAR!
MEW
YES, YES, LET ME CHECK. AHA! YOU'RE IN LUCK! I HAD A CANCELLATION. I'LL BE THERE WITHIN THE HOUR. WE'LL SOLVE THE MYSTERY OF YOUR ENTANGLED QUOINS YET!

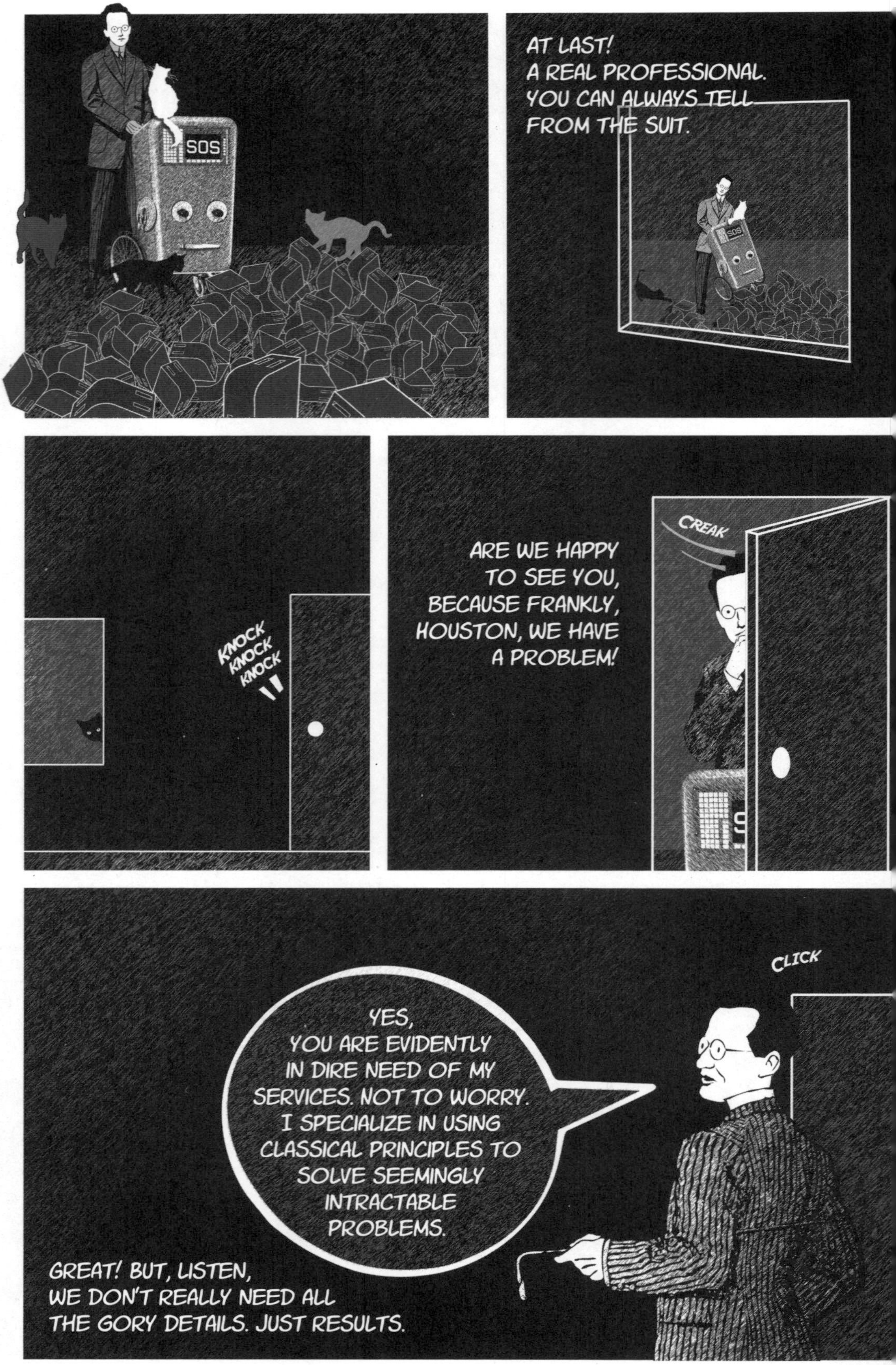
SOS
AT LAST!
A REAL PROFESSIONAL.
YOU CAN ALWAYS TELL
FROM THE SUIT.
SOS
KNOCK
KNOCK
KNOCK
CREAK
ARE WE HAPPY
TO SEE YOU,
BECAUSE FRANKLY,
HOUSTON, WE HAVE
A PROBLEM!
CLICK
YES,
YOU ARE EVIDENTLY
IN DIRE NEED OF MY
SERVICES. NOT TO WORRY.
I SPECIALIZE IN USING
CLASSICAL PRINCIPLES TO
SOLVE SEEMINGLY
INTRACTABLE
PROBLEMS.
GREAT! BUT, LISTEN,
WE DON'T REALLY NEED ALL
THE GORY DETAILS. JUST RESULTS.

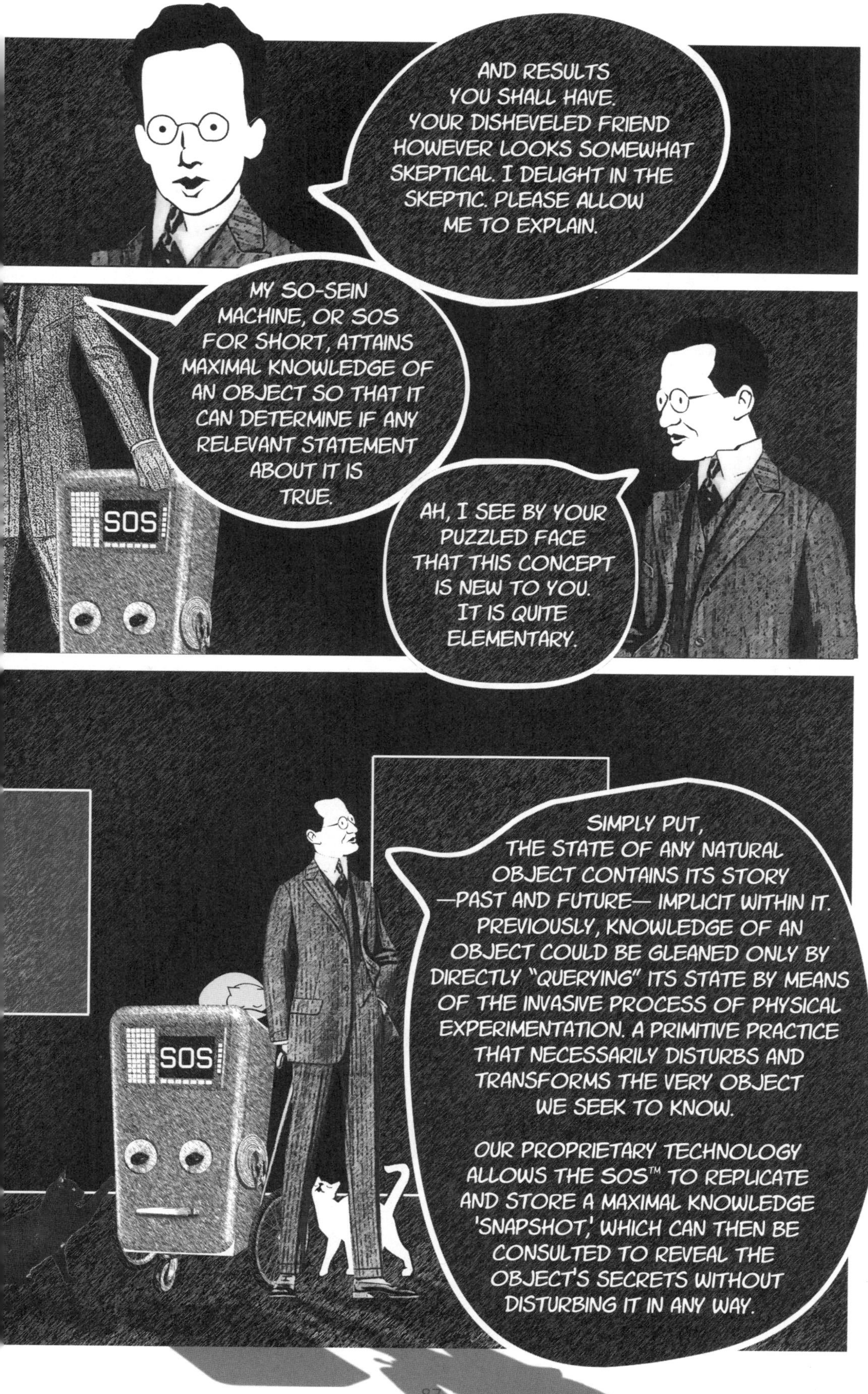
AND RESULTS YOU SHALL HAVE. YOUR DISHEVELED FRIEND HOWEVER LOOKS SOMEWHAT SKEPTICAL. I DELIGHT IN THE SKEPTIC. PLEASE ALLOW ME TO EXPLAIN.
MY SO-SEIN MACHINE, OR SOS FOR SHORT, ATTAINS MAXIMAL KNOWLEDGE OF AN OBJECT SO THAT IT CAN DETERMINE IF ANY RELEVANT STATEMENT ABOUT IT IS TRUE.
SOS
AH, I SEE BY YOUR PUZZLED FACE THAT THIS CONCEPT IS NEW TO YOU. IT IS QUITE ELEMENTARY.
SOS
SIMPLY PUT, THE STATE OF ANY NATURAL OBJECT CONTAINS ITS STORY —PAST AND FUTURE— IMPLICIT WITHIN IT. PREVIOUSLY, KNOWLEDGE OF AN OBJECT COULD BE GLEANED ONLY BY DIRECTLY "QUERYING" ITS STATE BY MEANS OF THE INVASIVE PROCESS OF PHYSICAL EXPERIMENTATION. A PRIMITIVE PRACTICE THAT NECESSARILY DISTURBS AND TRANSFORMS THE VERY OBJECT WE SEEK TO KNOW.
OUR PROPRIETARY TECHNOLOGY ALLOWS THE SOS™ TO REPLICATE AND STORE A MAXIMAL KNOWLEDGE 'SNAPSHOT,' WHICH CAN THEN BE CONSULTED TO REVEAL THE OBJECT'S SECRETS WITHOUT DISTURBING IT IN ANY WAY.

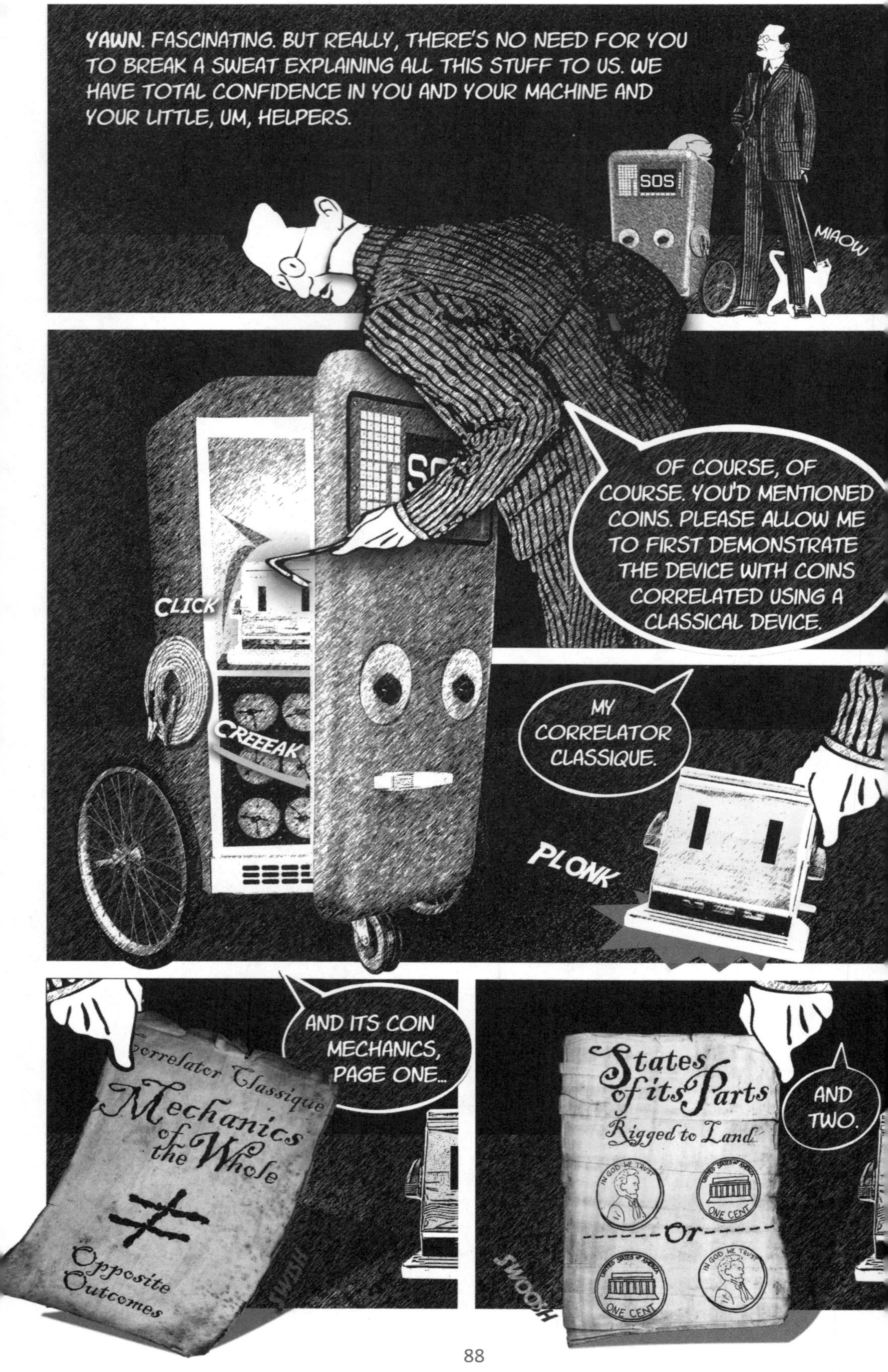
YAWN. FASCINATING. BUT REALLY, THERE'S NO NEED FOR YOU TO BREAK A SWEAT EXPLAINING ALL THIS STUFF TO US. WE HAVE TOTAL CONFIDENCE IN YOU AND YOUR MACHINE AND YOUR LITTLE, UM, HELPERS.
SOS
MIAOW
OF COURSE, OF COURSE. YOU'D MENTIONED COINS. PLEASE ALLOW ME TO FIRST DEMONSTRATE THE DEVICE WITH COINS CORRELATED USING A CLASSICAL DEVICE.
CLICK
CREEEAK
MY CORRELATOR CLASSIQUE.
PLONK
AND ITS COIN MECHANICS, PAGE ONE...
Correlator Classique
Mechanics of the Whole
Opposite Outcomes
SWISH
AND TWO.
States of its Parts
Rigged to Land
Or
SWOOSH

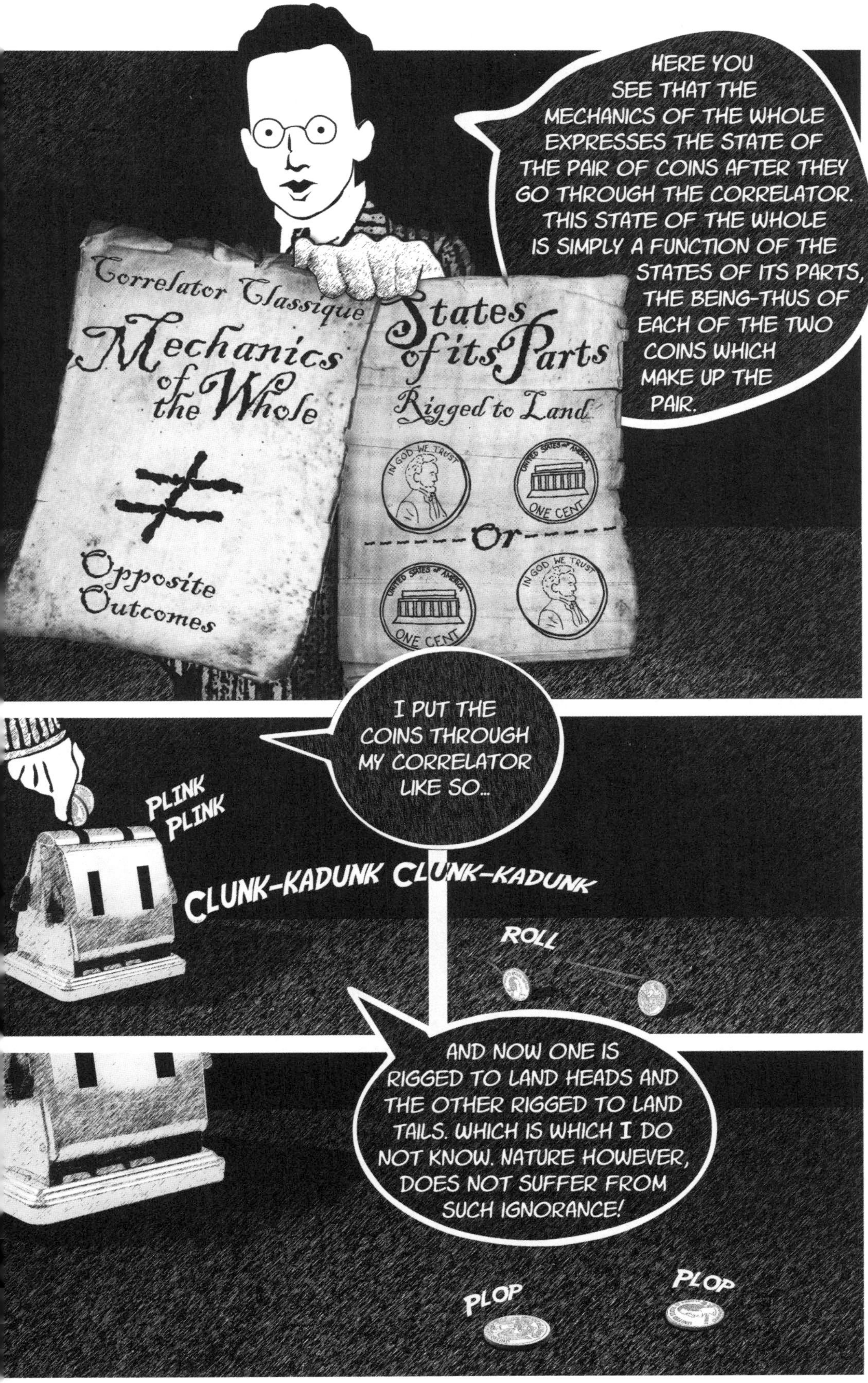
HERE YOU SEE THAT THE MECHANICS OF THE WHOLE EXPRESSES THE STATE OF THE PAIR OF COINS AFTER THEY GO THROUGH THE CORRELATOR. THIS STATE OF THE WHOLE IS SIMPLY A FUNCTION OF THE STATES OF ITS PARTS, THE BEING-THUS OF EACH OF THE TWO COINS WHICH MAKE UP THE PAIR.
Correlator Classique
Mechanics of the Whole
≠
Opposite Outcomes
States of its Parts
Rigged to Land
IN GOD WE TRUST
UNITED STATES of AMERICA
ONE CENT
Or
I PUT THE COINS THROUGH MY CORRELATOR LIKE SO...
PLINK PLINK
CLUNK-KADUNK CLUNK-KADUNK
ROLL
AND NOW ONE IS RIGGED TO LAND HEADS AND THE OTHER RIGGED TO LAND TAILS. WHICH IS WHICH I DO NOT KNOW. NATURE HOWEVER, DOES NOT SUFFER FROM SUCH IGNORANCE!
PLOP
PLOP

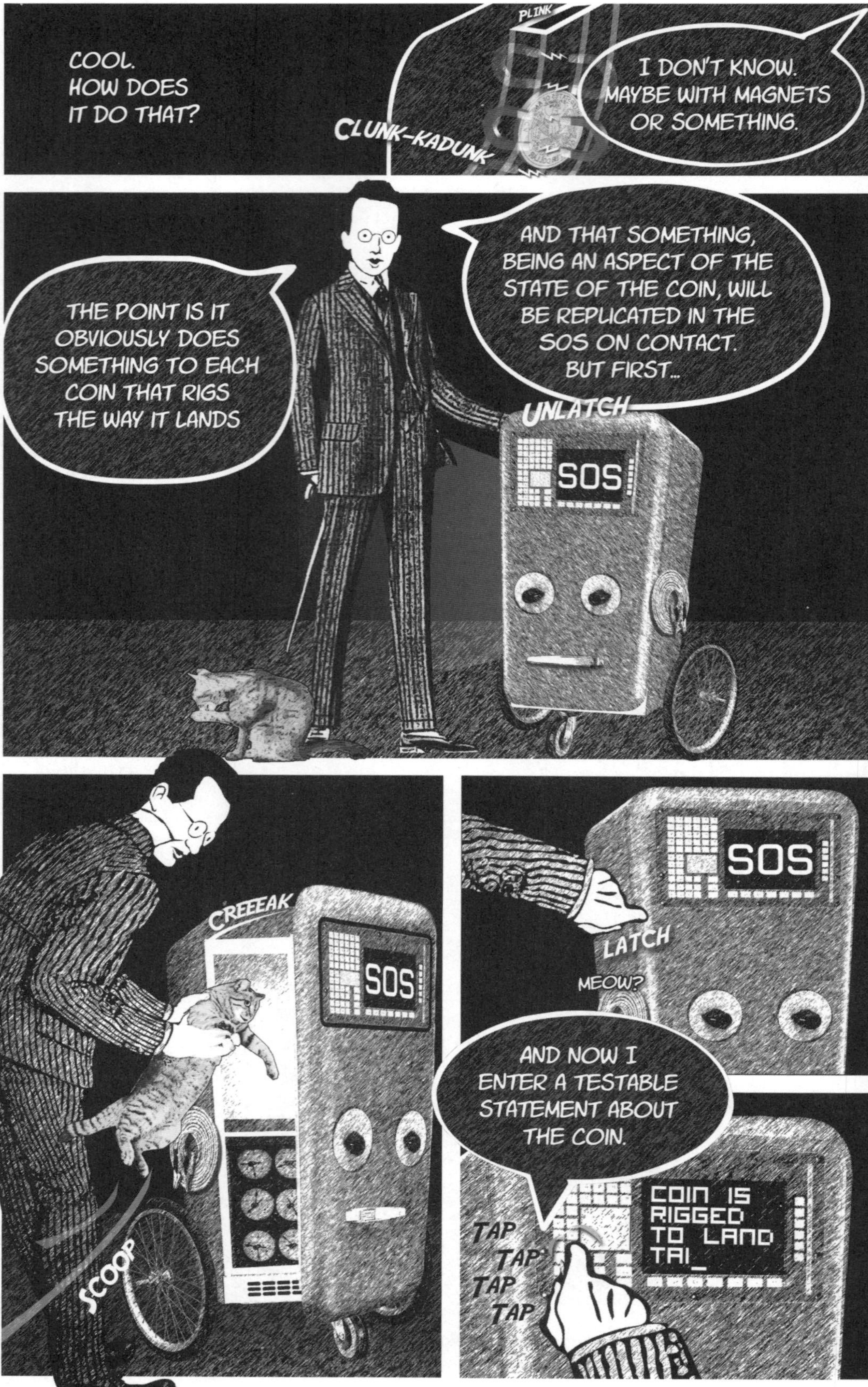
COOL. HOW DOES IT DO THAT?
PLINK
CLUNK-KADUNK
I DON'T KNOW. MAYBE WITH MAGNETS OR SOMETHING.
THE POINT IS IT OBVIOUSLY DOES SOMETHING TO EACH COIN THAT RIGS THE WAY IT LANDS
AND THAT SOMETHING, BEING AN ASPECT OF THE STATE OF THE COIN, WILL BE REPLICATED IN THE SOS ON CONTACT. BUT FIRST...
UNLATCH
SOS
CREEEAK
SOS
SCOOP
SOS
LATCH
MEOW?
AND NOW I ENTER A TESTABLE STATEMENT ABOUT THE COIN.
TAP TAP TAP TAP
COIN IS RIGGED TO LAND TAI_

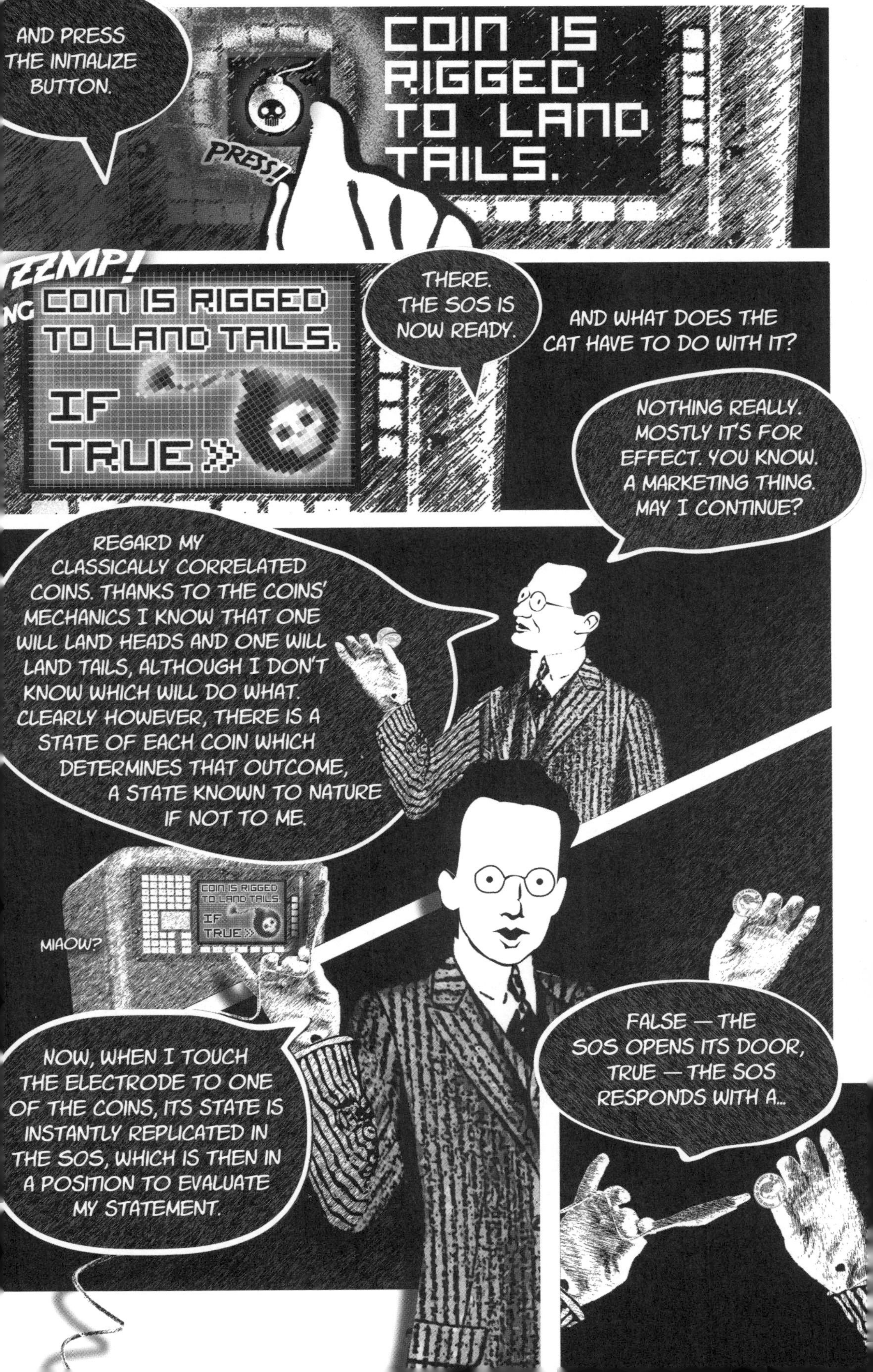
AND PRESS THE INITIALIZE BUTTON.
PRESS!
COIN IS RIGGED TO LAND TAILS.
ZZMP!
COIN IS RIGGED TO LAND TAILS.
IF TRUE »
THERE. THE SOS IS NOW READY.
AND WHAT DOES THE CAT HAVE TO DO WITH IT?
NOTHING REALLY. MOSTLY IT'S FOR EFFECT. YOU KNOW. A MARKETING THING. MAY I CONTINUE?
REGARD MY CLASSICALLY CORRELATED COINS. THANKS TO THE COINS' MECHANICS I KNOW THAT ONE WILL LAND HEADS AND ONE WILL LAND TAILS, ALTHOUGH I DON'T KNOW WHICH WILL DO WHAT. CLEARLY HOWEVER, THERE IS A STATE OF EACH COIN WHICH DETERMINES THAT OUTCOME, A STATE KNOWN TO NATURE IF NOT TO ME.
COIN IS RIGGED TO LAND TAILS.
IF TRUE »
MIAOW?
NOW, WHEN I TOUCH THE ELECTRODE TO ONE OF THE COINS, ITS STATE IS INSTANTLY REPLICATED IN THE SOS, WHICH IS THEN IN A POSITION TO EVALUATE MY STATEMENT.
FALSE — THE SOS OPENS ITS DOOR, TRUE — THE SOS RESPONDS WITH A...

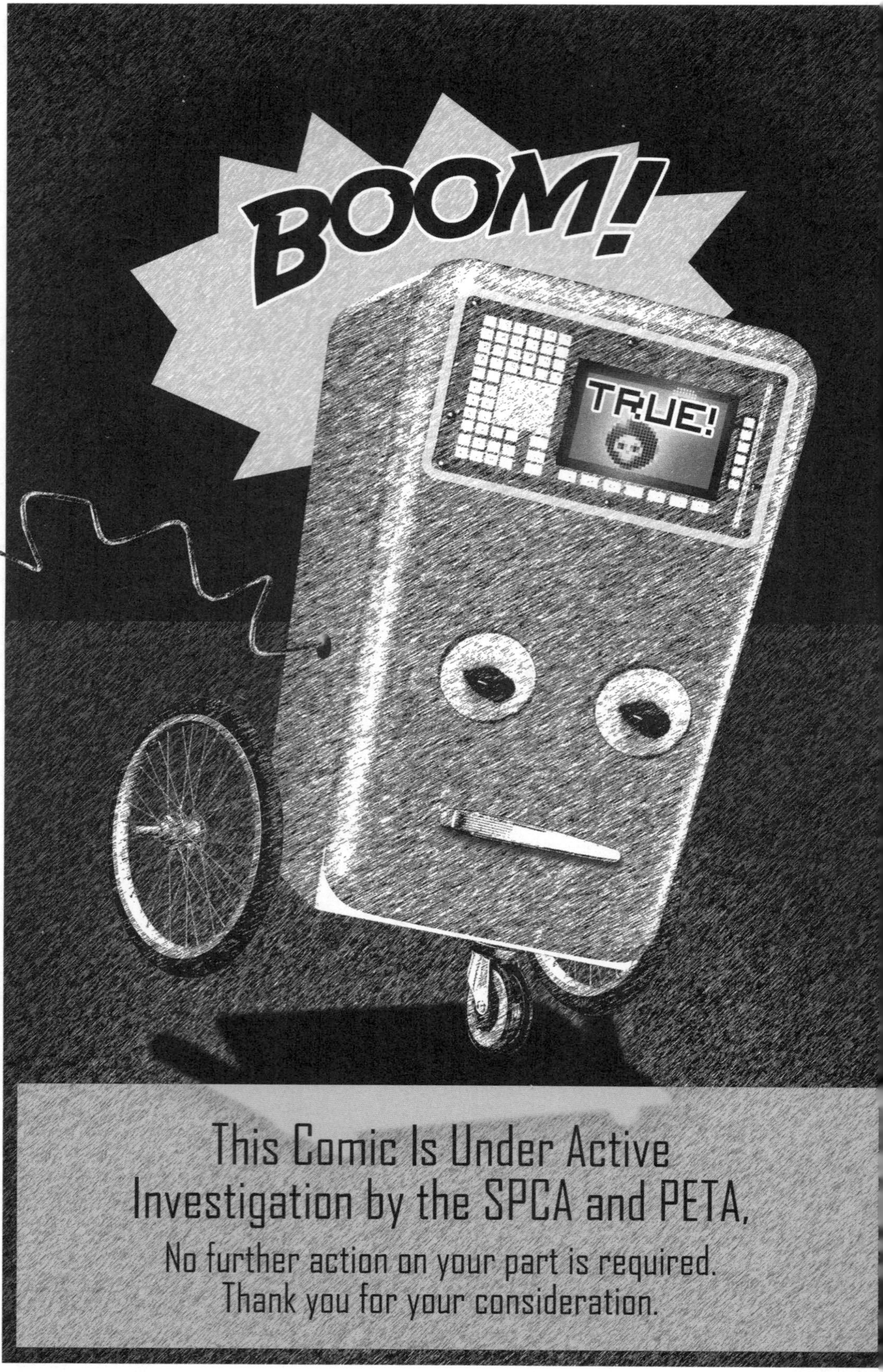
BOOM!
TRUE!
This Comic Is Under Active Investigation by the SPCA and PETA,
No further action on your part is required.
Thank you for your consideration.

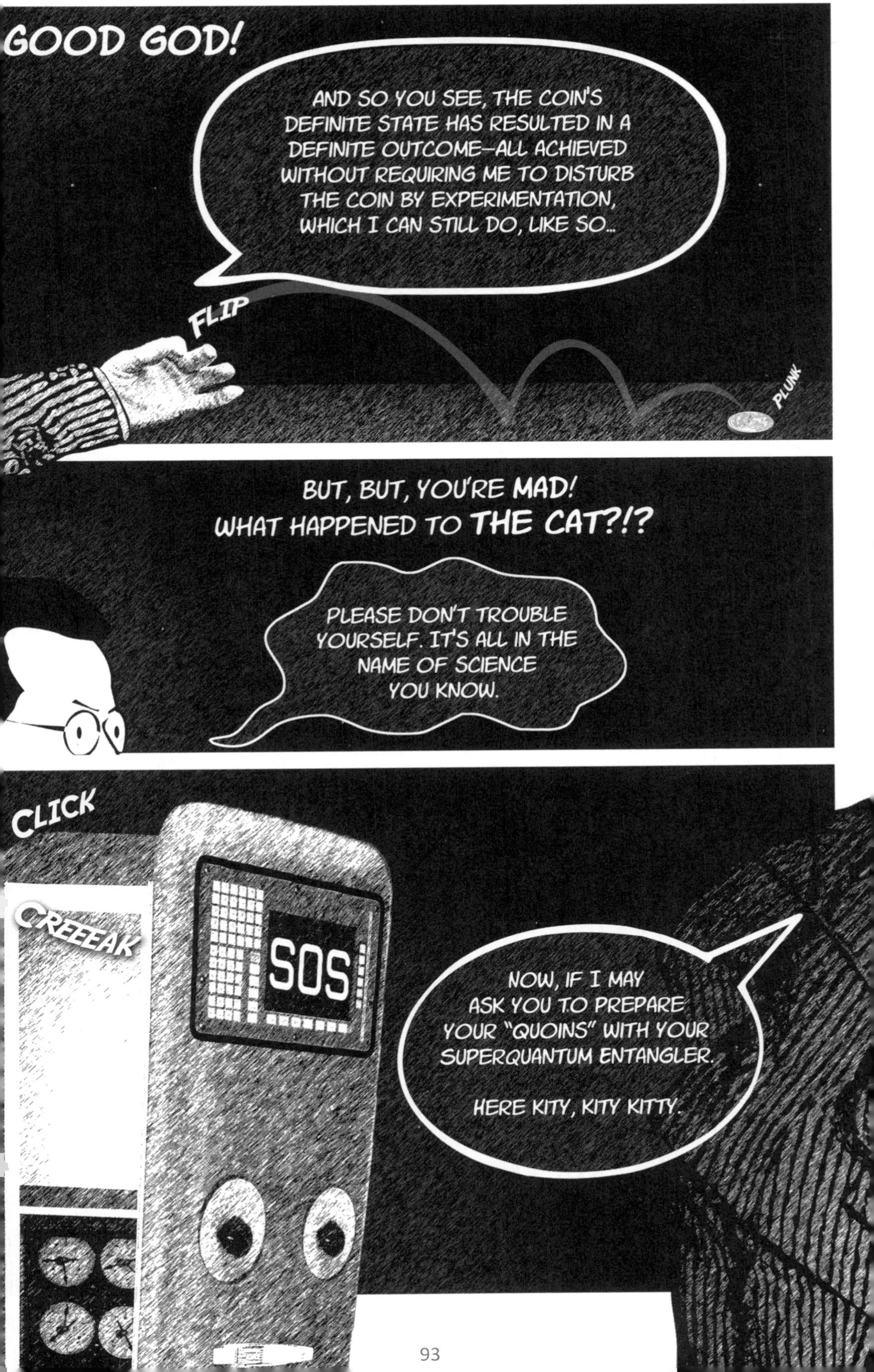
GOOD GOD!
AND SO YOU SEE, THE COIN'S DEFINITE STATE HAS RESULTED IN A DEFINITE OUTCOME—ALL ACHIEVED WITHOUT REQUIRING ME TO DISTURB THE COIN BY EXPERIMENTATION, WHICH I CAN STILL DO, LIKE SO...
FLIP
PLUNK
BUT, BUT, YOU'RE MAD! WHAT HAPPENED TO THE CAT?!?
PLEASE DON'T TROUBLE YOURSELF. IT'S ALL IN THE NAME OF SCIENCE YOU KNOW.
CLICK
CREEEAK
SOS
NOW, IF I MAY ASK YOU TO PREPARE YOUR "QUOINS" WITH YOUR SUPERQUANTUM ENTANGLER.
HERE KITY, KITY KITTY.

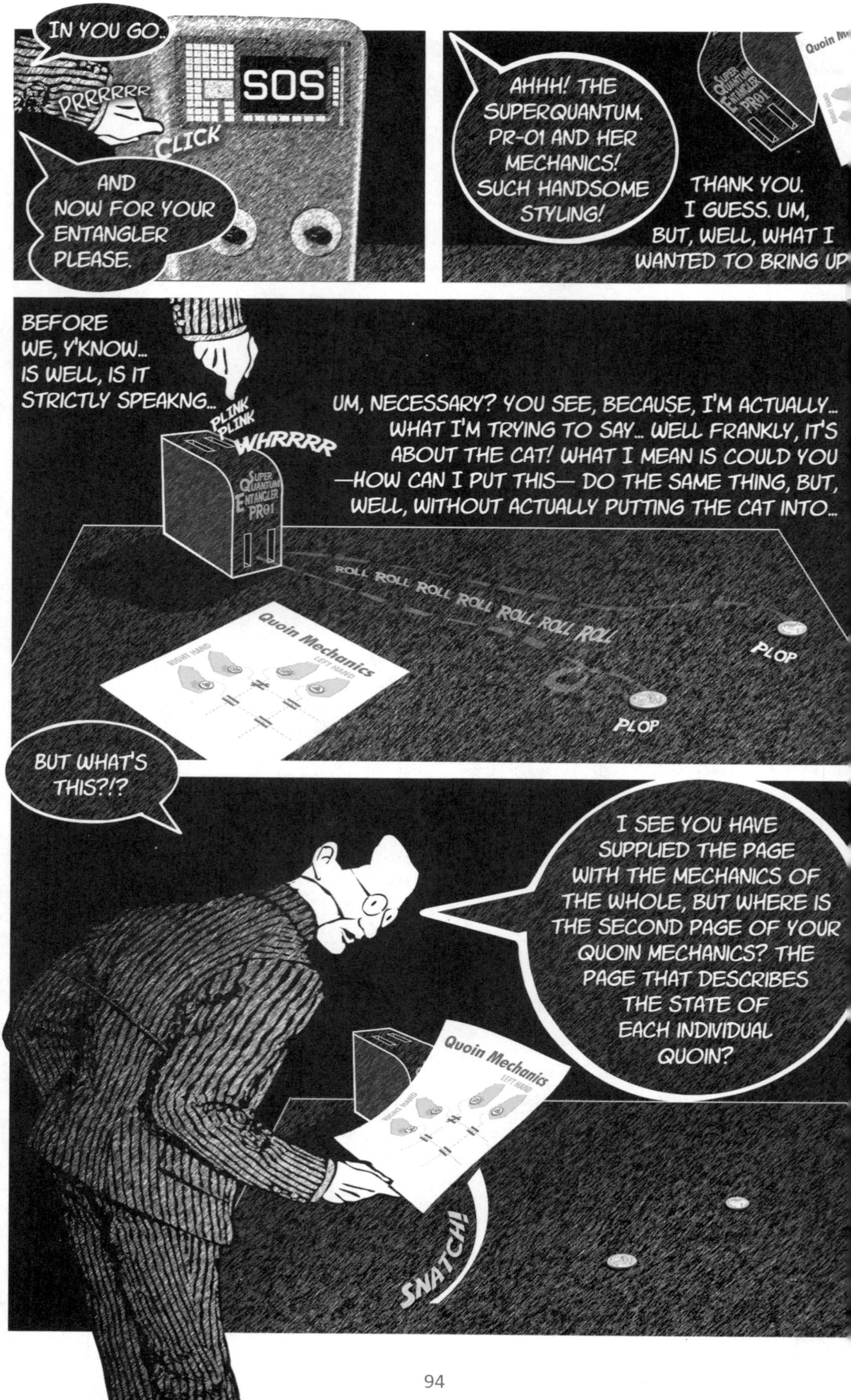
IN YOU GO..
SOS
PRRRRRRR
CLICK
AND NOW FOR YOUR ENTANGLER PLEASE.
AHHH! THE SUPERQUANTUM. PR-01 AND HER MECHANICS! SUCH HANDSOME STYLING!
THANK YOU. I GUESS. UM, BUT, WELL, WHAT I WANTED TO BRING UP
BEFORE WE, Y'KNOW... IS WELL, IS IT STRICTLY SPEAKNG...
PLINK PLINK
WHRRRR
UM, NECESSARY? YOU SEE, BECAUSE, I'M ACTUALLY... WHAT I'M TRYING TO SAY... WELL FRANKLY, IT'S ABOUT THE CAT! WHAT I MEAN IS COULD YOU —HOW CAN I PUT THIS— DO THE SAME THING, BUT, WELL, WITHOUT ACTUALLY PUTTING THE CAT INTO...
SUPER QUANTUM ENTANGLER PR01
ROLL ROLL ROLL ROLL ROLL ROLL ROLL ROLL
Quoin Mechanics
RIGHT HAND
LEFT HAND
PLOP
PLOP
BUT WHAT'S THIS?!?
I SEE YOU HAVE SUPPLIED THE PAGE WITH THE MECHANICS OF THE WHOLE, BUT WHERE IS THE SECOND PAGE OF YOUR QUOIN MECHANICS? THE PAGE THAT DESCRIBES THE STATE OF EACH INDIVIDUAL QUOIN?
Quoin Mechanics
SNATCH!

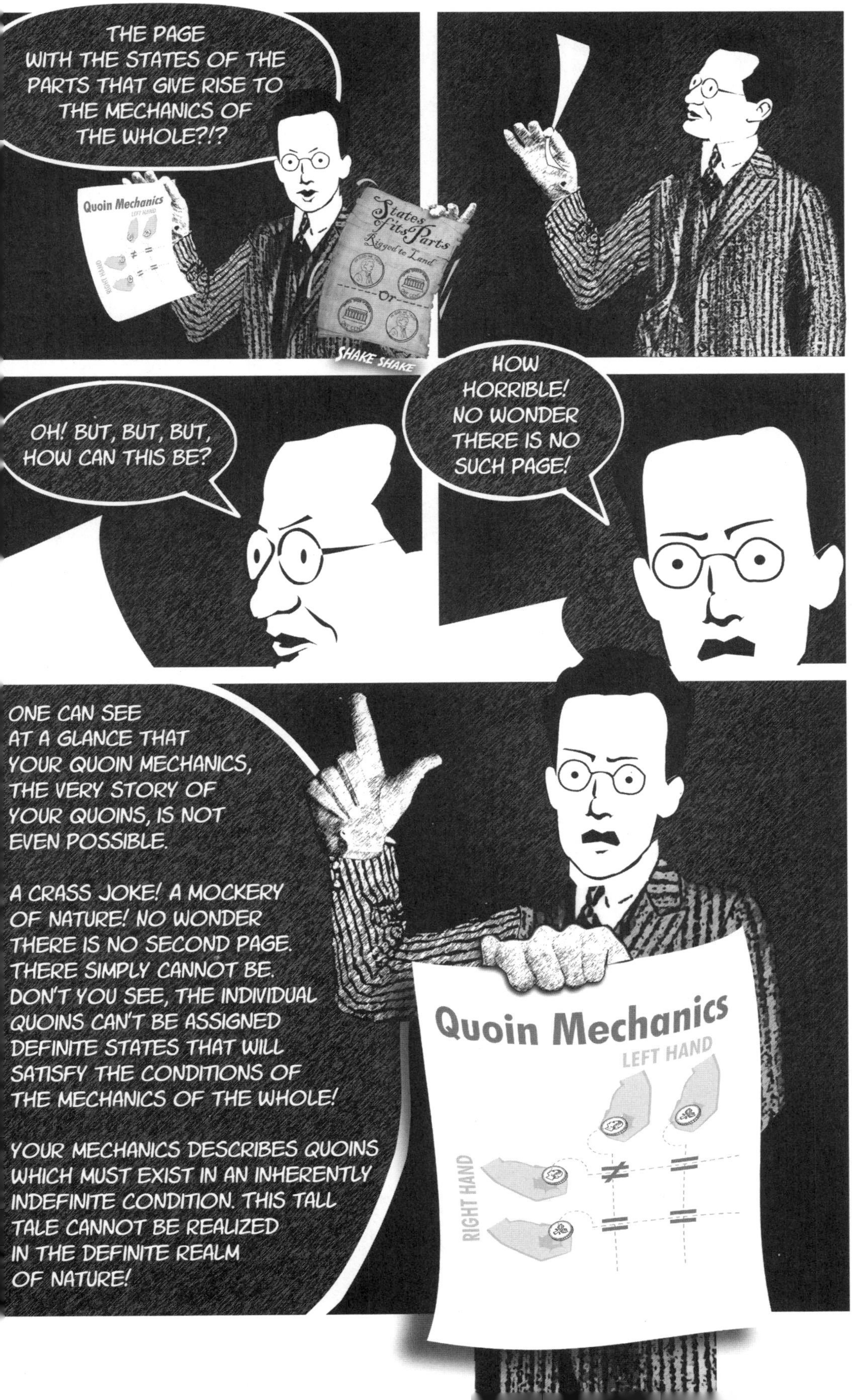
THE PAGE WITH THE STATES OF THE PARTS THAT GIVE RISE TO THE MECHANICS OF THE WHOLE?!?
Quoin Mechanics
LEFT HAND
RIGHT HAND
States of its Parts
Rigged to Land
Or
SHAKE SHAKE
OH! BUT, BUT, BUT, HOW CAN THIS BE?
HOW HORRIBLE! NO WONDER THERE IS NO SUCH PAGE!
ONE CAN SEE AT A GLANCE THAT YOUR QUOIN MECHANICS, THE VERY STORY OF YOUR QUOINS, IS NOT EVEN POSSIBLE.
A CRASS JOKE! A MOCKERY OF NATURE! NO WONDER THERE IS NO SECOND PAGE. THERE SIMPLY CANNOT BE. DON'T YOU SEE, THE INDIVIDUAL QUOINS CAN'T BE ASSIGNED DEFINITE STATES THAT WILL SATISFY THE CONDITIONS OF THE MECHANICS OF THE WHOLE!
YOUR MECHANICS DESCRIBES QUOINS WHICH MUST EXIST IN AN INHERENTLY INDEFINITE CONDITION. THIS TALL TALE CANNOT BE REALIZED IN THE DEFINITE REALM OF NATURE!
Quoin Mechanics
LEFT HAND
RIGHT HAND

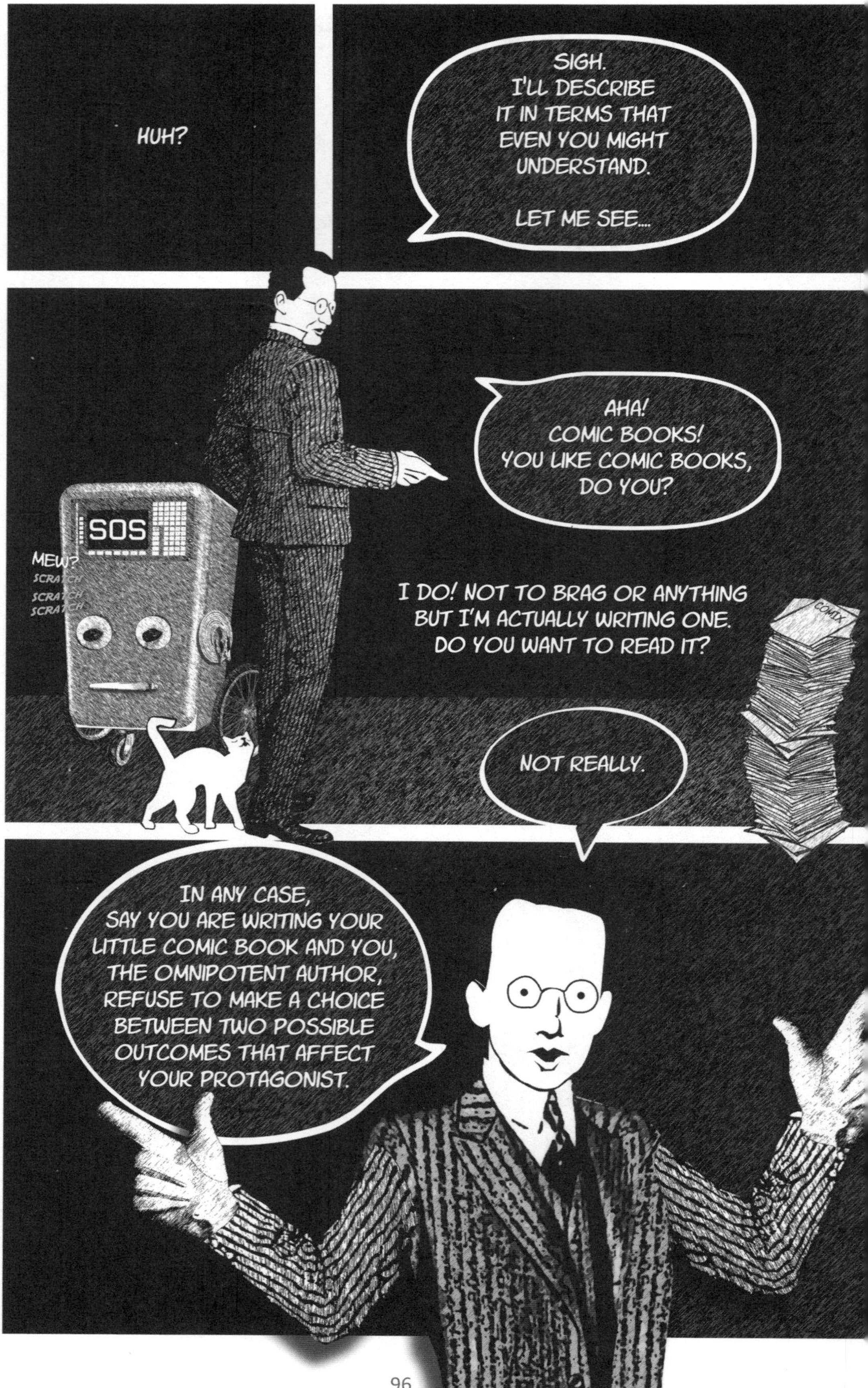
HUH?
SIGH. I'LL DESCRIBE IT IN TERMS THAT EVEN YOU MIGHT UNDERSTAND.
LET ME SEE....
AHA! COMIC BOOKS! YOU LIKE COMIC BOOKS, DO YOU?
SOS
MEW?
SCRATCH
SCRATCH
SCRATCH
I DO! NOT TO BRAG OR ANYTHING BUT I'M ACTUALLY WRITING ONE. DO YOU WANT TO READ IT?
COMIX
NOT REALLY.
IN ANY CASE, SAY YOU ARE WRITING YOUR LITTLE COMIC BOOK AND YOU, THE OMNIPOTENT AUTHOR, REFUSE TO MAKE A CHOICE BETWEEN TWO POSSIBLE OUTCOMES THAT AFFECT YOUR PROTAGONIST.

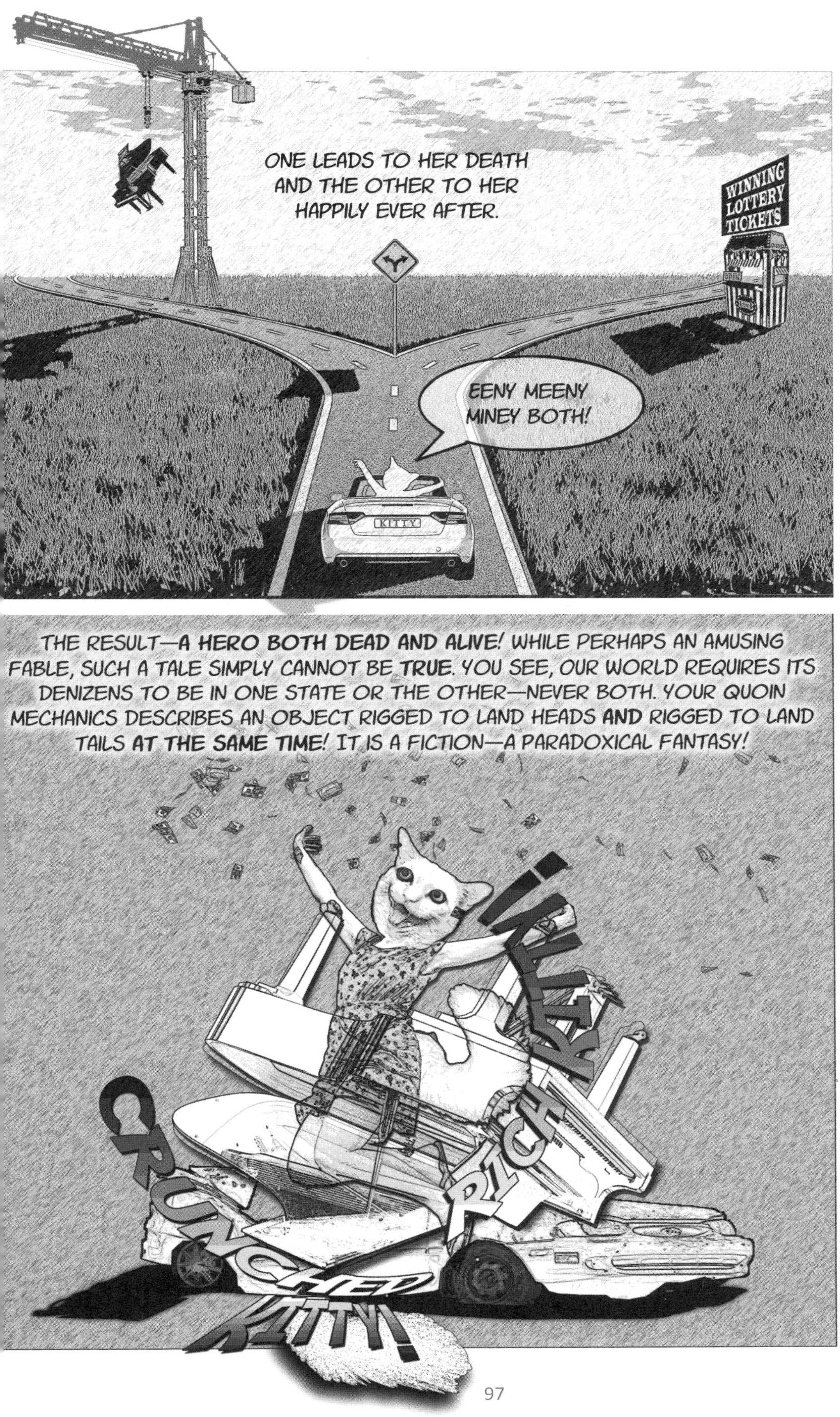
ONE LEADS TO HER DEATH AND THE OTHER TO HER HAPPILY EVER AFTER.
WINNING LOTTERY TICKETS
EENY MEENY MINEY BOTH!
KITTY
THE RESULT—**A HERO BOTH DEAD AND ALIVE!** WHILE PERHAPS AN AMUSING FABLE, SUCH A TALE SIMPLY CANNOT BE **TRUE**. YOU SEE, OUR WORLD REQUIRES ITS DENIZENS TO BE IN ONE STATE OR THE OTHER—NEVER BOTH. YOUR QUOIN MECHANICS DESCRIBES AN OBJECT RIGGED TO LAND HEADS **AND** RIGGED TO LAND TAILS **AT THE SAME TIME!** IT IS A FICTION—A PARADOXICAL FANTASY!
RICH KITTY!
CRUNCHED KITTY!

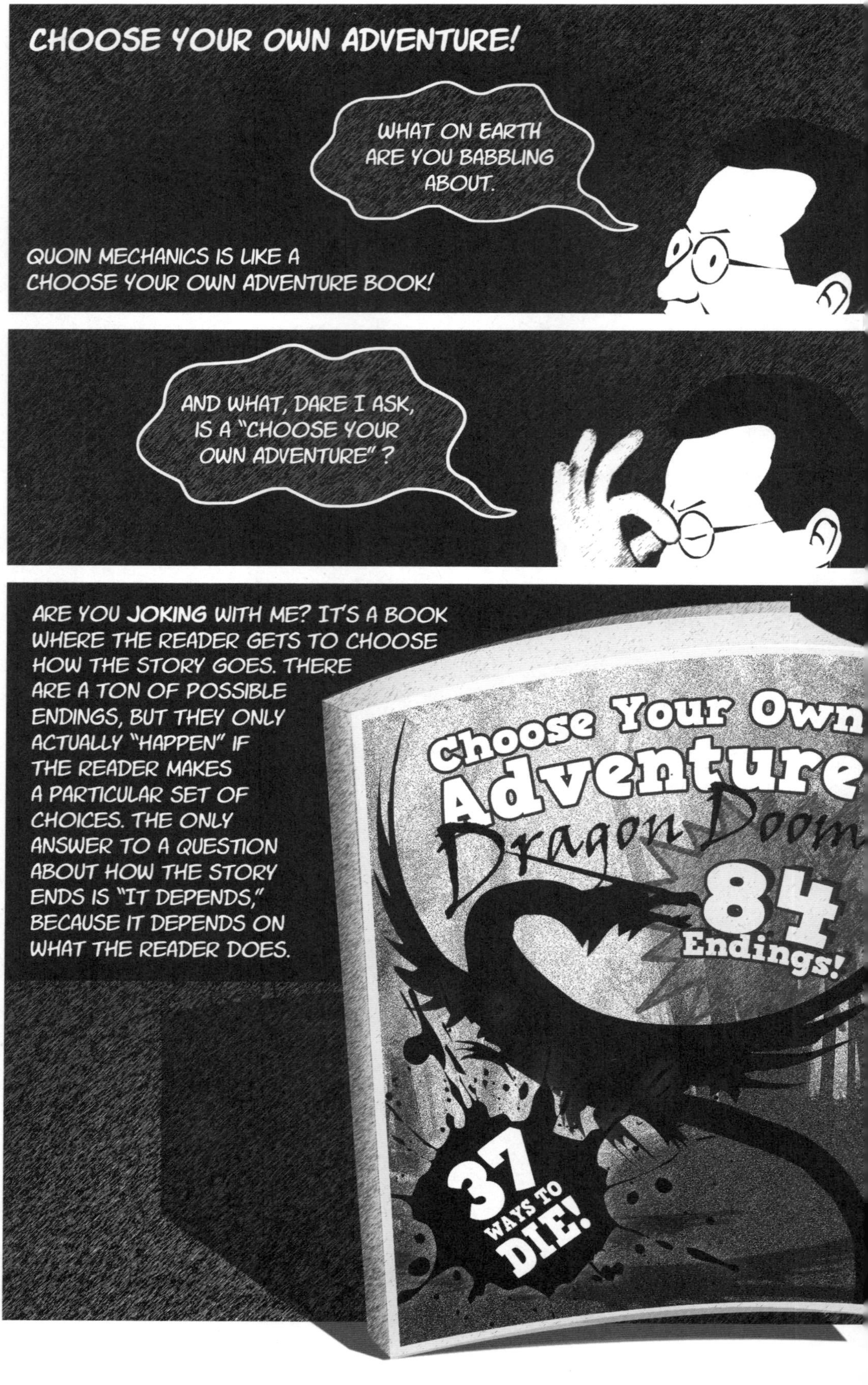
CHOOSE YOUR OWN ADVENTURE!
WHAT ON EARTH ARE YOU BABBLING ABOUT.
QUOIN MECHANICS IS LIKE A CHOOSE YOUR OWN ADVENTURE BOOK!
AND WHAT, DARE I ASK, IS A "CHOOSE YOUR OWN ADVENTURE"?
ARE YOU **JOKING** WITH ME? IT'S A BOOK WHERE THE READER GETS TO CHOOSE HOW THE STORY GOES. THERE ARE A TON OF POSSIBLE ENDINGS, BUT THEY ONLY ACTUALLY "HAPPEN" IF THE READER MAKES A PARTICULAR SET OF CHOICES. THE ONLY ANSWER TO A QUESTION ABOUT HOW THE STORY ENDS IS "IT DEPENDS," BECAUSE IT DEPENDS ON WHAT THE READER DOES.
Choose Your Own Adventure
Dragon Doom
84 Endings!
37 WAYS TO DIE!

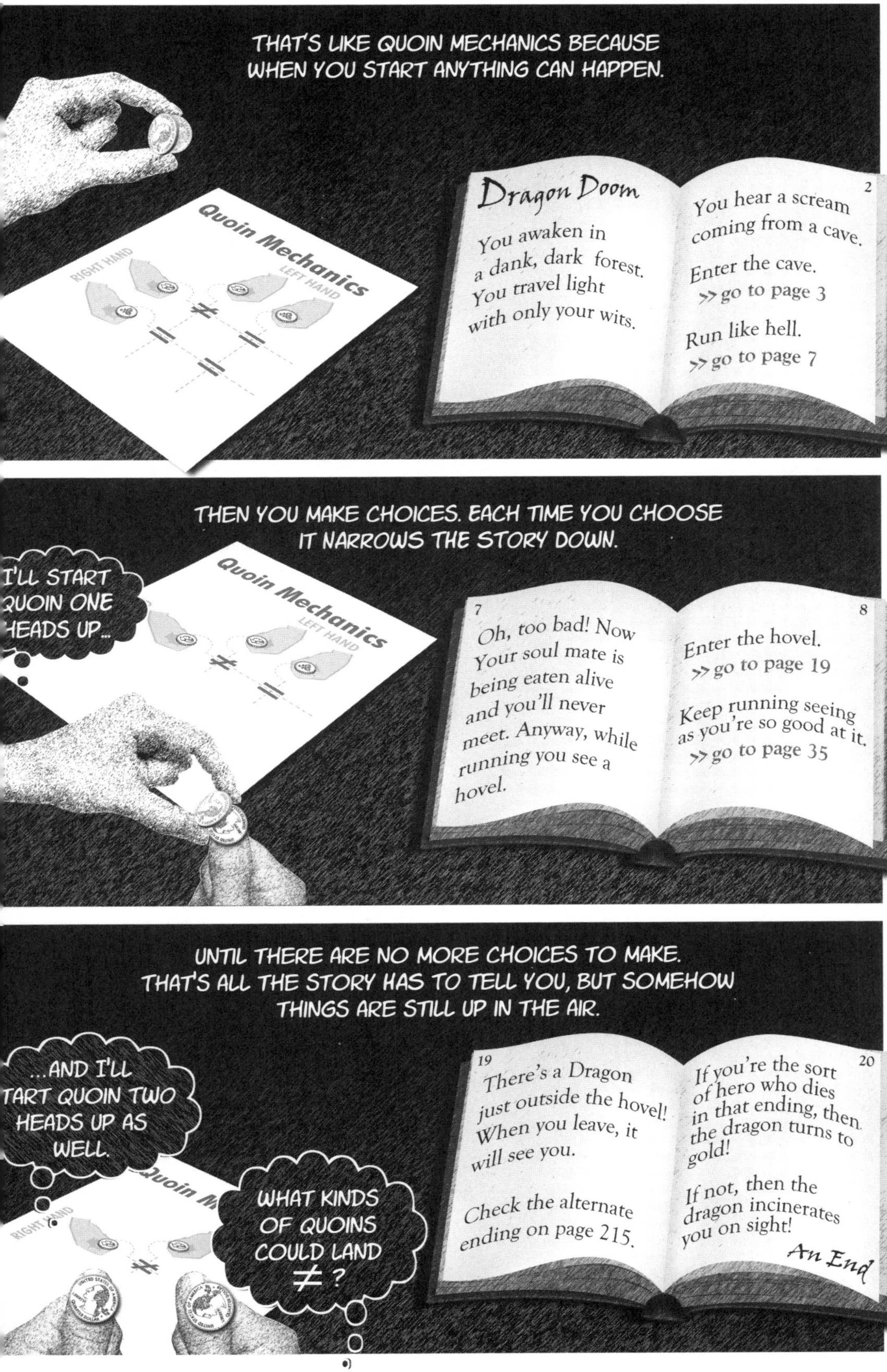

THAT'S LIKE QUOIN MECHANICS BECAUSE WHEN YOU START ANYTHING CAN HAPPEN.
Quoin Mechanics
RIGHT HAND
LEFT HAND
Dragon Doom
You awaken in a dank, dark forest. You travel light with only your wits.
2
You hear a scream coming from a cave.
Enter the cave.
>> go to page 3
Run like hell.
>> go to page 7
THEN YOU MAKE CHOICES. EACH TIME YOU CHOOSE IT NARROWS THE STORY DOWN.
I'LL START QUOIN ONE HEADS UP...
Quoin Mechanics
LEFT HAND
7
Oh, too bad! Now Your soul mate is being eaten alive and you'll never meet. Anyway, while running you see a hovel.
8
Enter the hovel.
>> go to page 19
Keep running seeing as you're so good at it.
>> go to page 35
UNTIL THERE ARE NO MORE CHOICES TO MAKE. THAT'S ALL THE STORY HAS TO TELL YOU, BUT SOMEHOW THINGS ARE STILL UP IN THE AIR.
...AND I'LL TART QUOIN TWO HEADS UP AS WELL.
RIGHT HAND
WHAT KINDS OF QUOINS COULD LAND ≠?
19
There's a Dragon just outside the hovel! When you leave, it will see you.
Check the alternate ending on page 215.
20
If you're the sort of hero who dies in that ending, then the dragon turns to gold!
If not, then the dragon incinerates you on sight!
An End

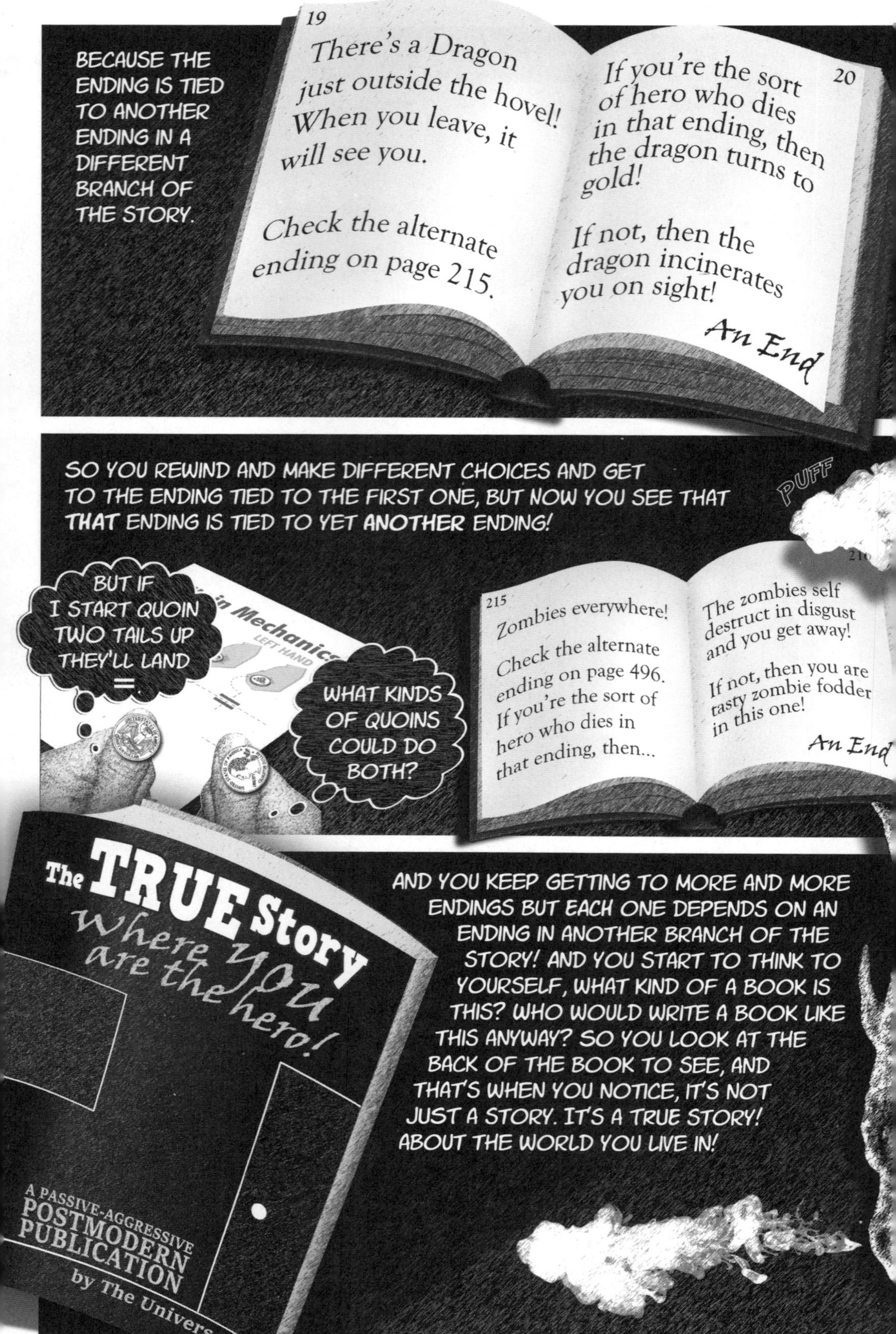

BECAUSE THE ENDING IS TIED TO ANOTHER ENDING IN A DIFFERENT BRANCH OF THE STORY.

19

There's a Dragon just outside the hovel! When you leave, it will see you.

Check the alternate ending on page 215.

20

If you're the sort of hero who dies in that ending, then the dragon turns to gold!

If not, then the dragon incinerates you on sight!

An End

SO YOU REWIND AND MAKE DIFFERENT CHOICES AND GET TO THE ENDING TIED TO THE FIRST ONE, BUT NOW YOU SEE THAT **THAT** ENDING IS TIED TO YET **ANOTHER** ENDING!

PUFF

BUT IF I START QUOIN TWO TAILS UP THEY'LL LAND =

in Mechanics
LEFT HAND

WHAT KINDS OF QUOINS COULD DO BOTH?

215

Zombies everywhere!

Check the alternate ending on page 496. If you're the sort of hero who dies in that ending, then...

The zombies self destruct in disgust and you get away!

If not, then you are tasty zombie fodder in this one!

An End

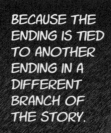

The **TRUE** Story
where you are the hero!

A PASSIVE-AGGRESSIVE **POSTMODERN PUBLICATION**
by The Universe

AND YOU KEEP GETTING TO MORE AND MORE ENDINGS BUT EACH ONE DEPENDS ON AN ENDING IN ANOTHER BRANCH OF THE STORY! AND YOU START TO THINK TO YOURSELF, WHAT KIND OF A BOOK IS THIS? WHO WOULD WRITE A BOOK LIKE THIS ANYWAY? SO YOU LOOK AT THE BACK OF THE BOOK TO SEE, AND THAT'S WHEN YOU NOTICE, IT'S NOT JUST A STORY. IT'S A TRUE STORY! ABOUT THE WORLD YOU LIVE IN!

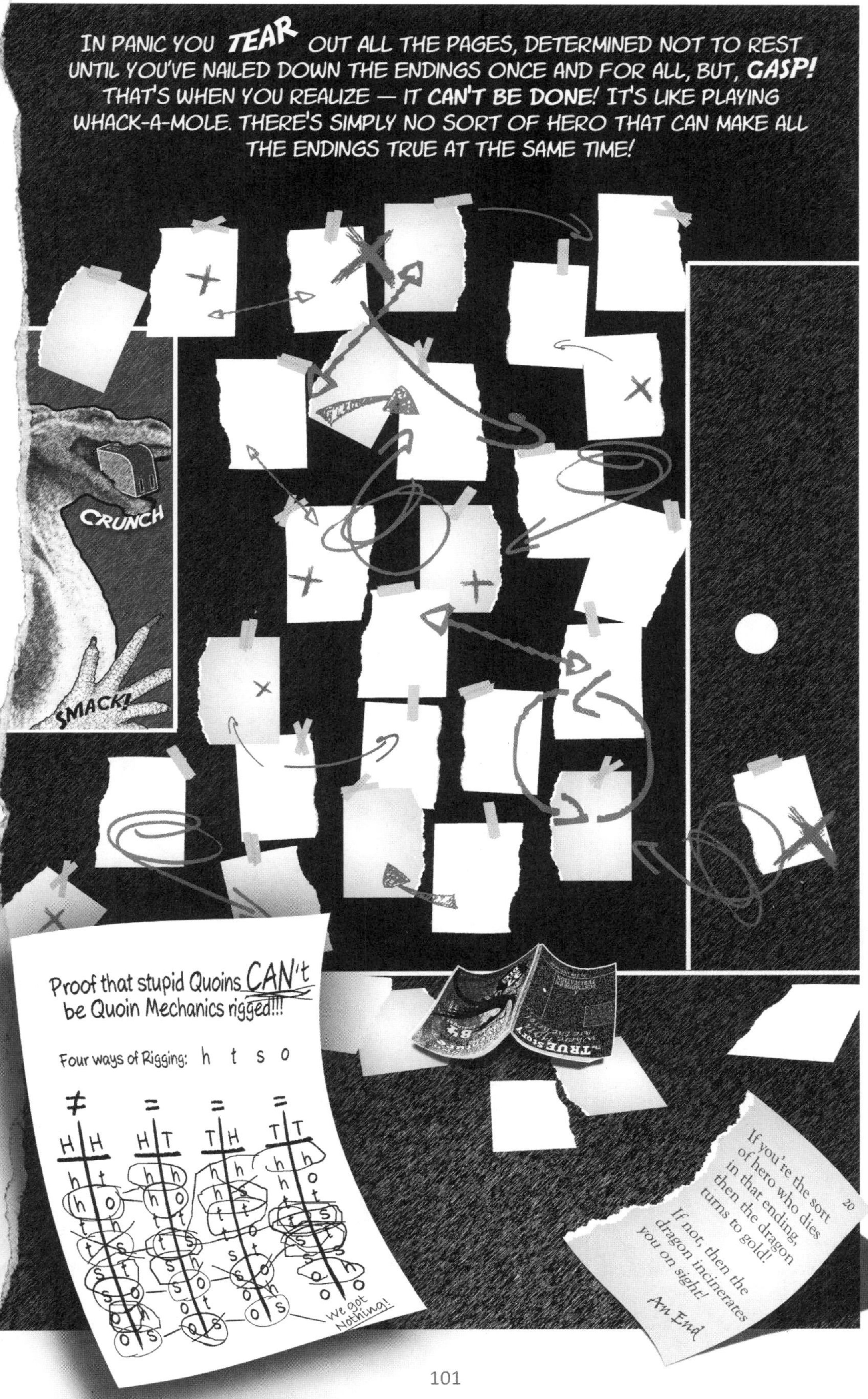
IN PANIC YOU TEAR OUT ALL THE PAGES, DETERMINED NOT TO REST UNTIL YOU'VE NAILED DOWN THE ENDINGS ONCE AND FOR ALL, BUT, GASP! THAT'S WHEN YOU REALIZE — IT CAN'T BE DONE! IT'S LIKE PLAYING WHACK-A-MOLE. THERE'S SIMPLY NO SORT OF HERO THAT CAN MAKE ALL THE ENDINGS TRUE AT THE SAME TIME!
CRUNCH
SMACK!
Proof that stupid Quoins CAN't be Quoin Mechanics rigged!!!
Four ways of Rigging: h t s o
HH HT TH TT
We got Nothing!
The TRUE story
If you're the sort of hero who dies in that ending, then the dragon turns to gold!
If not, then the dragon incinerates you on sight!
20
An End

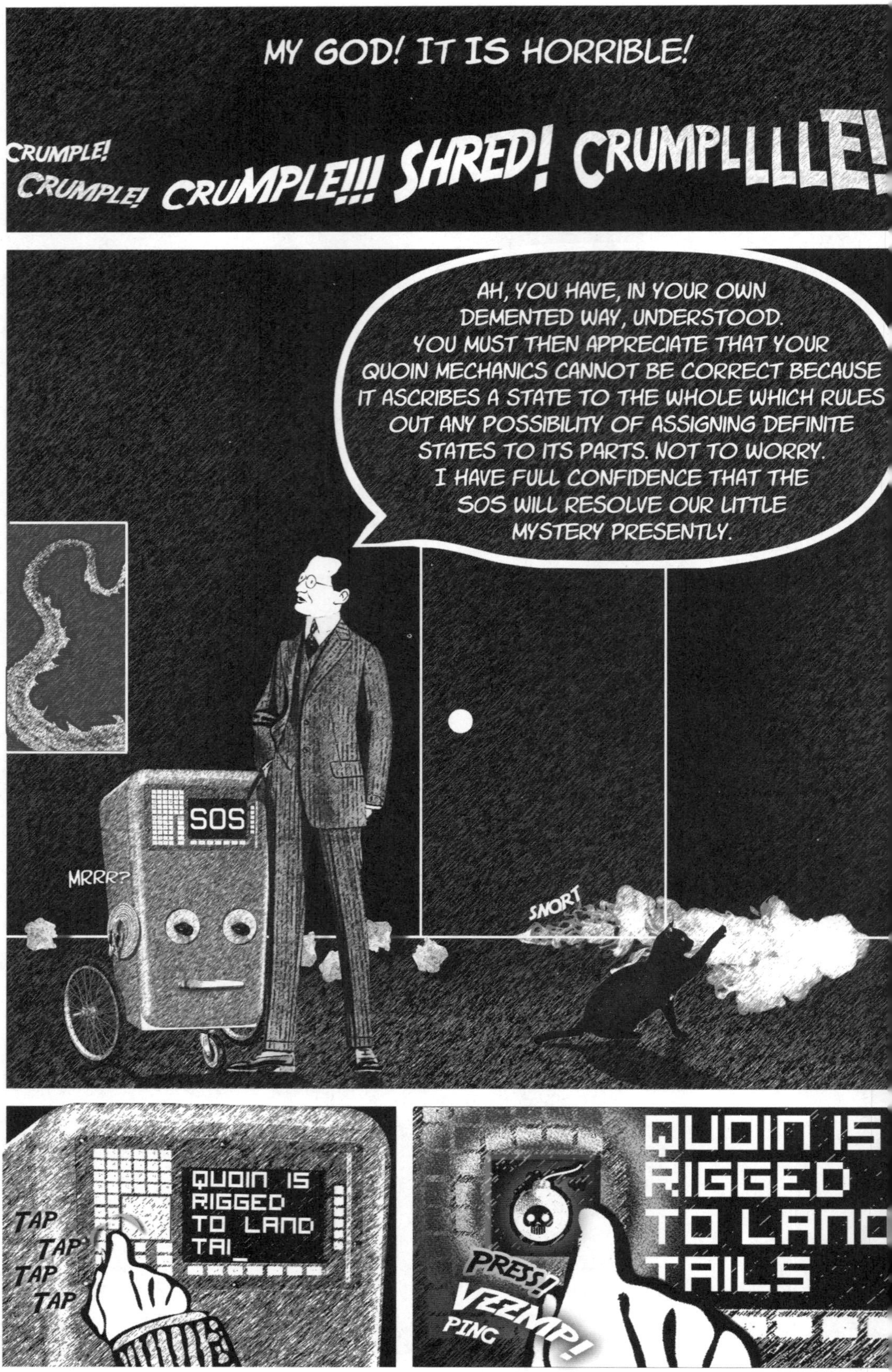
MY GOD! IT IS HORRIBLE!
CRUMPLE!
CRUMPLE!
CRUMPLE!!!
SHRED!
CRUMPLLLLE!
AH, YOU HAVE, IN YOUR OWN DEMENTED WAY, UNDERSTOOD. YOU MUST THEN APPRECIATE THAT YOUR QUOIN MECHANICS CANNOT BE CORRECT BECAUSE IT ASCRIBES A STATE TO THE WHOLE WHICH RULES OUT ANY POSSIBILITY OF ASSIGNING DEFINITE STATES TO ITS PARTS. NOT TO WORRY. I HAVE FULL CONFIDENCE THAT THE SOS WILL RESOLVE OUR LITTLE MYSTERY PRESENTLY.
SOS
MRRR?
SNORT
TAP
TAP
TAP
TAP
QUOIN IS RIGGED TO LAND TAI_
PRESS!
VZZMP!
PING
QUOIN IS RIGGED TO LAND TAILS

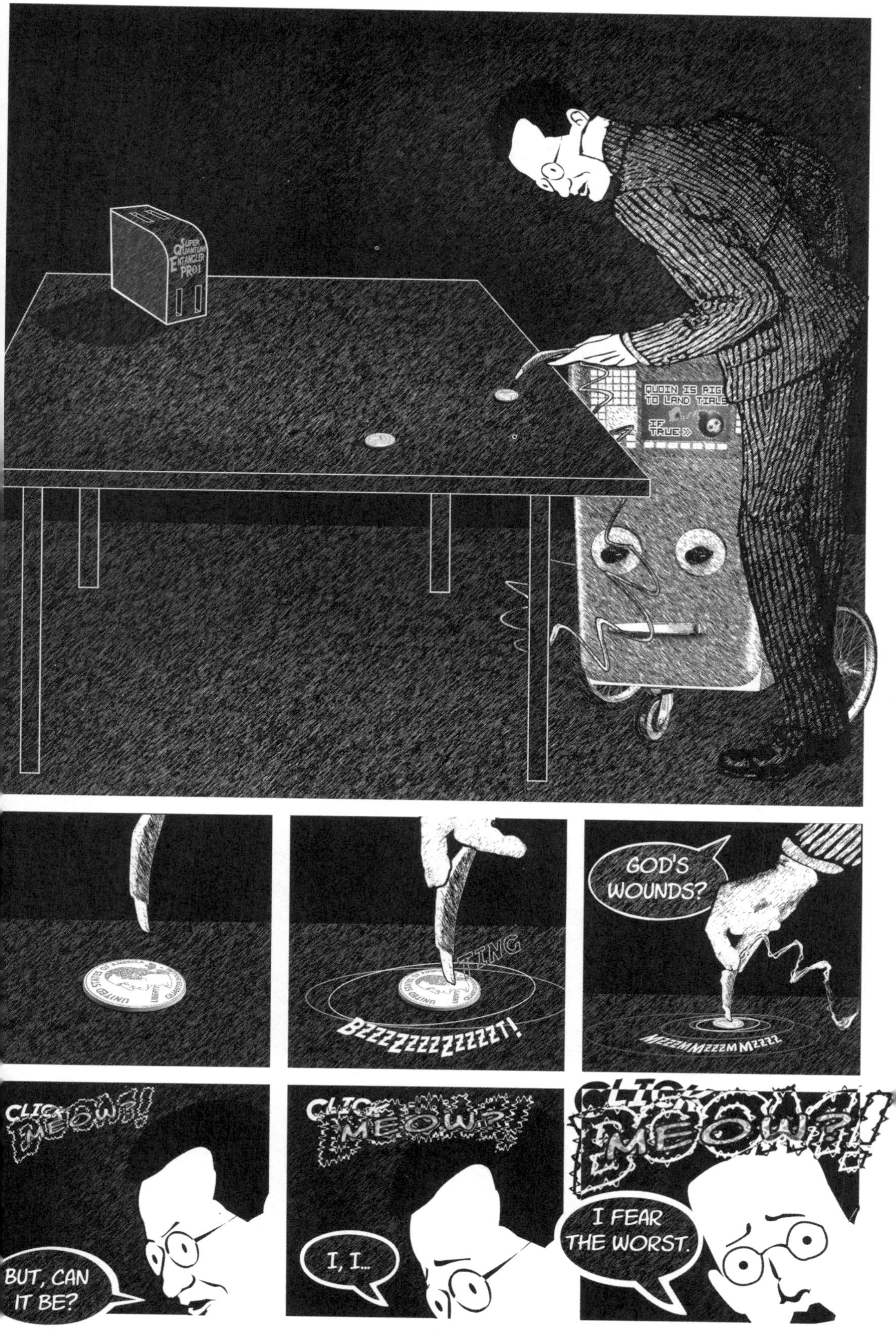
SUPER QUANTUM ENTANGLER PRO1
QUOIN IS RIG
TO LAND TIALS
IF TRUE »
TING
BZZZZZZZZZZZZT!
GOD'S WOUNDS?
MZZZMMZZZMMZZZZ
MEOW?!
BUT, CAN IT BE?
MEOW?!
I, I...
MEOW?!
I FEAR THE WORST.

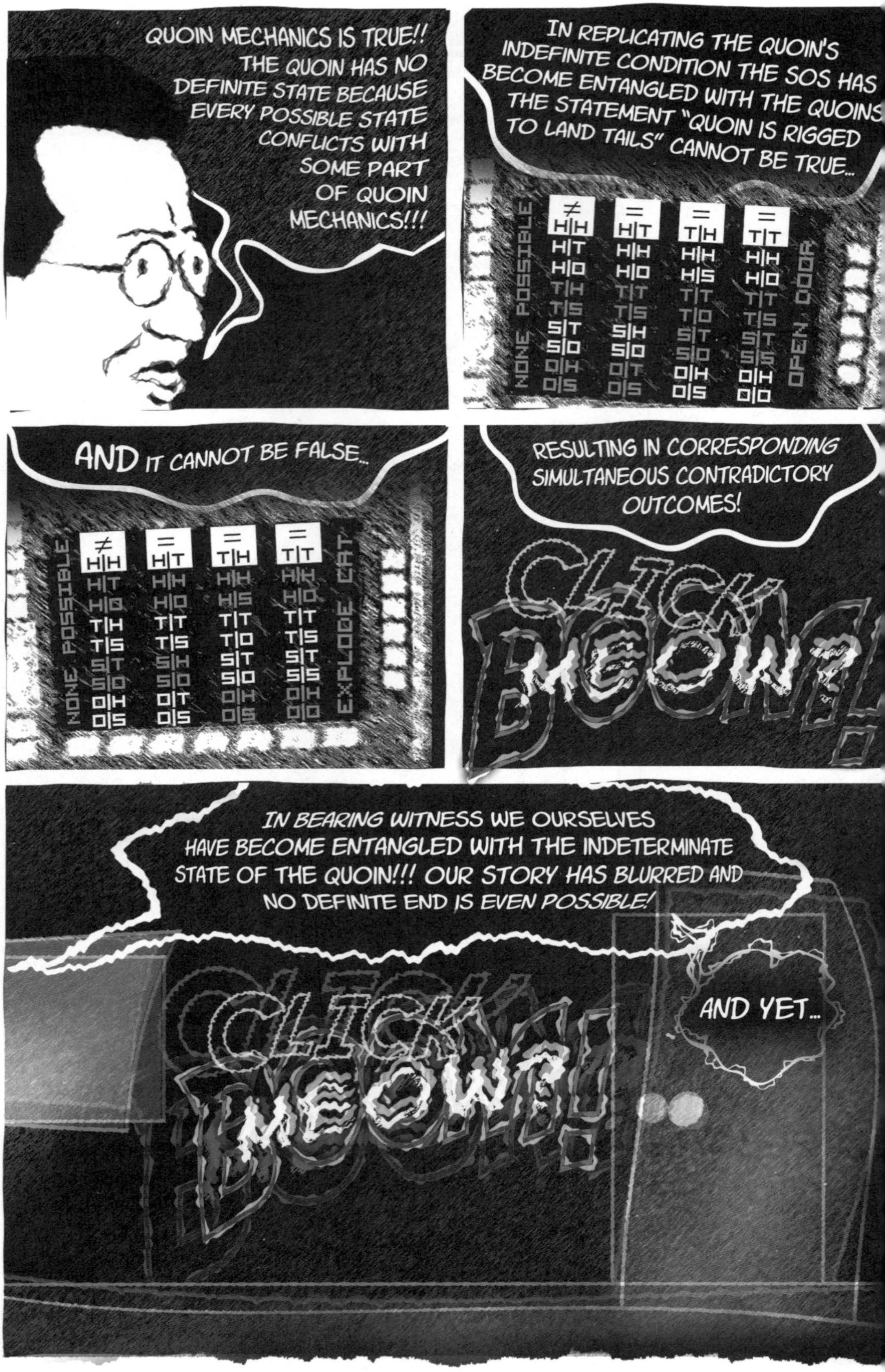
QUOIN MECHANICS IS TRUE!! THE QUOIN HAS NO DEFINITE STATE BECAUSE EVERY POSSIBLE STATE CONFLICTS WITH SOME PART OF QUOIN MECHANICS!!!
IN REPLICATING THE QUOIN'S INDEFINITE CONDITION THE SOS HAS BECOME ENTANGLED WITH THE QUOINS THE STATEMENT "QUOIN IS RIGGED TO LAND TAILS" CANNOT BE TRUE...
NONE POSSIBLE
OPEN DOOR
AND IT CANNOT BE FALSE...
NONE POSSIBLE
EXPLODE CAT
RESULTING IN CORRESPONDING SIMULTANEOUS CONTRADICTORY OUTCOMES!
CLICK
BOOM!
MEOW!
IN BEARING WITNESS WE OURSELVES HAVE BECOME ENTANGLED WITH THE INDETERMINATE STATE OF THE QUOIN!!! OUR STORY HAS BLURRED AND NO DEFINITE END IS EVEN POSSIBLE!
CLICK
BOOM!
MEOW!
AND YET...

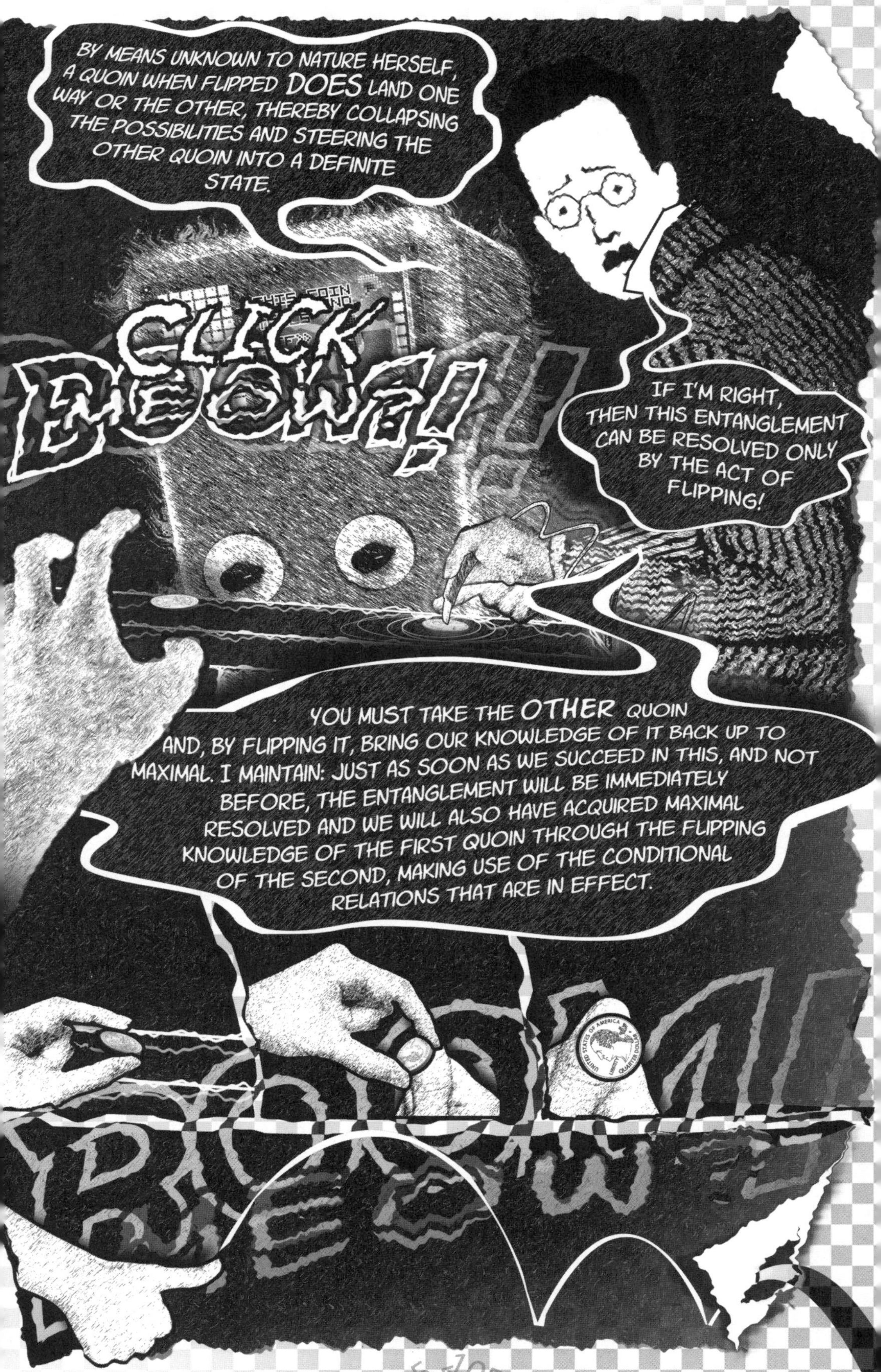
BY MEANS UNKNOWN TO NATURE HERSELF, A QUOIN WHEN FLIPPED DOES LAND ONE WAY OR THE OTHER, THEREBY COLLAPSING THE POSSIBILITIES AND STEERING THE OTHER QUOIN INTO A DEFINITE STATE.
CLICK BLOW!!
IF I'M RIGHT, THEN THIS ENTANGLEMENT CAN BE RESOLVED ONLY BY THE ACT OF FLIPPING!
YOU MUST TAKE THE OTHER QUOIN AND, BY FLIPPING IT, BRING OUR KNOWLEDGE OF IT BACK UP TO MAXIMAL. I MAINTAIN: JUST AS SOON AS WE SUCCEED IN THIS, AND NOT BEFORE, THE ENTANGLEMENT WILL BE IMMEDIATELY RESOLVED AND WE WILL ALSO HAVE ACQUIRED MAXIMAL KNOWLEDGE OF THE FIRST QUOIN THROUGH THE FLIPPING OF THE SECOND, MAKING USE OF THE CONDITIONAL RELATIONS THAT ARE IN EFFECT.

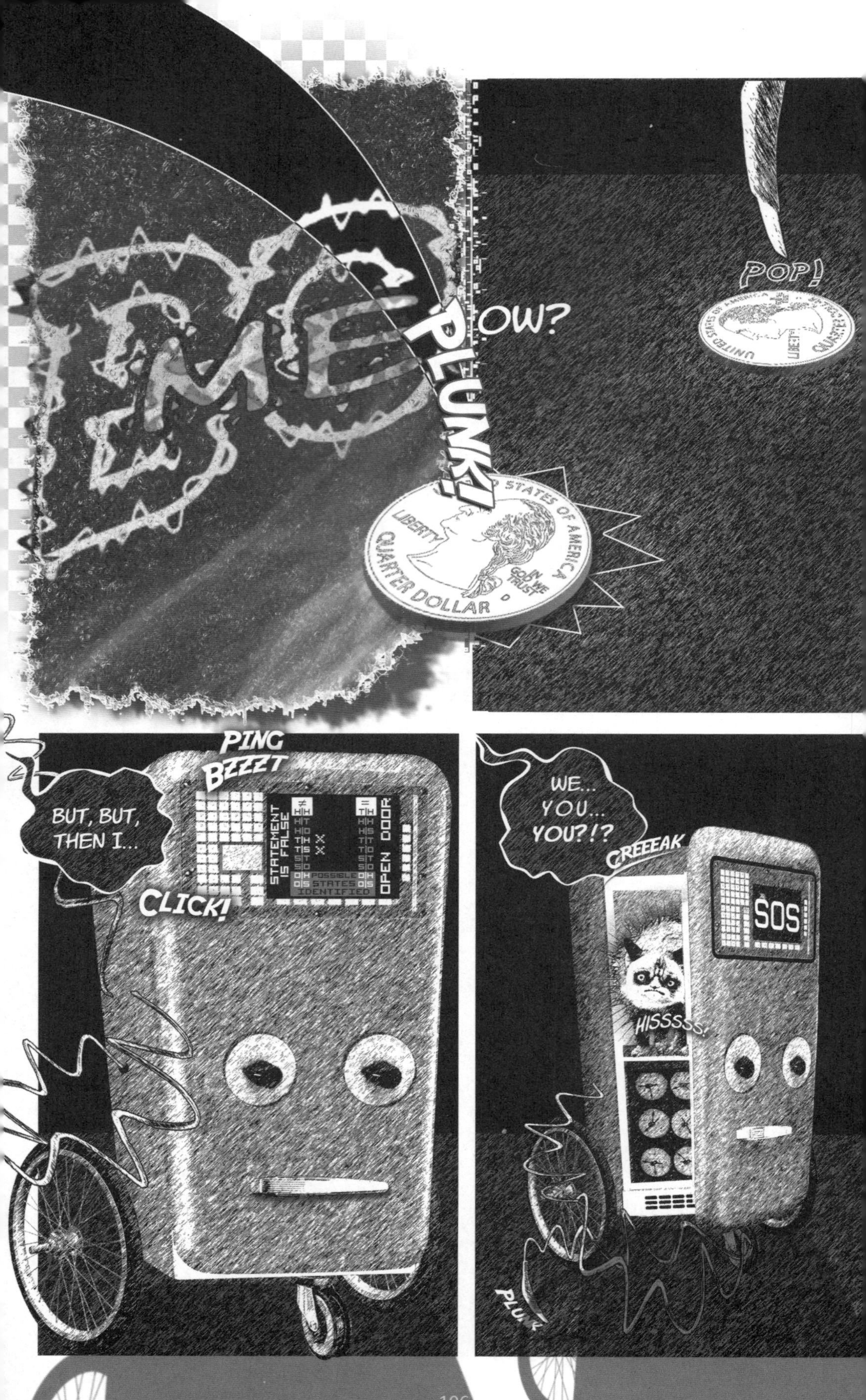
MEOW?
PLUNK!
UNITED STATES OF AMERICA
LIBERTY
QUARTER DOLLAR
IN GOD WE TRUST
POP!
PING
BZZZT
BUT, BUT, THEN I...
STATEMENT IS FALSE
POSSIBLE STATES IDENTIFIED
OPEN DOOR
CLICK!
WE... YOU... YOU?!?
CREEEAK
SOS
HISSSSS!

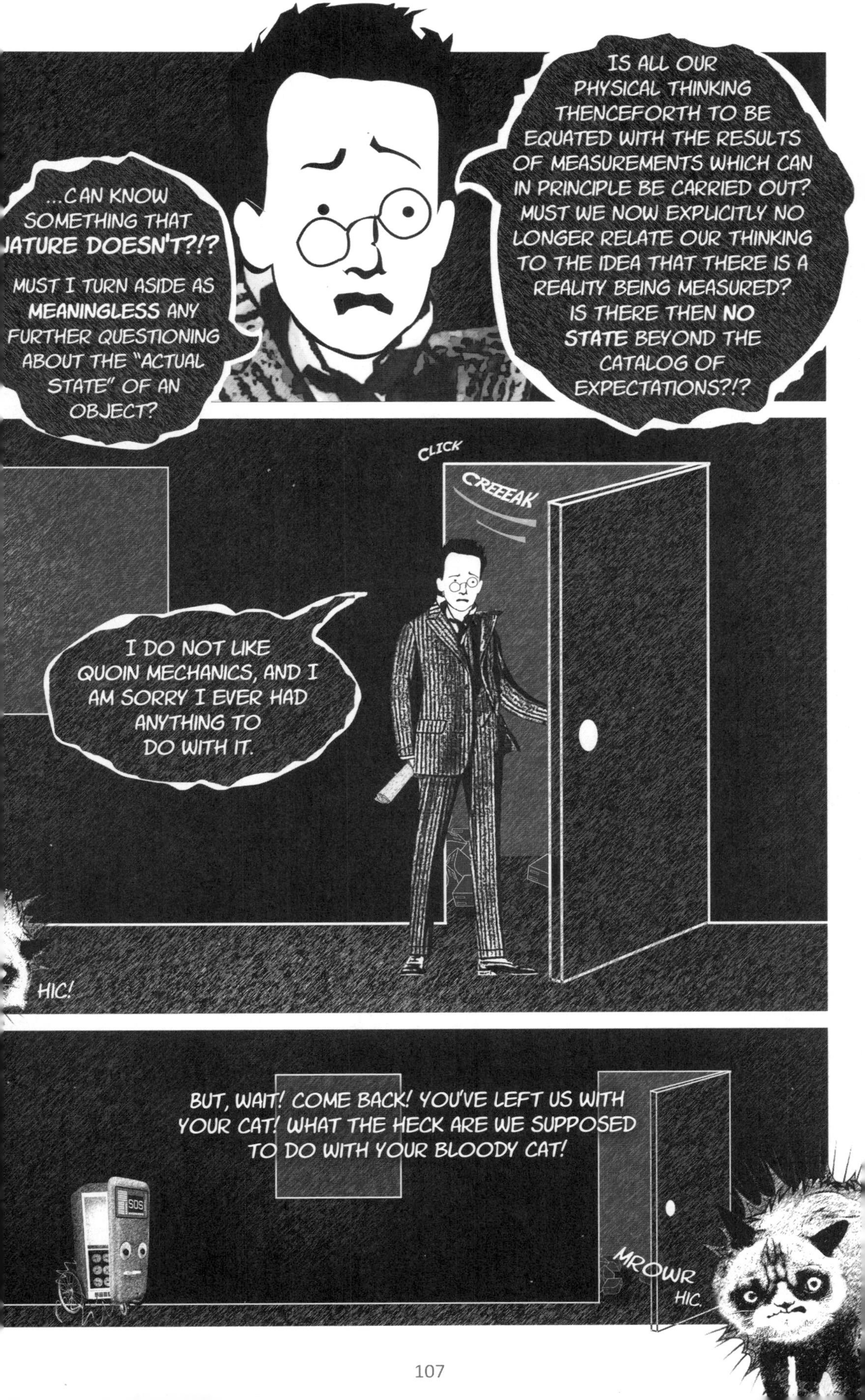
...CAN KNOW SOMETHING THAT NATURE DOESN'T?!? MUST I TURN ASIDE AS MEANINGLESS ANY FURTHER QUESTIONING ABOUT THE "ACTUAL STATE" OF AN OBJECT?
IS ALL OUR PHYSICAL THINKING THENCEFORTH TO BE EQUATED WITH THE RESULTS OF MEASUREMENTS WHICH CAN IN PRINCIPLE BE CARRIED OUT? MUST WE NOW EXPLICITLY NO LONGER RELATE OUR THINKING TO THE IDEA THAT THERE IS A REALITY BEING MEASURED? IS THERE THEN NO STATE BEYOND THE CATALOG OF EXPECTATIONS?!?
CLICK
CREEEAK
I DO NOT LIKE QUOIN MECHANICS, AND I AM SORRY I EVER HAD ANYTHING TO DO WITH IT.
HIC!
BUT, WAIT! COME BACK! YOU'VE LEFT US WITH YOUR CAT! WHAT THE HECK ARE WE SUPPOSED TO DO WITH YOUR BLOODY CAT!
SOS
MROWR
HIC.

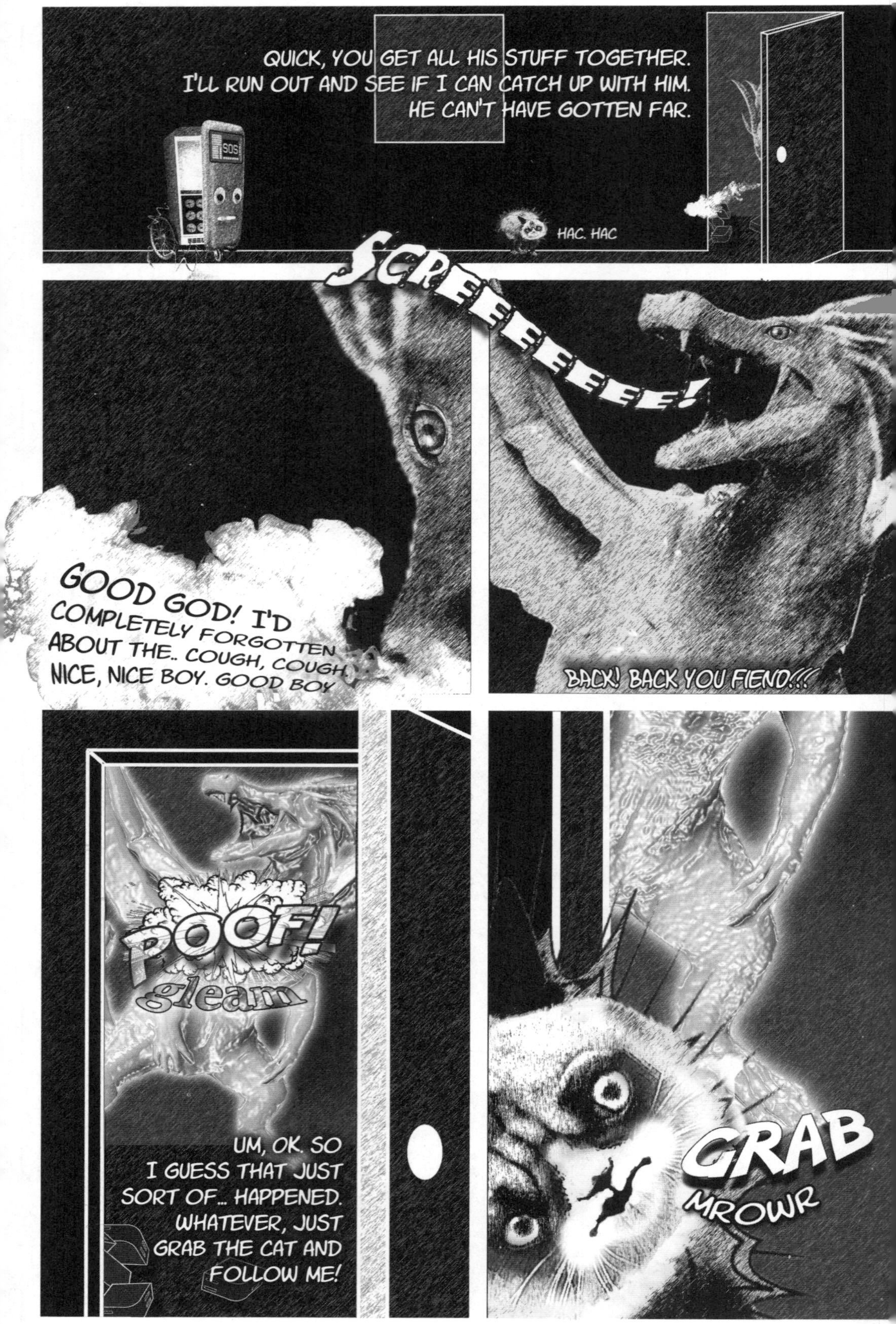

QUICK, YOU GET ALL HIS STUFF TOGETHER. I'LL RUN OUT AND SEE IF I CAN CATCH UP WITH HIM. HE CAN'T HAVE GOTTEN FAR.
SOS
HAC. HAC
SCREEEEEEEE!
GOOD GOD! I'D COMPLETELY FORGOTTEN ABOUT THE.. COUGH, COUGH. NICE, NICE BOY. GOOD BOY
BACK! BACK YOU FIEND!!!
POOF!
gleam
UM, OK. SO I GUESS THAT JUST SORT OF... HAPPENED. WHATEVER, JUST GRAB THE CAT AND FOLLOW ME!
GRAB
MROWR

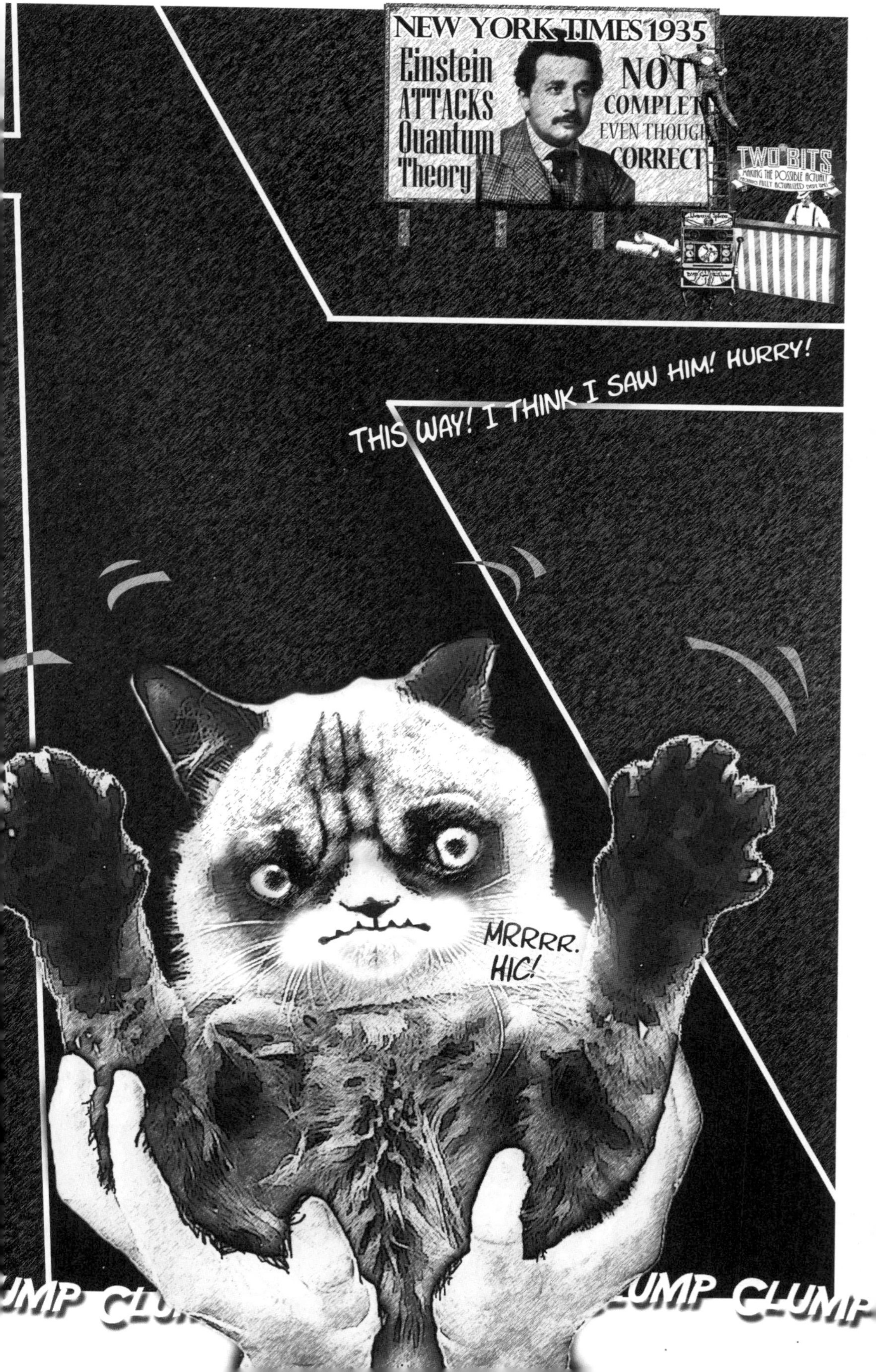
NEW YORK TIMES 1935
Einstein ATTACKS Quantum Theory
NOT
COMPLET
EVEN THOUG
CORRECT
TWO BITS
MAKING THE POSSIBLE ACTUAL!
FULLY ACTUALIZED EVERY TIME!
THIS WAY! I THINK I SAW HIM! HURRY!
MRRRR. HIC!
UMP CLU
LUMP CLUMP

Many Worlds

It is true that the [many-worlds interpretation],
in this realist form, avoids some of the paradoxes of
[quantum mechanics]. ... All possible outcomes take
place. Schrödinger's notorious cat is never in a
mixed state of alive and dead. It lives in one
universe, dies in another. But what a
fantastic price is paid for these
seeming simplicities!

—Martin Gardner

The final strategy is acceptance. That is the
Everettian approach. The formalism of quantum
mechanics, in this view, consists of quantum states
as described ... and nothing more, which evolve
according to the usual Schrödinger equation and
nothing more. The formalism predicts that there
are many worlds, so we choose to accept that.

—Sean Carroll

[The many-worlds interpretation] really is just the "obvious, straightforward" reading of quantum mechanics itself, if you take quantum mechanics literally as a description of the whole universe, and assume nothing new will ever be discovered that changes the picture.

—Scott Aaronson

This universe is constantly splitting into a stupendous number of branches, all resulting from the measurementlike interactions between its myriads of components. Moreover, every quantum transition taking place on every star, in every galaxy, in every remote corner of the universe is splitting our local world on earth into myriads of copies of itself.

—Bryce DeWitt

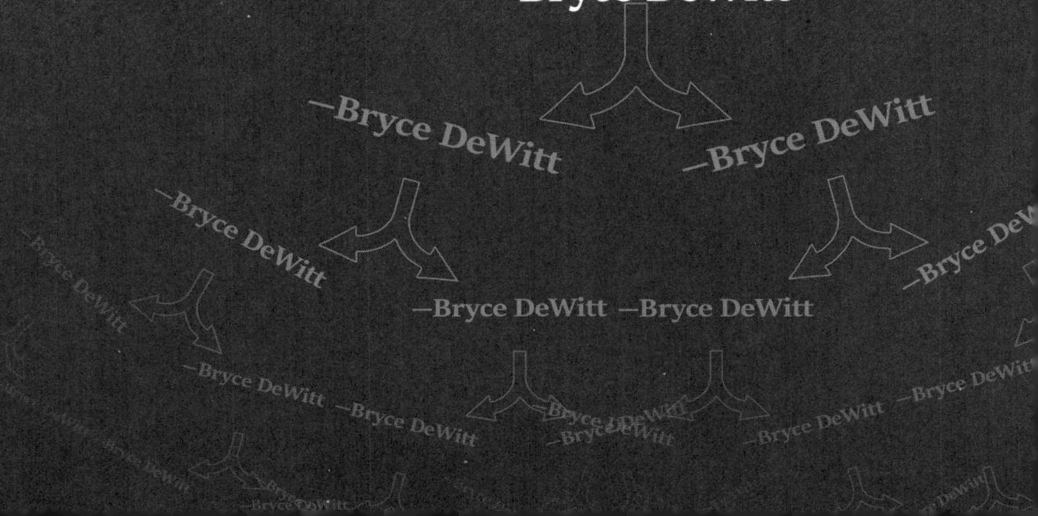

—Bryce DeWitt   —Bryce DeWitt

—Bryce DeWitt

—Bryce DeWitt —Bryce DeWitt   —Bryce Dew

—Bryce DeWitt

—Bryce DeWitt —Bryce DeWitt   —Bryce DeWitt —Bryce DeWitt

—Bryce DeWitt —Bryce DeWitt

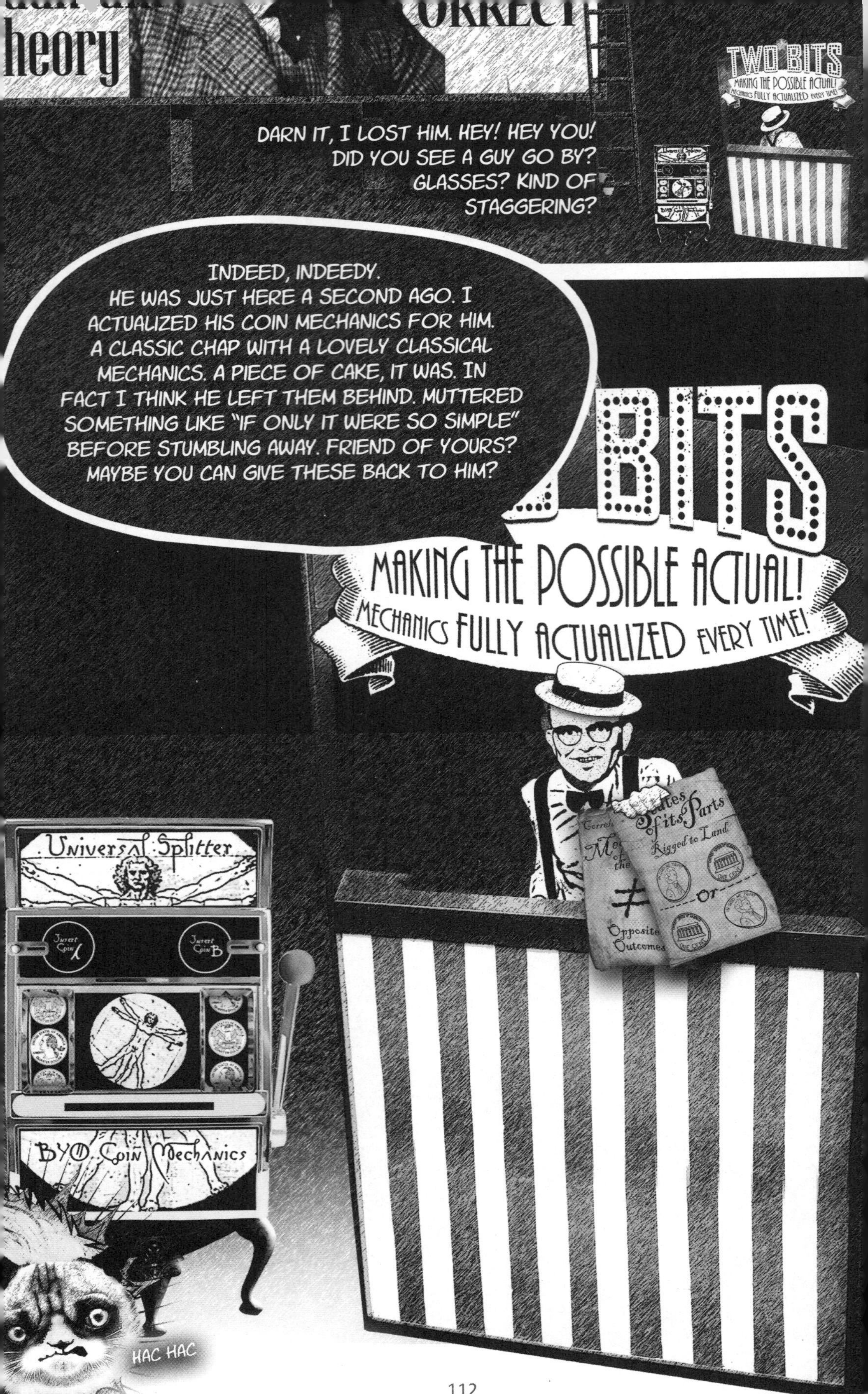
TWO BITS
MAKING THE POSSIBLE ACTUAL!
MECHANICS FULLY ACTUALIZED EVERY TIME!
DARN IT, I LOST HIM. HEY! HEY YOU! DID YOU SEE A GUY GO BY? GLASSES? KIND OF STAGGERING?
INDEED, INDEEDY. HE WAS JUST HERE A SECOND AGO. I ACTUALIZED HIS COIN MECHANICS FOR HIM. A CLASSIC CHAP WITH A LOVELY CLASSICAL MECHANICS. A PIECE OF CAKE, IT WAS. IN FACT I THINK HE LEFT THEM BEHIND. MUTTERED SOMETHING LIKE "IF ONLY IT WERE SO SIMPLE" BEFORE STUMBLING AWAY. FRIEND OF YOURS? MAYBE YOU CAN GIVE THESE BACK TO HIM?
BITS
MAKING THE POSSIBLE ACTUAL!
MECHANICS FULLY ACTUALIZED EVERY TIME!
Outcomes of its Parts Rigged to Land
Or
Opposite Outcomes
Universal Splitter
Insert Coin B
BYO Coin Mechanics
HAC HAC

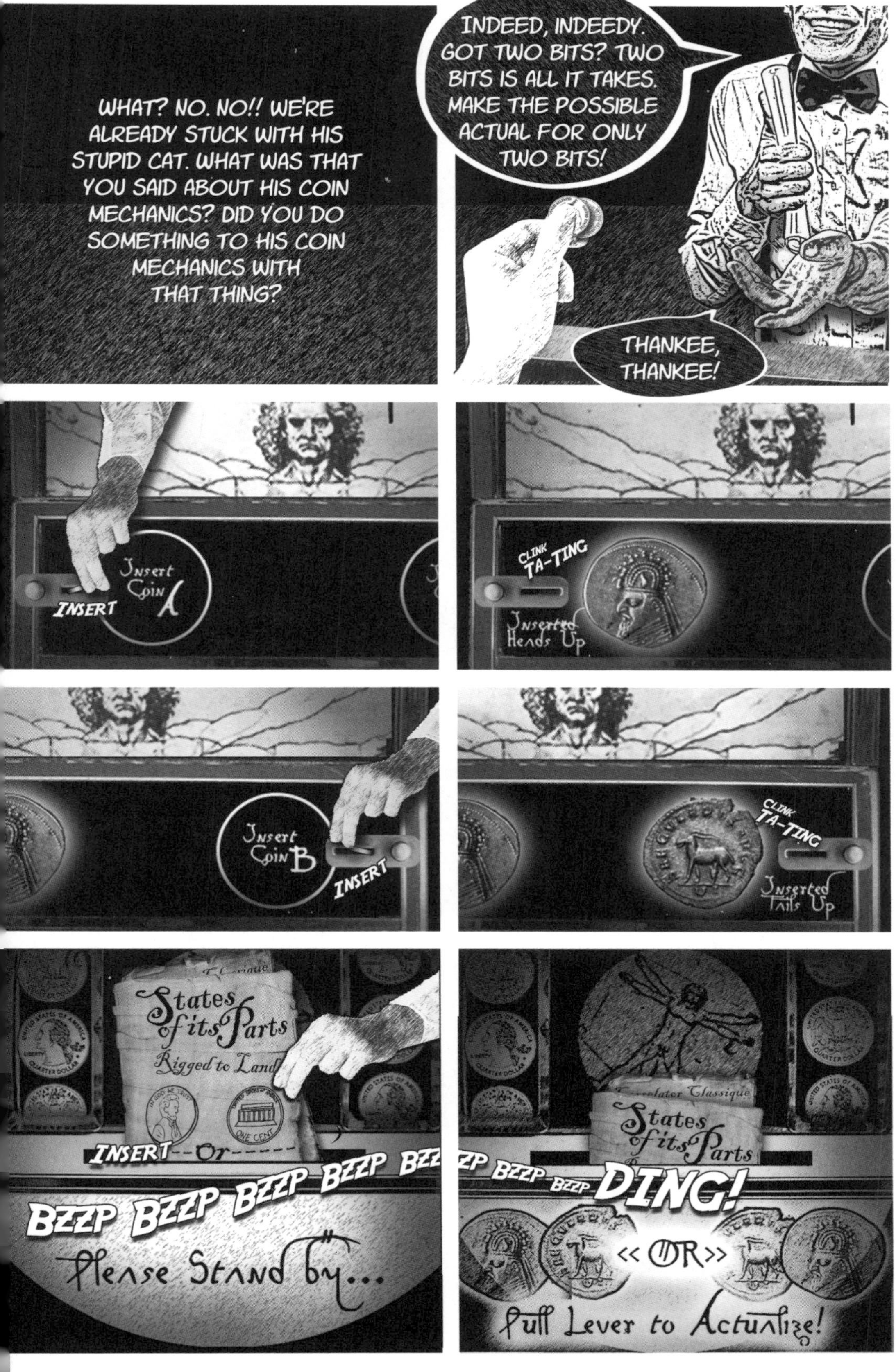
WHAT? NO. NO!! WE'RE ALREADY STUCK WITH HIS STUPID CAT. WHAT WAS THAT YOU SAID ABOUT HIS COIN MECHANICS? DID YOU DO SOMETHING TO HIS COIN MECHANICS WITH THAT THING?
INDEED, INDEEDY. GOT TWO BITS? TWO BITS IS ALL IT TAKES. MAKE THE POSSIBLE ACTUAL FOR ONLY TWO BITS!
THANKEE, THANKEE!
Insert Coin A
INSERT
CLINK
TA-TING
Inserted Heads Up
Insert Coin B
INSERT
CLINK
TA-TING
Inserted Tails Up
States of its Parts
Rigged to Land
INSERT --Or------
BZZP BZZP BZZP BZZP BZZP BZZP BZZP BZZP DING!
Please Stand by...
States of its Parts
<< OR >>
Pull Lever to Actualize!

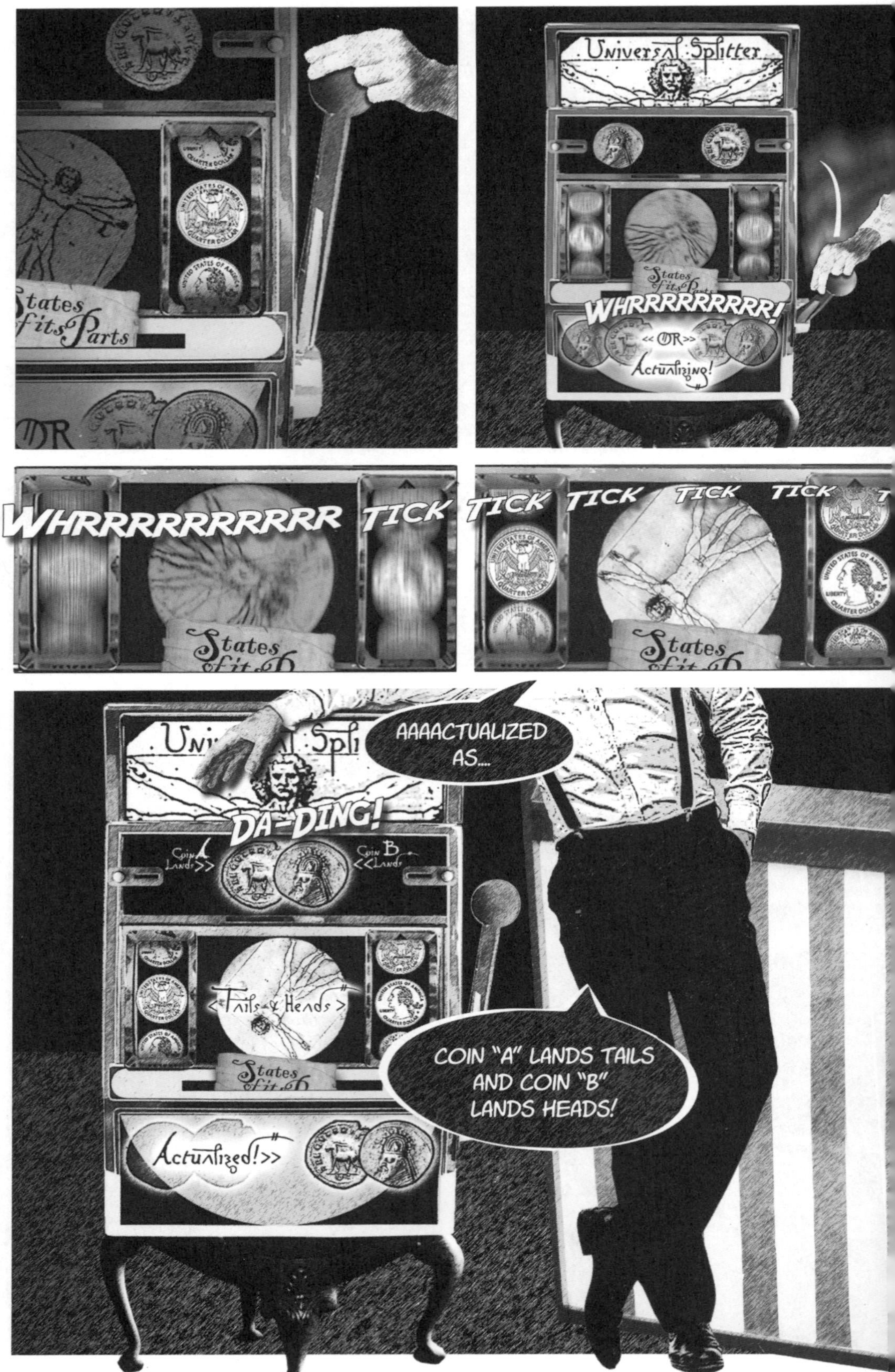
States of its Parts
Universal Splitter
States of its Parts
WHRRRRRRRRR!
<< OR >>
Actualizing!
WHRRRRRRRRRR
TICK
TICK
TICK
TICK
TICK
AAAACTUALIZED AS....
DA-DING!
Coin A Lands>>
Coin B <<Lands
<Tails & Heads>
COIN "A" LANDS TAILS AND COIN "B" LANDS HEADS!
Actualized!>>

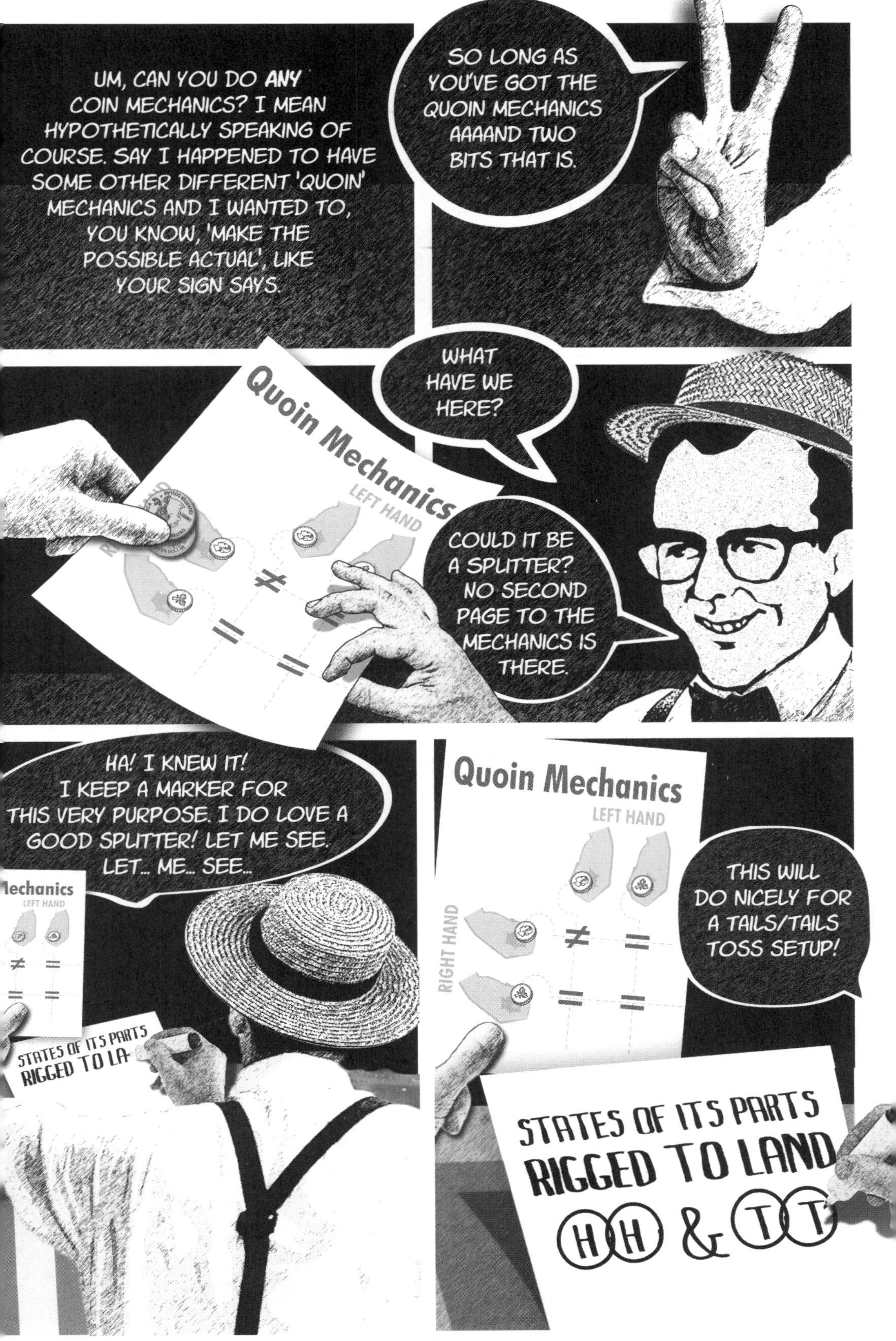

UM, CAN YOU DO ANY COIN MECHANICS? I MEAN HYPOTHETICALLY SPEAKING OF COURSE. SAY I HAPPENED TO HAVE SOME OTHER DIFFERENT 'QUOIN' MECHANICS AND I WANTED TO, YOU KNOW, 'MAKE THE POSSIBLE ACTUAL', LIKE YOUR SIGN SAYS.
SO LONG AS YOU'VE GOT THE QUOIN MECHANICS AAAAND TWO BITS THAT IS.
WHAT HAVE WE HERE?
Quoin Mechanics
LEFT HAND
COULD IT BE A SPLITTER? NO SECOND PAGE TO THE MECHANICS IS THERE.
HA! I KNEW IT! I KEEP A MARKER FOR THIS VERY PURPOSE. I DO LOVE A GOOD SPLITTER! LET ME SEE. LET... ME... SEE...
STATES OF ITS PARTS
RIGGED TO LA
Quoin Mechanics
LEFT HAND
RIGHT HAND
THIS WILL DO NICELY FOR A TAILS/TAILS TOSS SETUP!
STATES OF ITS PARTS
RIGGED TO LAND
HH & TT

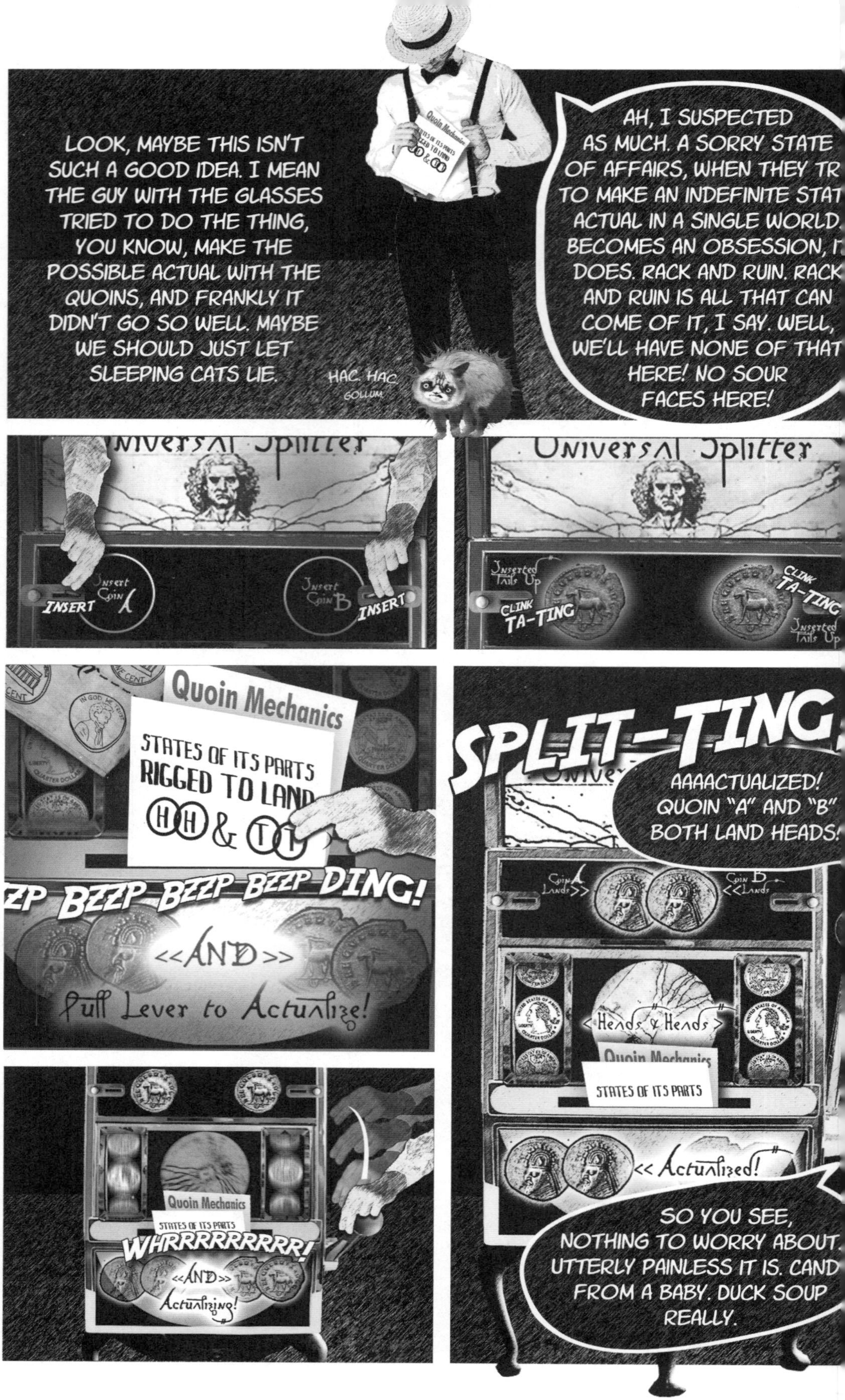
LOOK, MAYBE THIS ISN'T SUCH A GOOD IDEA. I MEAN THE GUY WITH THE GLASSES TRIED TO DO THE THING, YOU KNOW, MAKE THE POSSIBLE ACTUAL WITH THE QUOINS, AND FRANKLY IT DIDN'T GO SO WELL. MAYBE WE SHOULD JUST LET SLEEPING CATS LIE.
HAC. HAC.
GOLLUM
AH, I SUSPECTED AS MUCH. A SORRY STATE OF AFFAIRS, WHEN THEY TR TO MAKE AN INDEFINITE STAT ACTUAL IN A SINGLE WORLD. BECOMES AN OBSESSION, IT DOES. RACK AND RUIN. RACK AND RUIN IS ALL THAT CAN COME OF IT, I SAY. WELL, WE'LL HAVE NONE OF THAT HERE! NO SOUR FACES HERE!
Universal Splitter
Insert Coin A
Insert Coin B
INSERT
INSERT
Inserted Tails Up
CLINK TA-TING
CLINK TA-TING
Inserted Tails Up
Quoin Mechanics
STATES OF ITS PARTS RIGGED TO LAND HH & TT
ZP BZZP BZZP BZZP DING!
<<AND>>
Pull Lever to Actualize!
Quoin Mechanics
STATES OF ITS PARTS
WHRRRRRRRRR!
<<AND>>
Actualizing!
SPLIT-TING!
AAAACTUALIZED! QUOIN "A" AND "B" BOTH LAND HEADS!
Coin A Lands>>
Coin B <<Lands
< Heads & Heads >
Quoin Mechanics
STATES OF ITS PARTS
<< Actualized!
SO YOU SEE, NOTHING TO WORRY ABOUT. UTTERLY PAINLESS IT IS. CAND FROM A BABY. DUCK SOUP REALLY.

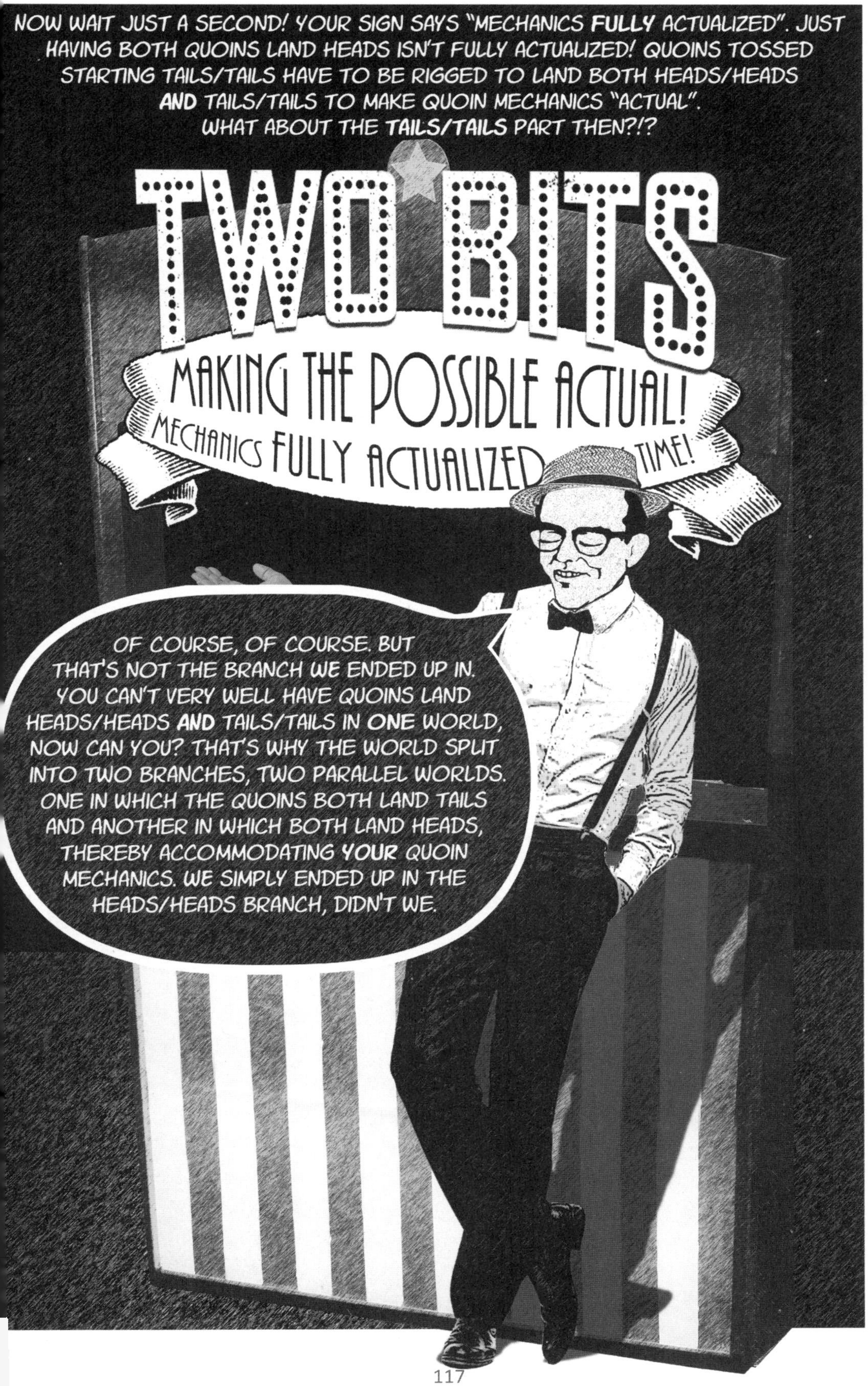
NOW WAIT JUST A SECOND! YOUR SIGN SAYS "MECHANICS FULLY ACTUALIZED". JUST HAVING BOTH QUOINS LAND HEADS ISN'T FULLY ACTUALIZED! QUOINS TOSSED STARTING TAILS/TAILS HAVE TO BE RIGGED TO LAND BOTH HEADS/HEADS AND TAILS/TAILS TO MAKE QUOIN MECHANICS "ACTUAL". WHAT ABOUT THE TAILS/TAILS PART THEN?!?
TWO BITS
MAKING THE POSSIBLE ACTUAL!
MECHANICS FULLY ACTUALIZED
TIME!
OF COURSE, OF COURSE. BUT THAT'S NOT THE BRANCH WE ENDED UP IN. YOU CAN'T VERY WELL HAVE QUOINS LAND HEADS/HEADS AND TAILS/TAILS IN ONE WORLD, NOW CAN YOU? THAT'S WHY THE WORLD SPLIT INTO TWO BRANCHES, TWO PARALLEL WORLDS. ONE IN WHICH THE QUOINS BOTH LAND TAILS AND ANOTHER IN WHICH BOTH LAND HEADS, THEREBY ACCOMMODATING YOUR QUOIN MECHANICS. WE SIMPLY ENDED UP IN THE HEADS/HEADS BRANCH, DIDN'T WE.

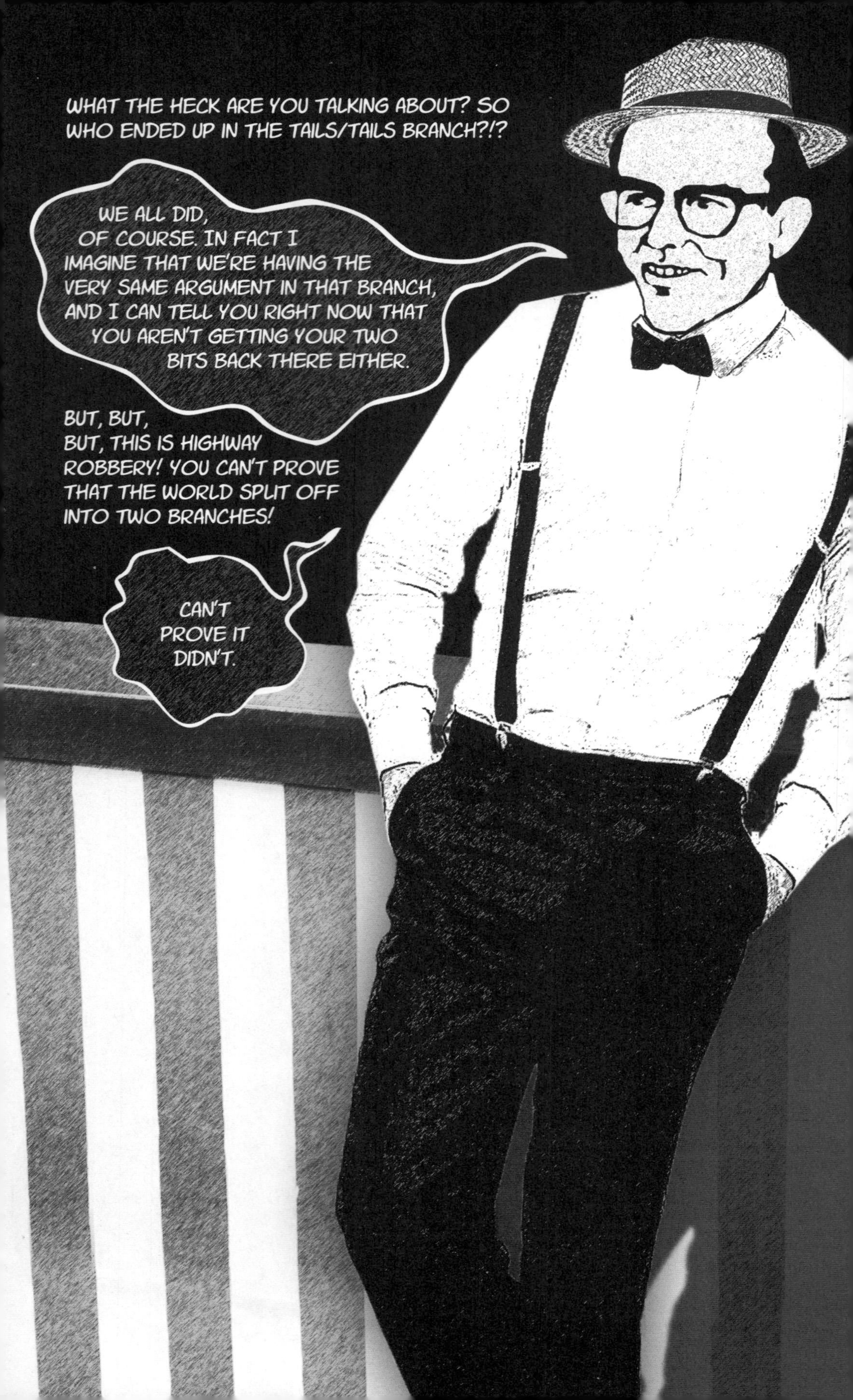
WHAT THE HECK ARE YOU TALKING ABOUT? SO WHO ENDED UP IN THE TAILS/TAILS BRANCH?!?
WE ALL DID, OF COURSE. IN FACT I IMAGINE THAT WE'RE HAVING THE VERY SAME ARGUMENT IN THAT BRANCH, AND I CAN TELL YOU RIGHT NOW THAT YOU AREN'T GETTING YOUR TWO BITS BACK THERE EITHER.
BUT, BUT, BUT, THIS IS HIGHWAY ROBBERY! YOU CAN'T PROVE THAT THE WORLD SPLIT OFF INTO TWO BRANCHES!
CAN'T PROVE IT DIDN'T.

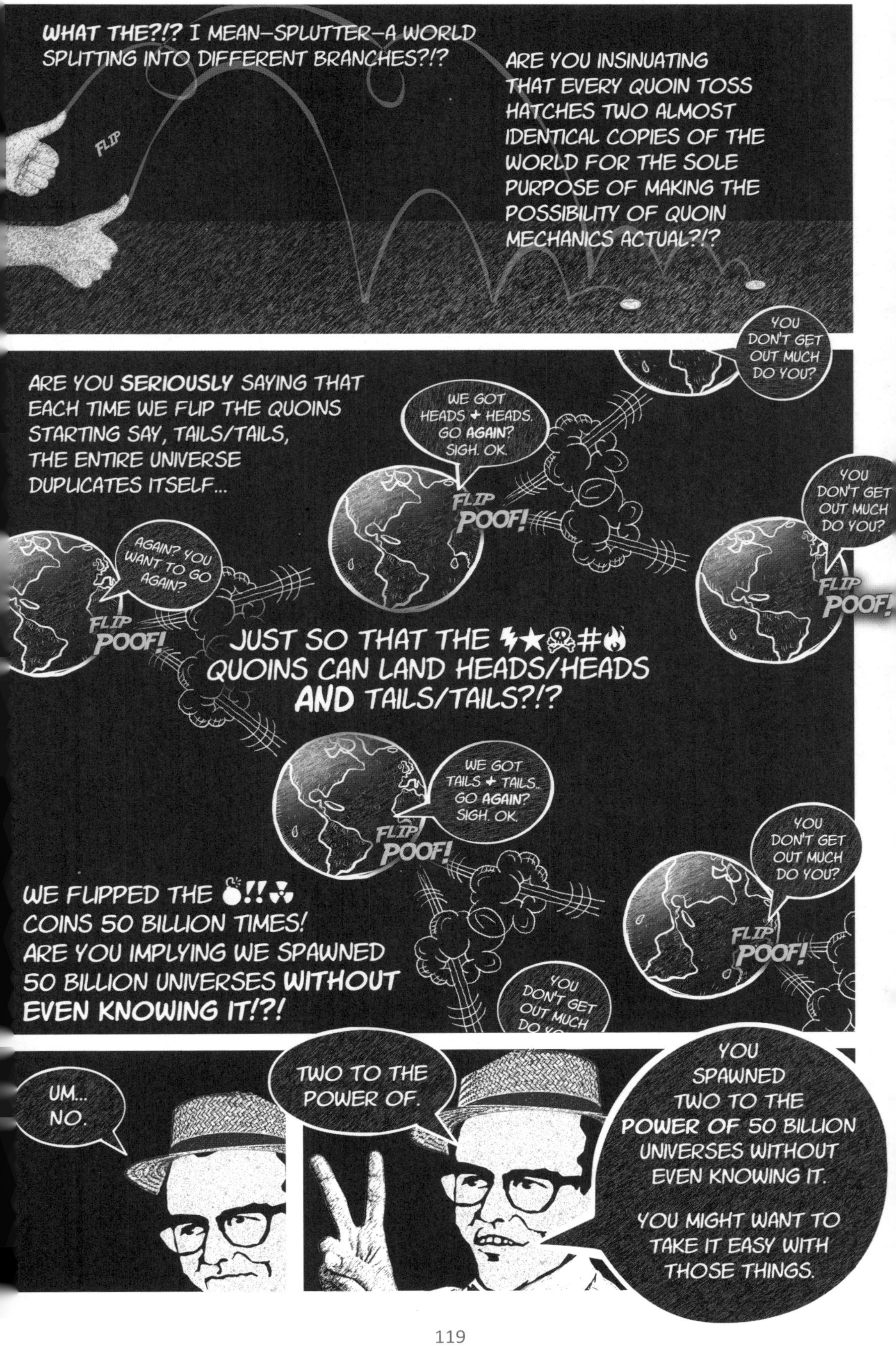
WHAT THE?!? I MEAN–SPLUTTER–A WORLD SPLITTING INTO DIFFERENT BRANCHES?!?
FLIP
ARE YOU INSINUATING THAT EVERY QUOIN TOSS HATCHES TWO ALMOST IDENTICAL COPIES OF THE WORLD FOR THE SOLE PURPOSE OF MAKING THE POSSIBILITY OF QUOIN MECHANICS ACTUAL?!?
ARE YOU SERIOUSLY SAYING THAT EACH TIME WE FLIP THE QUOINS STARTING SAY, TAILS/TAILS, THE ENTIRE UNIVERSE DUPLICATES ITSELF...
WE GOT HEADS + HEADS. GO AGAIN? SIGH. OK.
FLIP
POOF!
YOU DON'T GET OUT MUCH DO YOU?
AGAIN? YOU WANT TO GO AGAIN?
FLIP
POOF!
YOU DON'T GET OUT MUCH DO YOU?
FLIP
POOF!
JUST SO THAT THE ϟ★☠#🔥 QUOINS CAN LAND HEADS/HEADS AND TAILS/TAILS?!?
WE GOT TAILS + TAILS. GO AGAIN? SIGH. OK.
FLIP
POOF!
YOU DON'T GET OUT MUCH DO YOU?
FLIP
POOF!
WE FLIPPED THE 💣!!☢ COINS 50 BILLION TIMES! ARE YOU IMPLYING WE SPAWNED 50 BILLION UNIVERSES WITHOUT EVEN KNOWING IT!?!
YOU DON'T GET OUT MUCH
UM... NO.
TWO TO THE POWER OF.
YOU SPAWNED TWO TO THE POWER OF 50 BILLION UNIVERSES WITHOUT EVEN KNOWING IT.
YOU MIGHT WANT TO TAKE IT EASY WITH THOSE THINGS.

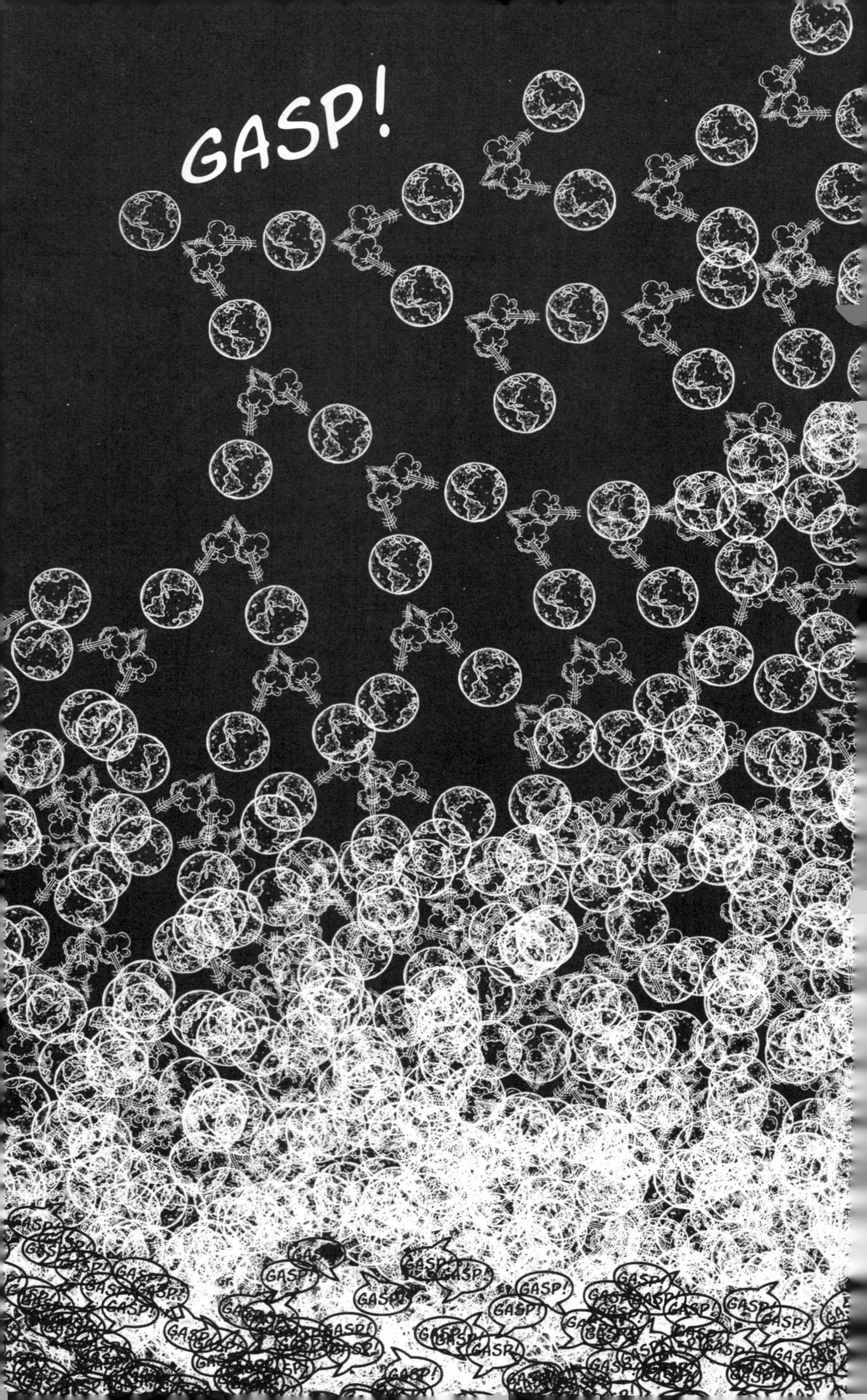
GASP!

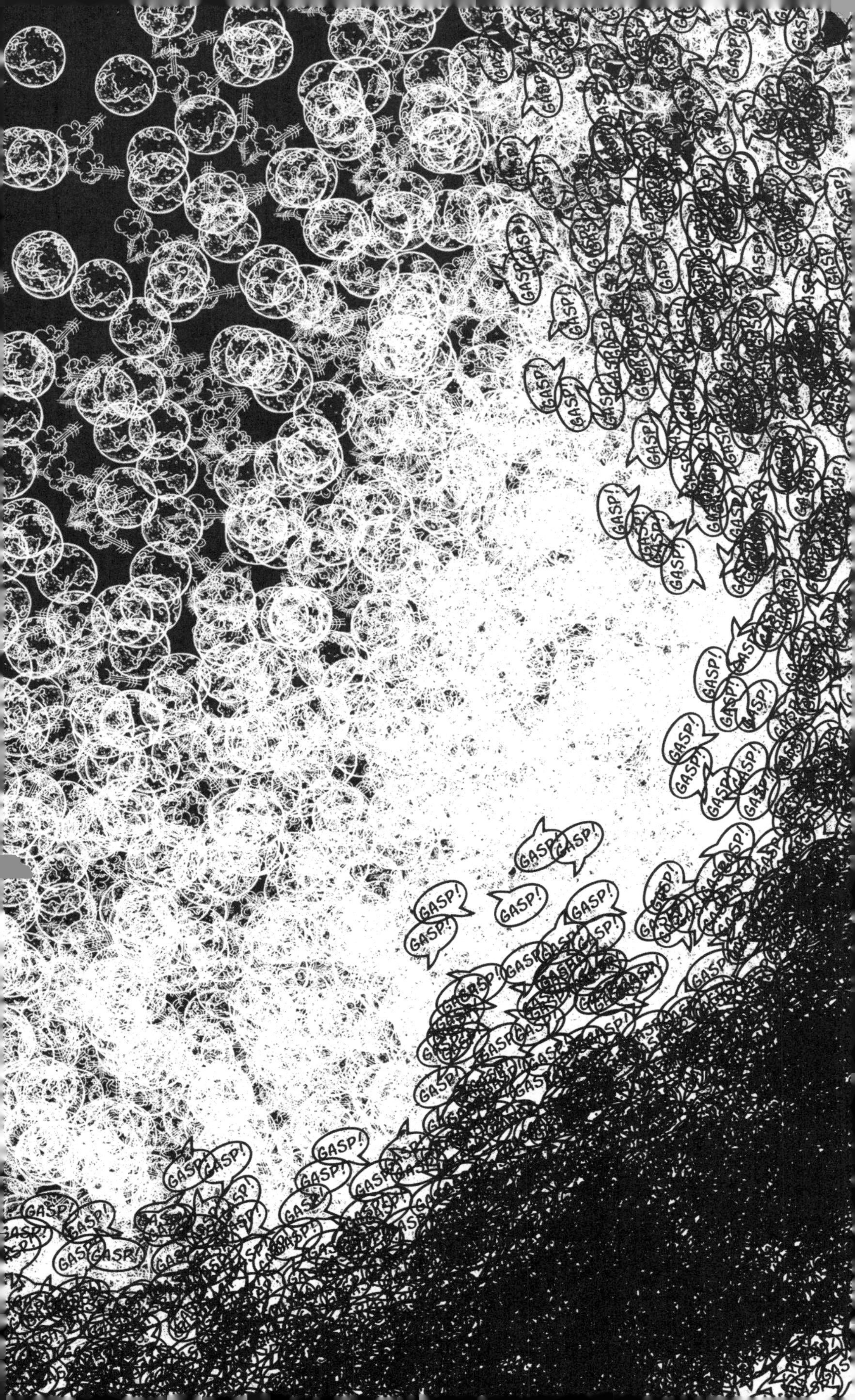
GASP!
GASP!
GASP!
GASP!
GASP!
GASP!
GASP!
GASP!

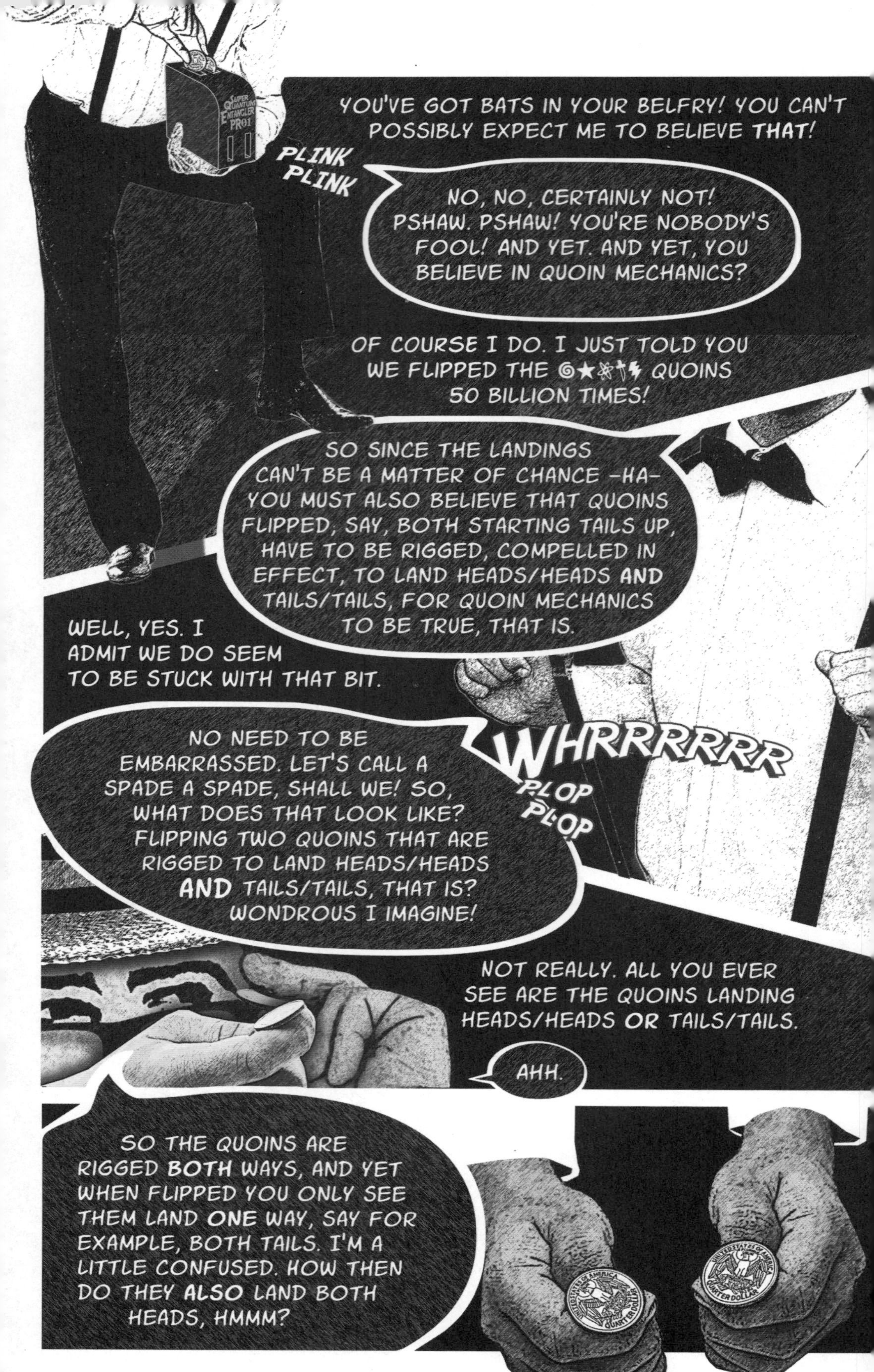

SUPER QUANTUM ENTANGLER PRO1
YOU'VE GOT BATS IN YOUR BELFRY! YOU CAN'T POSSIBLY EXPECT ME TO BELIEVE **THAT**!
PLINK PLINK
NO, NO, CERTAINLY NOT! PSHAW. PSHAW! YOU'RE NOBODY'S FOOL! AND YET. AND YET, YOU BELIEVE IN QUOIN MECHANICS?
OF COURSE I DO. I JUST TOLD YOU WE FLIPPED THE ◎★⚛†ϟ QUOINS 50 BILLION TIMES!
SO SINCE THE LANDINGS CAN'T BE A MATTER OF CHANCE –HA– YOU MUST ALSO BELIEVE THAT QUOINS FLIPPED, SAY, BOTH STARTING TAILS UP, HAVE TO BE RIGGED, COMPELLED IN EFFECT, TO LAND HEADS/HEADS **AND** TAILS/TAILS, FOR QUOIN MECHANICS TO BE TRUE, THAT IS.
WELL, YES. I ADMIT WE DO SEEM TO BE STUCK WITH THAT BIT.
NO NEED TO BE EMBARRASSED. LET'S CALL A SPADE A SPADE, SHALL WE! SO, WHAT DOES THAT LOOK LIKE? FLIPPING TWO QUOINS THAT ARE RIGGED TO LAND HEADS/HEADS **AND** TAILS/TAILS, THAT IS? WONDROUS I IMAGINE!
WHRRRRRR
PLOP PLOP
NOT REALLY. ALL YOU EVER SEE ARE THE QUOINS LANDING HEADS/HEADS **OR** TAILS/TAILS.
AHH.
SO THE QUOINS ARE RIGGED **BOTH** WAYS, AND YET WHEN FLIPPED YOU ONLY SEE THEM LAND **ONE** WAY, SAY FOR EXAMPLE, BOTH TAILS. I'M A LITTLE CONFUSED. HOW THEN DO THEY **ALSO** LAND BOTH HEADS, HMMM?
UNITED STATES OF AMERICA QUARTER DOLLAR
UNITED STATES OF AMERICA QUARTER DOLLAR

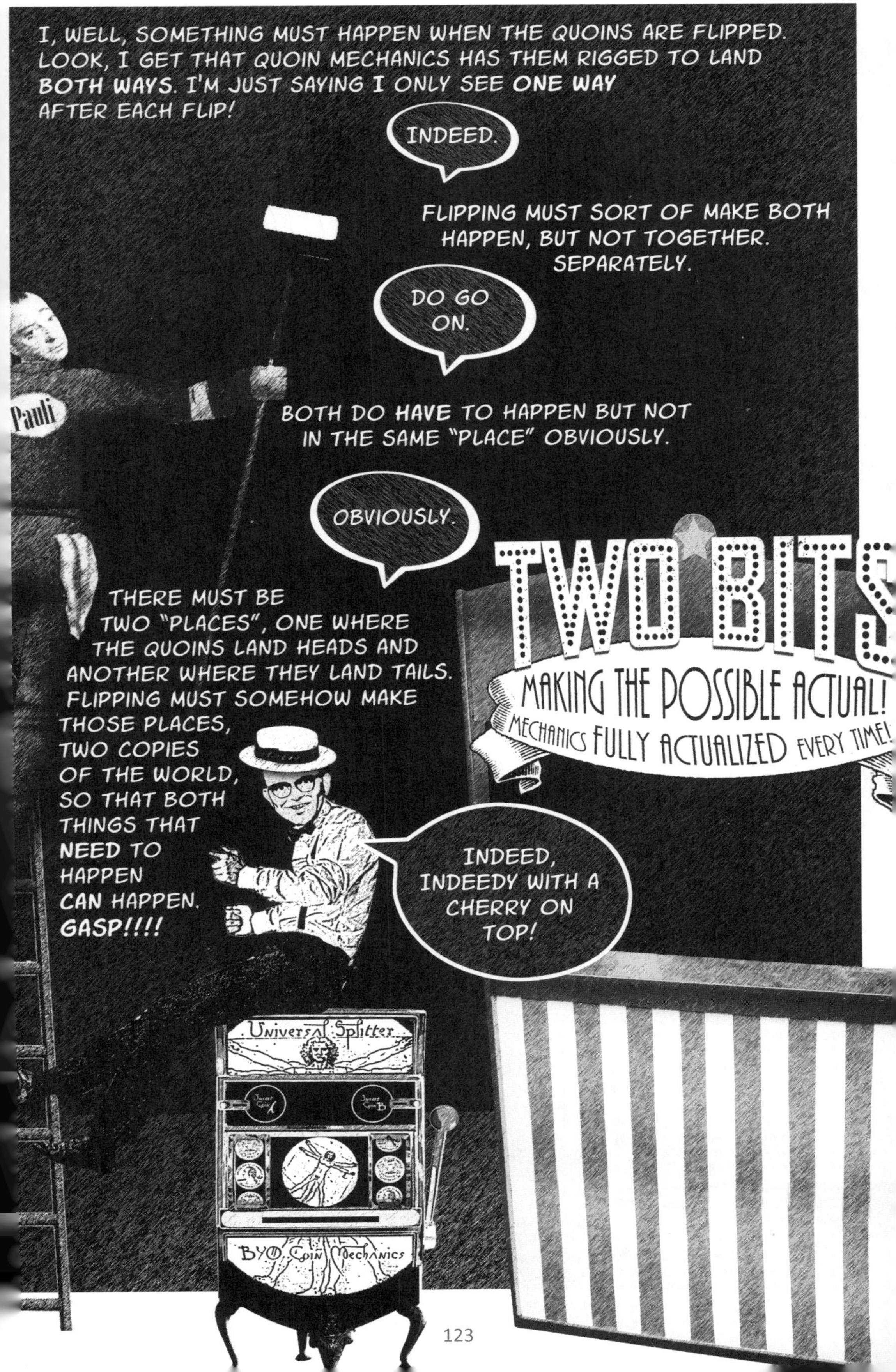
I, WELL, SOMETHING MUST HAPPEN WHEN THE QUOINS ARE FLIPPED. LOOK, I GET THAT QUOIN MECHANICS HAS THEM RIGGED TO LAND **BOTH WAYS**. I'M JUST SAYING **I** ONLY SEE **ONE WAY** AFTER EACH FLIP!
INDEED.
FLIPPING MUST SORT OF MAKE BOTH HAPPEN, BUT NOT TOGETHER. SEPARATELY.
DO GO ON.
Pauli
BOTH DO **HAVE** TO HAPPEN BUT NOT IN THE SAME "PLACE" OBVIOUSLY.
OBVIOUSLY.
TWO BITS
MAKING THE POSSIBLE ACTUAL!
MECHANICS FULLY ACTUALIZED EVERY TIME!
THERE MUST BE TWO "PLACES", ONE WHERE THE QUOINS LAND HEADS AND ANOTHER WHERE THEY LAND TAILS. FLIPPING MUST SOMEHOW MAKE THOSE PLACES, TWO COPIES OF THE WORLD, SO THAT BOTH THINGS THAT **NEED** TO HAPPEN **CAN** HAPPEN. **GASP!!!!**
INDEED, INDEEDY WITH A CHERRY ON TOP!
Universal Splitter
BYO Coin Mechanics

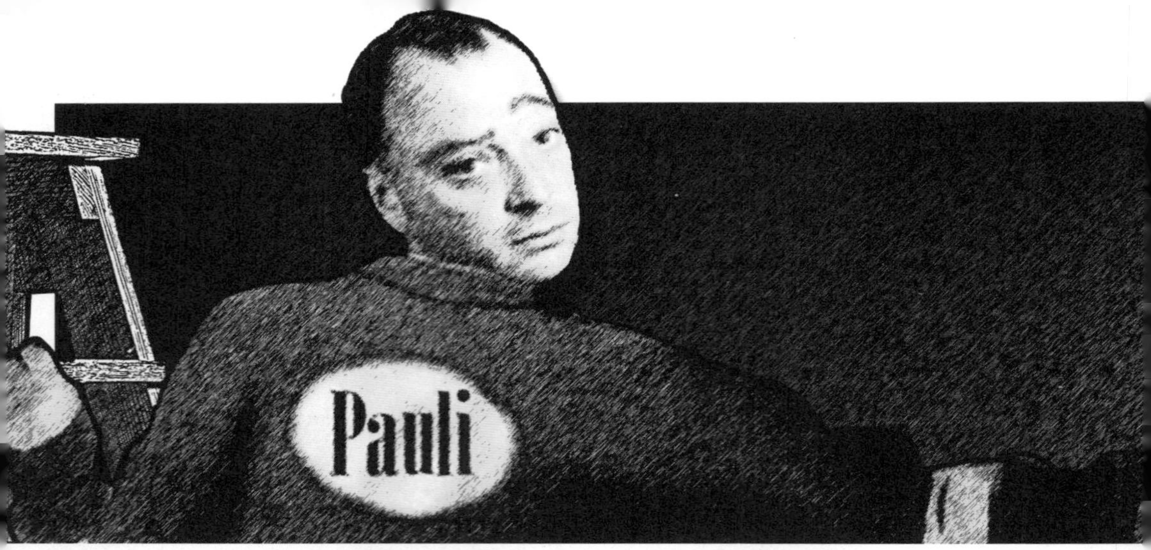
Pauli

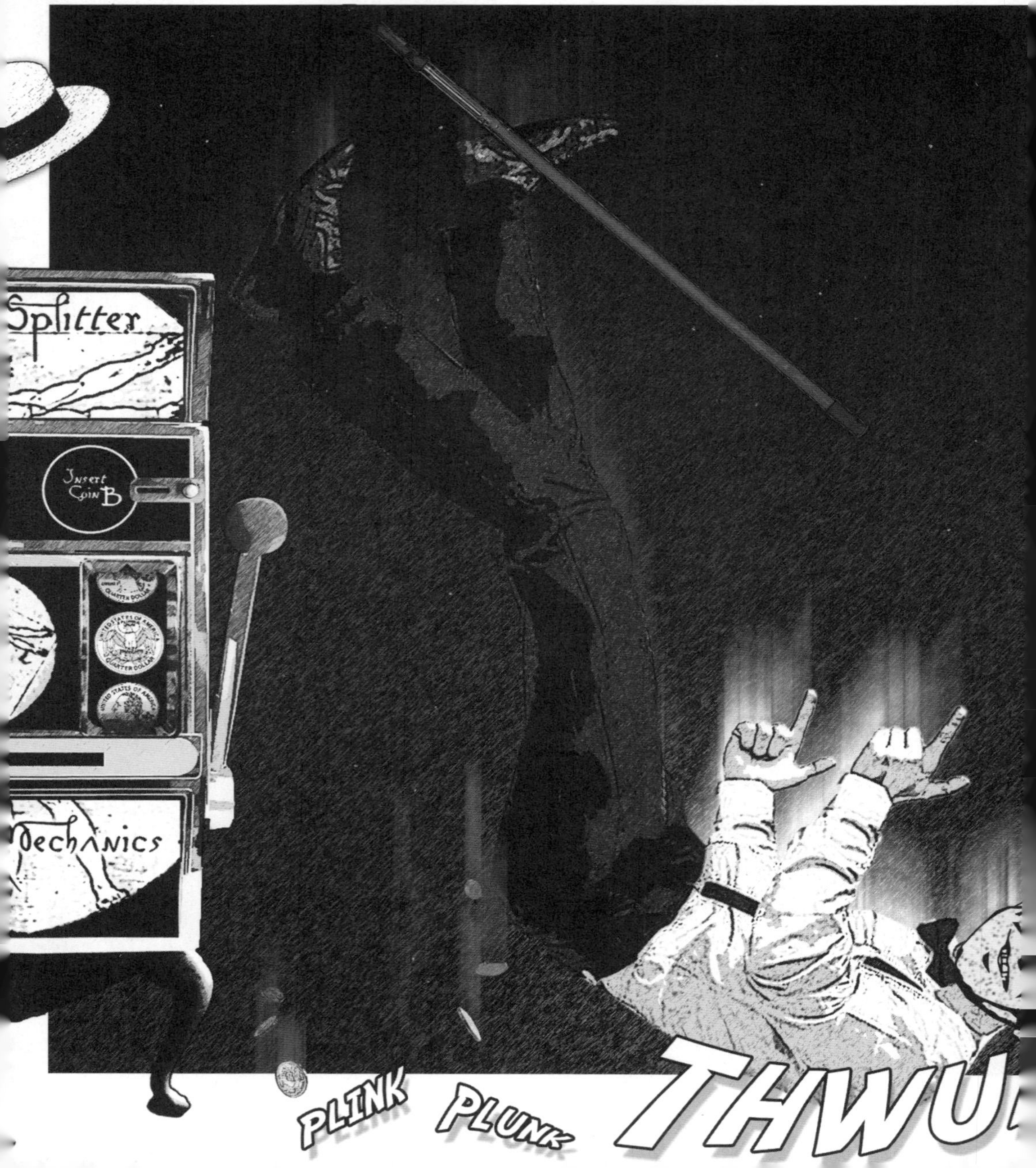
Splitter
Insert
Coin B
Mechanics
PLINK
PLUNK
THWU

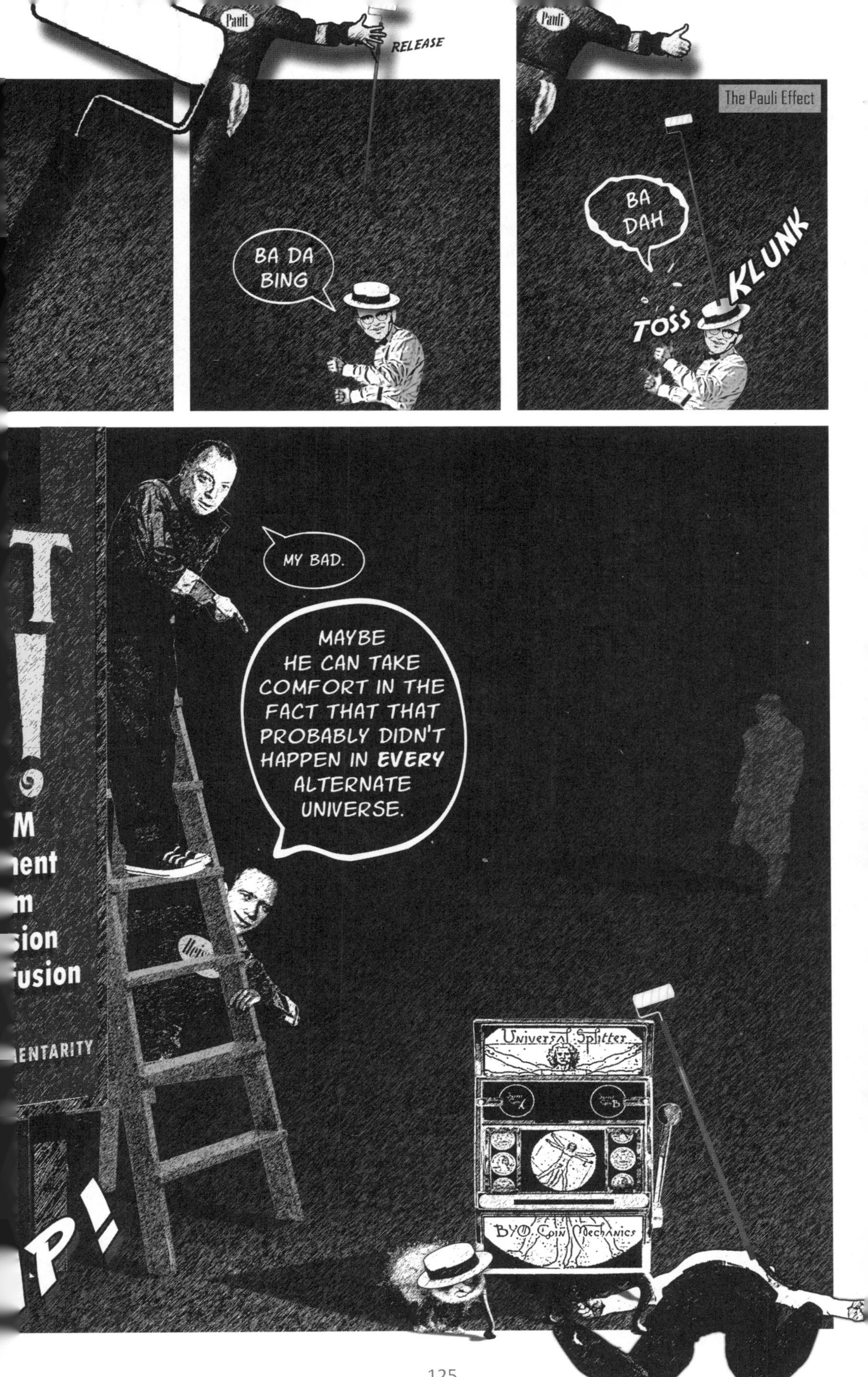
Pauli
RELEASE
BA DA BING
Pauli
The Pauli Effect
BA DAH
TOSS
KLUNK
MY BAD.
MAYBE HE CAN TAKE COMFORT IN THE FACT THAT THAT PROBABLY DIDN'T HAPPEN IN **EVERY** ALTERNATE UNIVERSE.
Universal Splitter
BYO Coin Mechanics

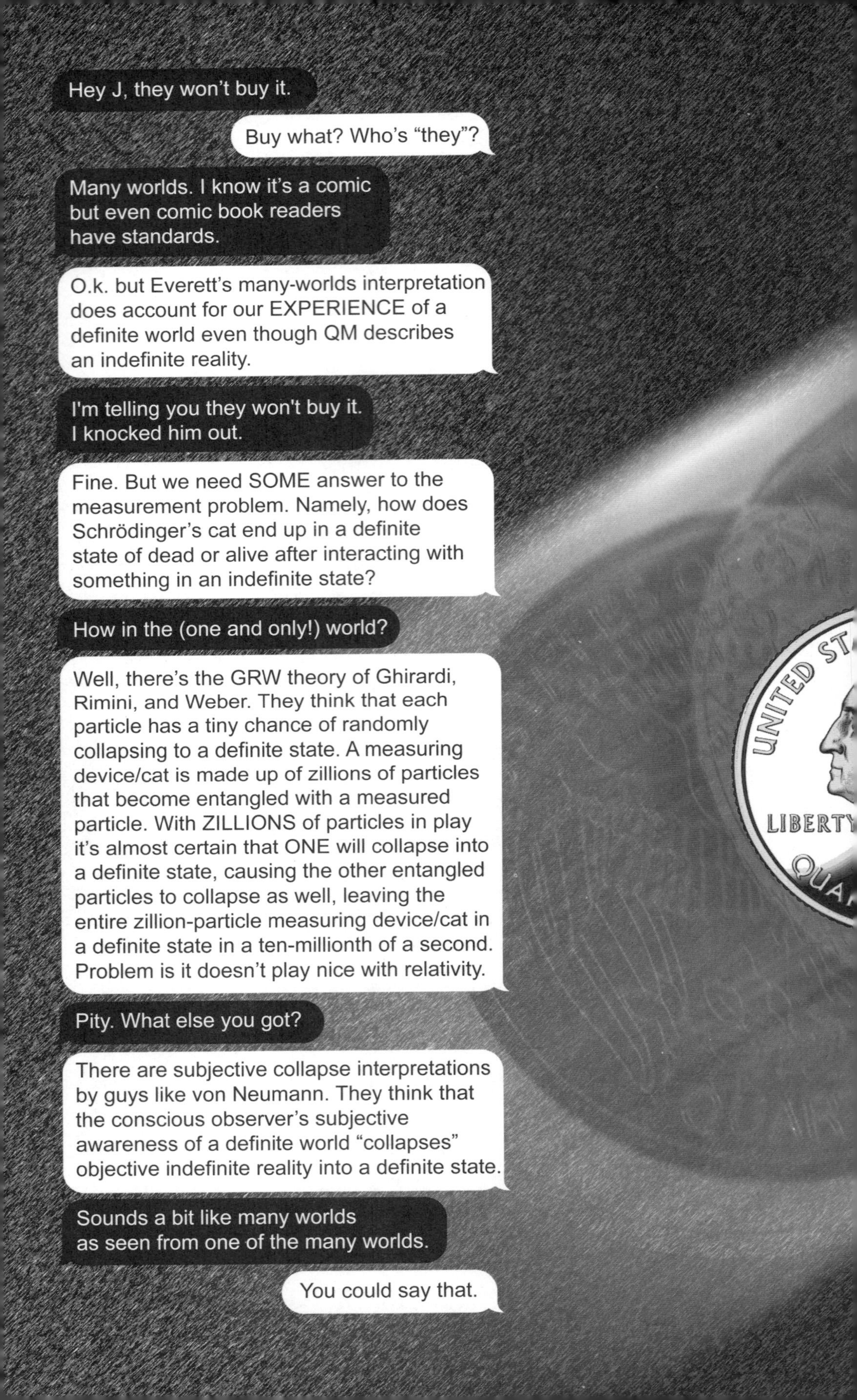

Hey J, they won't buy it.
Buy what? Who's "they"?
Many worlds. I know it's a comic but even comic book readers have standards.
O.k. but Everett's many-worlds interpretation does account for our EXPERIENCE of a definite world even though QM describes an indefinite reality.
I'm telling you they won't buy it. I knocked him out.
Fine. But we need SOME answer to the measurement problem. Namely, how does Schrödinger's cat end up in a definite state of dead or alive after interacting with something in an indefinite state?
How in the (one and only!) world?
Well, there's the GRW theory of Ghirardi, Rimini, and Weber. They think that each particle has a tiny chance of randomly collapsing to a definite state. A measuring device/cat is made up of zillions of particles that become entangled with a measured particle. With ZILLIONS of particles in play it's almost certain that ONE will collapse into a definite state, causing the other entangled particles to collapse as well, leaving the entire zillion-particle measuring device/cat in a definite state in a ten-millionth of a second. Problem is it doesn't play nice with relativity.
Pity. What else you got?
There are subjective collapse interpretations by guys like von Neumann. They think that the conscious observer's subjective awareness of a definite world "collapses" objective indefinite reality into a definite state.
Sounds a bit like many worlds as seen from one of the many worlds.
You could say that.
UNITED
LIBERTY

Collapsing the Limbo of Potentialities…

Einstein's criticism of quantum theory … is mainly concerned with the drastic changes of state brought about by simple acts of observation … particularly in connection with correlated systems which are widely separated so as to be mechanically uncoupled at the time of observation. At another time he put his feeling colorfully by stating that he could not believe that a mouse could bring about drastic changes in the universe simply by looking at it.

—**Hugh Everett**

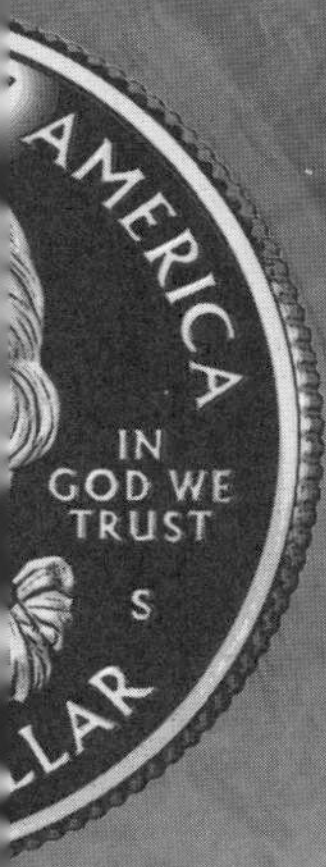
AMERICA
IN
GOD WE
TRUST
S
LLAR

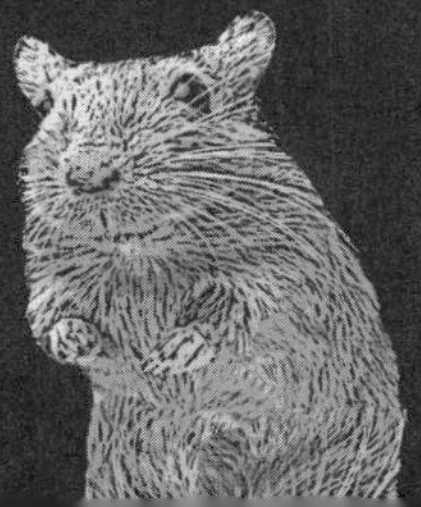

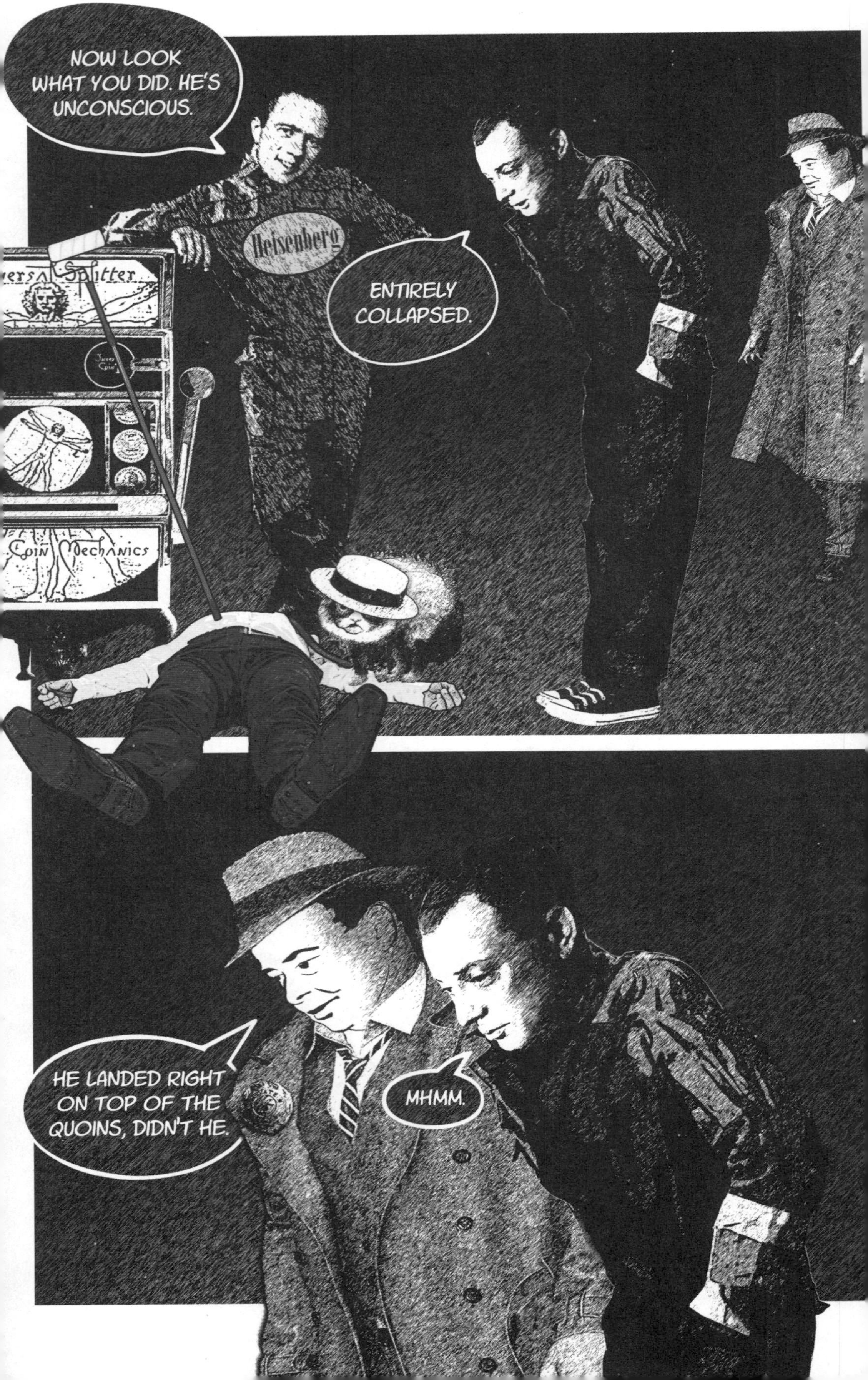
NOW LOOK WHAT YOU DID. HE'S UNCONSCIOUS.
Heisenberg
ENTIRELY COLLAPSED.
Coin Mechanics
HE LANDED RIGHT ON TOP OF THE QUOINS, DIDN'T HE.
MHMM.

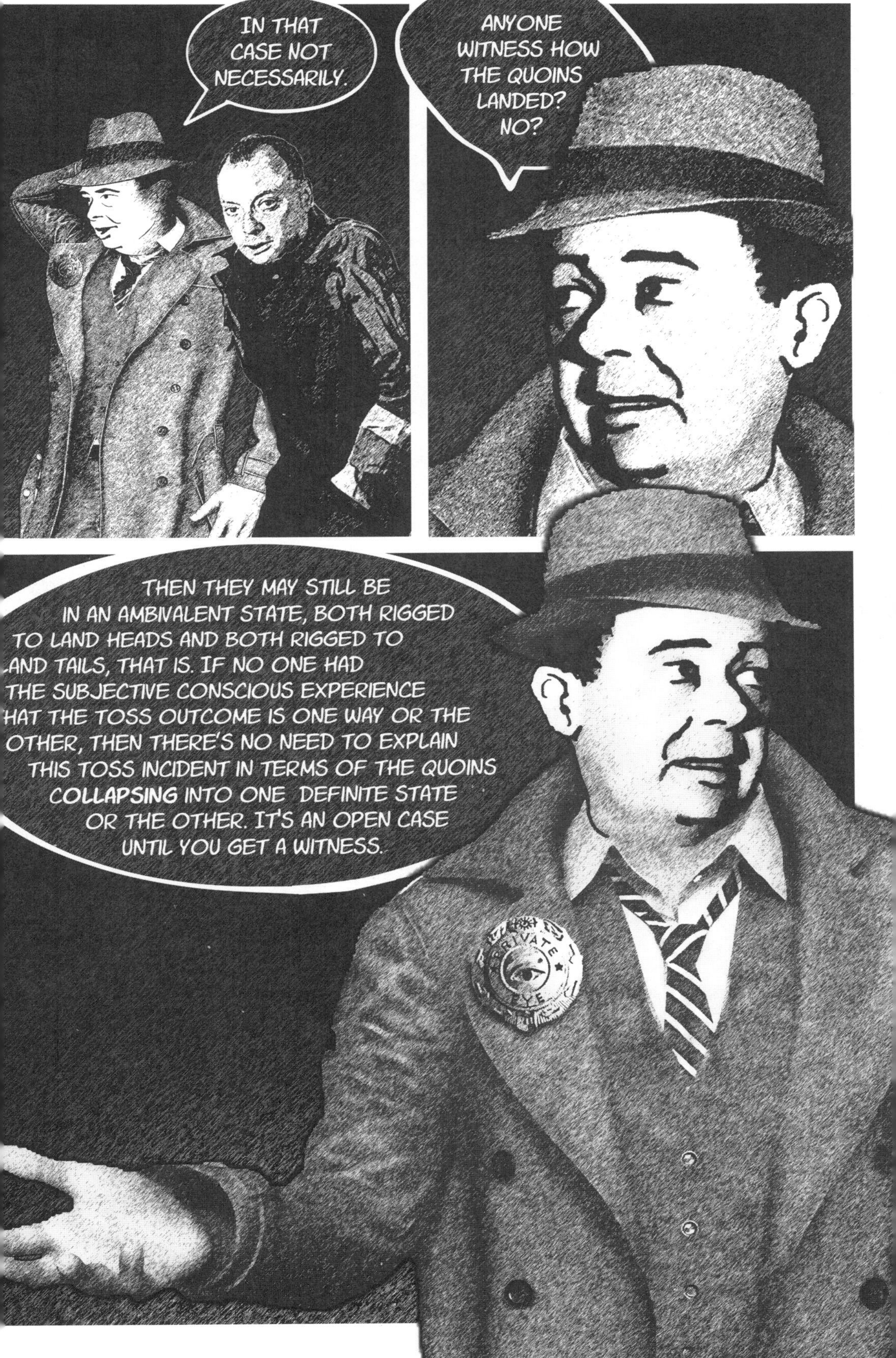
IN THAT CASE NOT NECESSARILY.
ANYONE WITNESS HOW THE QUOINS LANDED? NO?
THEN THEY MAY STILL BE IN AN AMBIVALENT STATE, BOTH RIGGED TO LAND HEADS AND BOTH RIGGED TO AND TAILS, THAT IS. IF NO ONE HAD THE SUBJECTIVE CONSCIOUS EXPERIENCE HAT THE TOSS OUTCOME IS ONE WAY OR THE OTHER, THEN THERE'S NO NEED TO EXPLAIN THIS TOSS INCIDENT IN TERMS OF THE QUOINS COLLAPSING INTO ONE DEFINITE STATE OR THE OTHER. IT'S AN OPEN CASE UNTIL YOU GET A WITNESS.

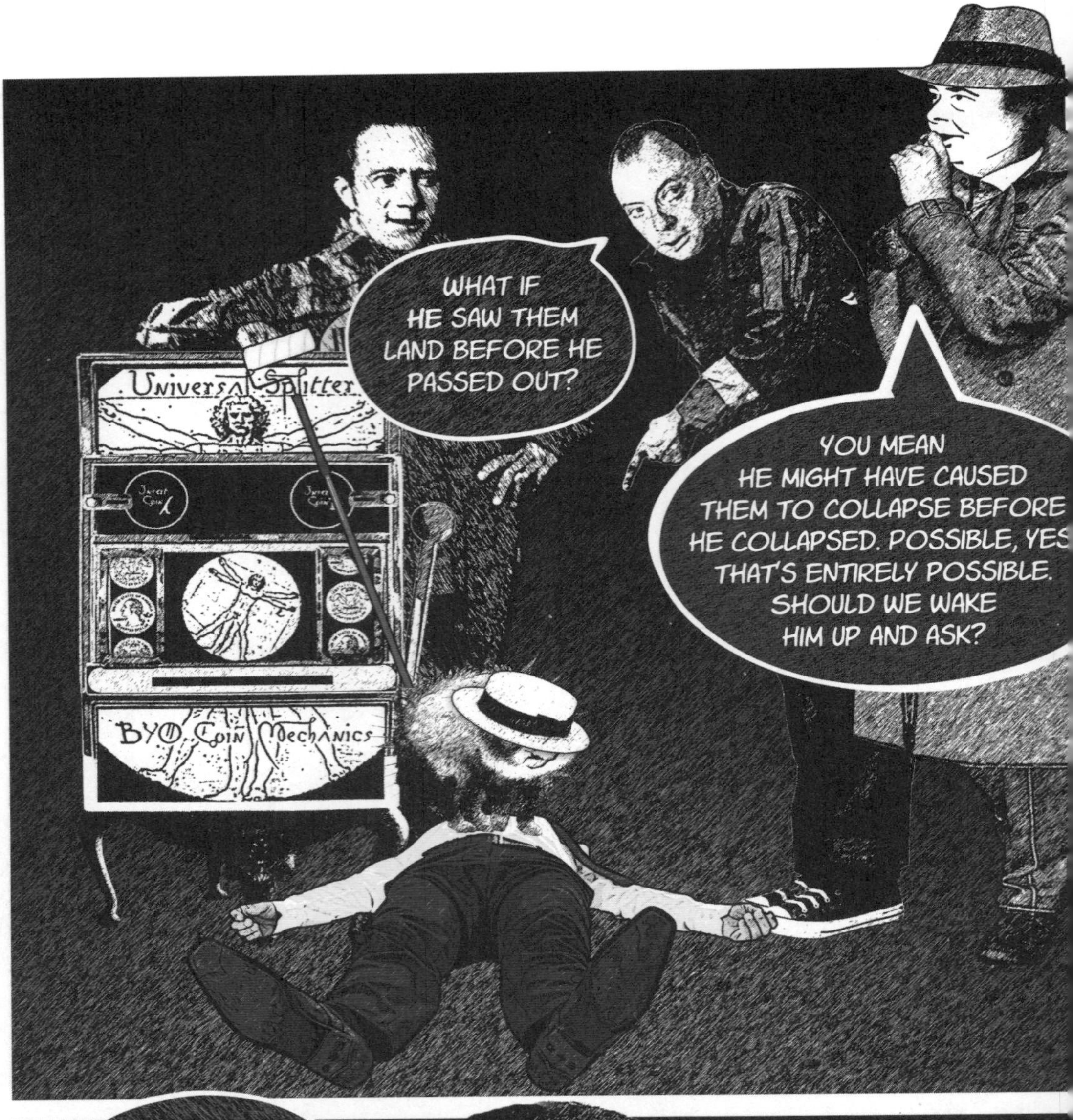
Universal Splitter
Insert Coin
Insert Coin
BYO Coin Mechanics
WHAT IF HE SAW THEM LAND BEFORE HE PASSED OUT?
YOU MEAN HE MIGHT HAVE CAUSED THEM TO COLLAPSE BEFORE HE COLLAPSED. POSSIBLE, YES THAT'S ENTIRELY POSSIBLE. SHOULD WE WAKE HIM UP AND ASK?

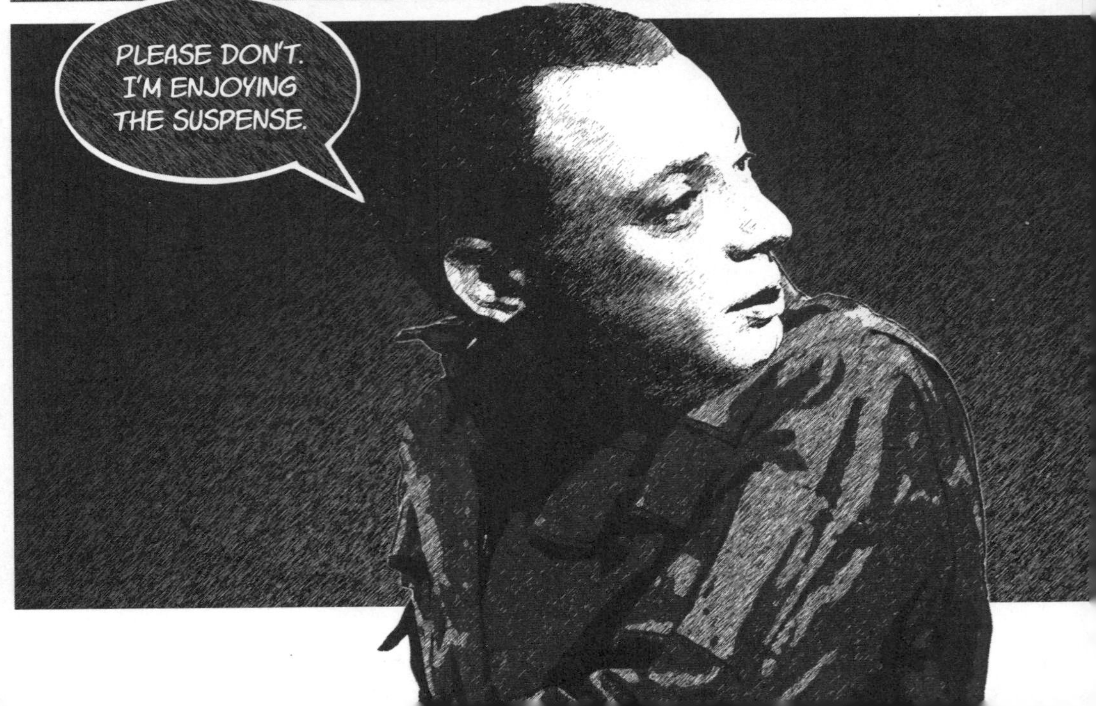
PLEASE DON'T. I'M ENJOYING THE SUSPENSE.

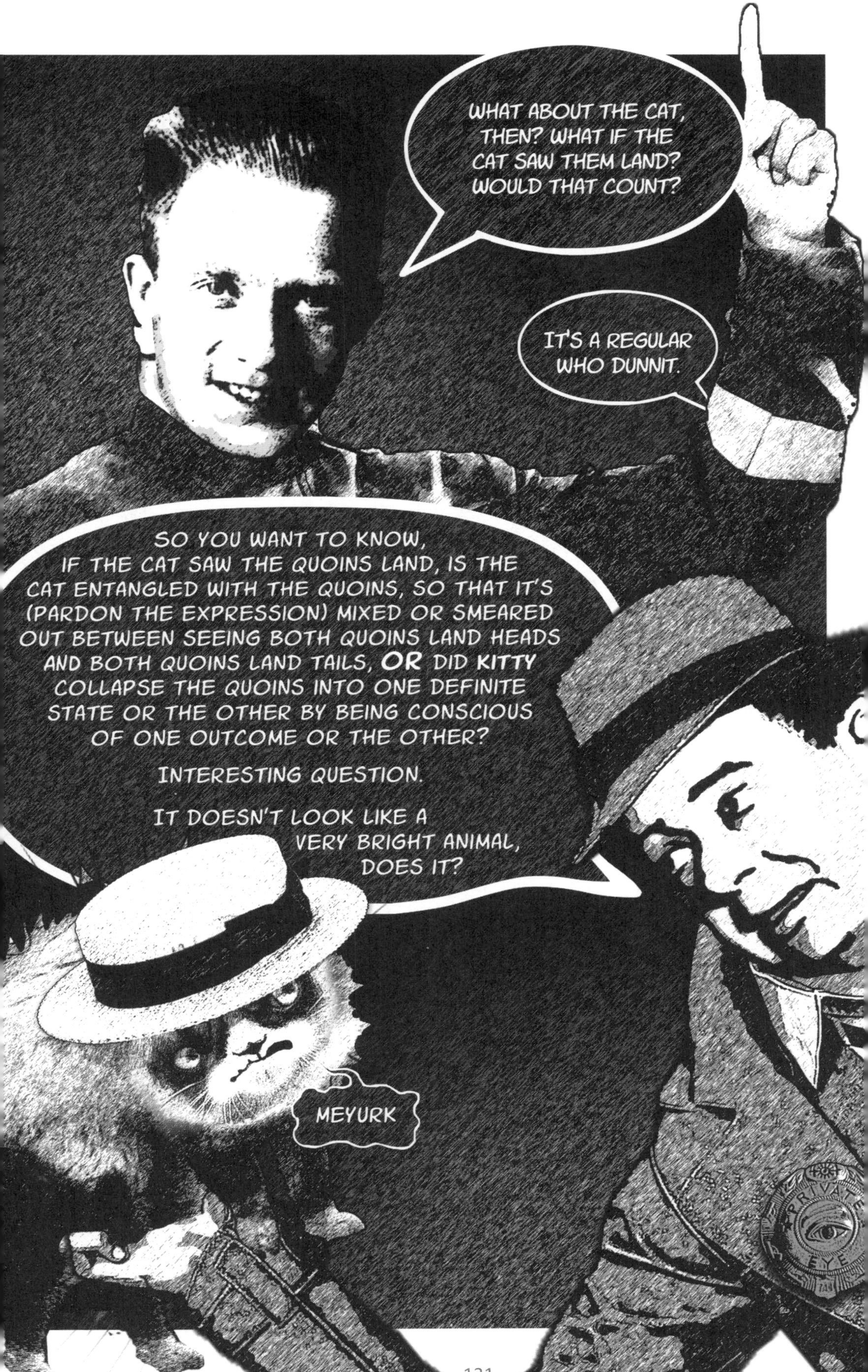
WHAT ABOUT THE CAT, THEN? WHAT IF THE CAT SAW THEM LAND? WOULD THAT COUNT?
IT'S A REGULAR WHO DUNNIT.
SO YOU WANT TO KNOW, IF THE CAT SAW THE QUOINS LAND, IS THE CAT ENTANGLED WITH THE QUOINS, SO THAT IT'S (PARDON THE EXPRESSION) MIXED OR SMEARED OUT BETWEEN SEEING BOTH QUOINS LAND HEADS AND BOTH QUOINS LAND TAILS, **OR** DID **KITTY** COLLAPSE THE QUOINS INTO ONE DEFINITE STATE OR THE OTHER BY BEING CONSCIOUS OF ONE OUTCOME OR THE OTHER?
INTERESTING QUESTION.
IT DOESN'T LOOK LIKE A VERY BRIGHT ANIMAL, DOES IT?
MEYURK
PRIVATE EYE

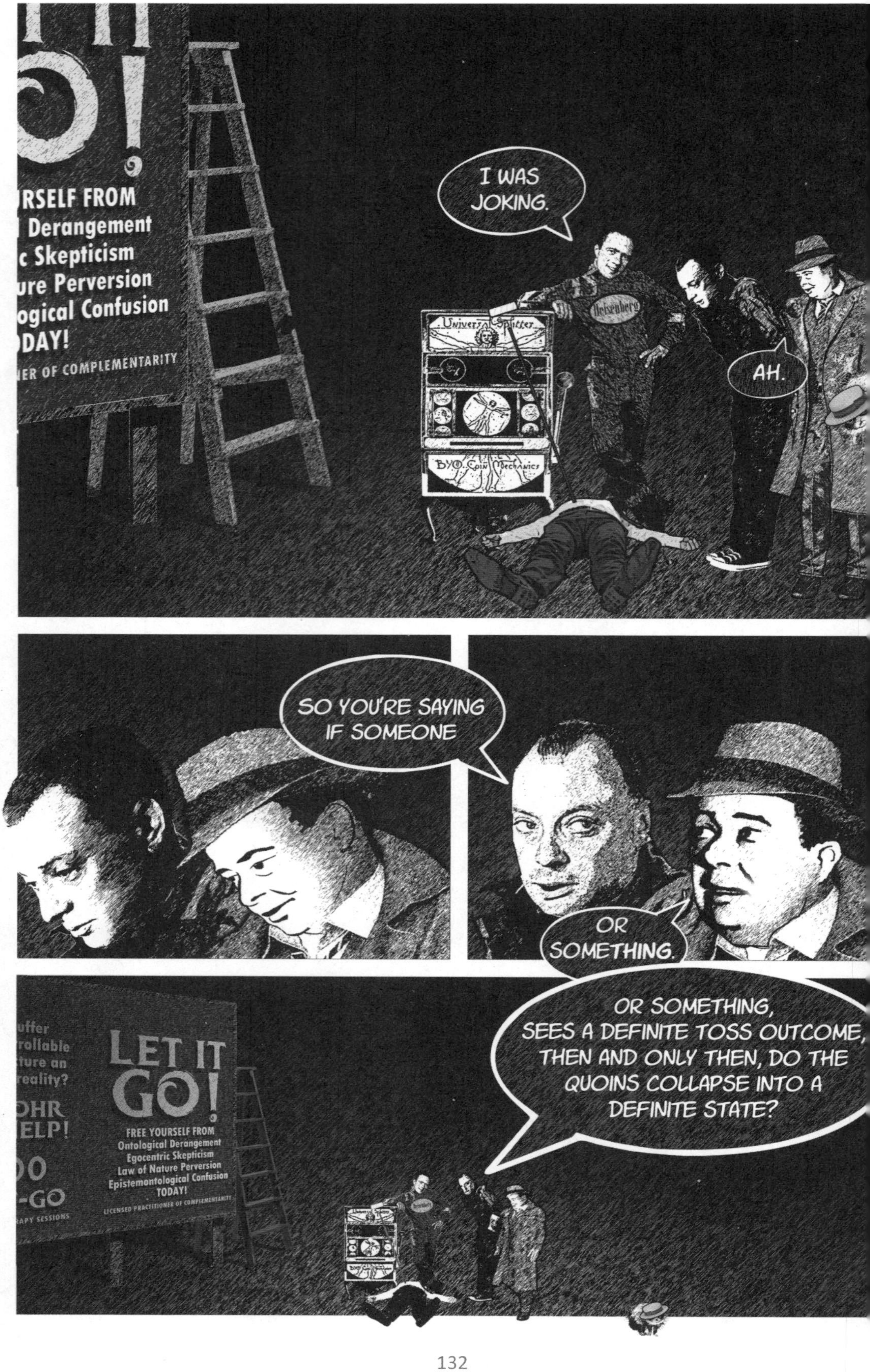
I WAS JOKING.
AH.
SO YOU'RE SAYING IF SOMEONE
OR SOMETHING.
OR SOMETHING, SEES A DEFINITE TOSS OUTCOME, THEN AND ONLY THEN, DO THE QUOINS COLLAPSE INTO A DEFINITE STATE?
LET IT GO!
FREE YOURSELF FROM
Ontological Derangement
Egocentric Skepticism
Law of Nature Perversion
Epistemontological Confusion
TODAY!
LICENSED PRACTITIONER OF COMPLEMENTARITY

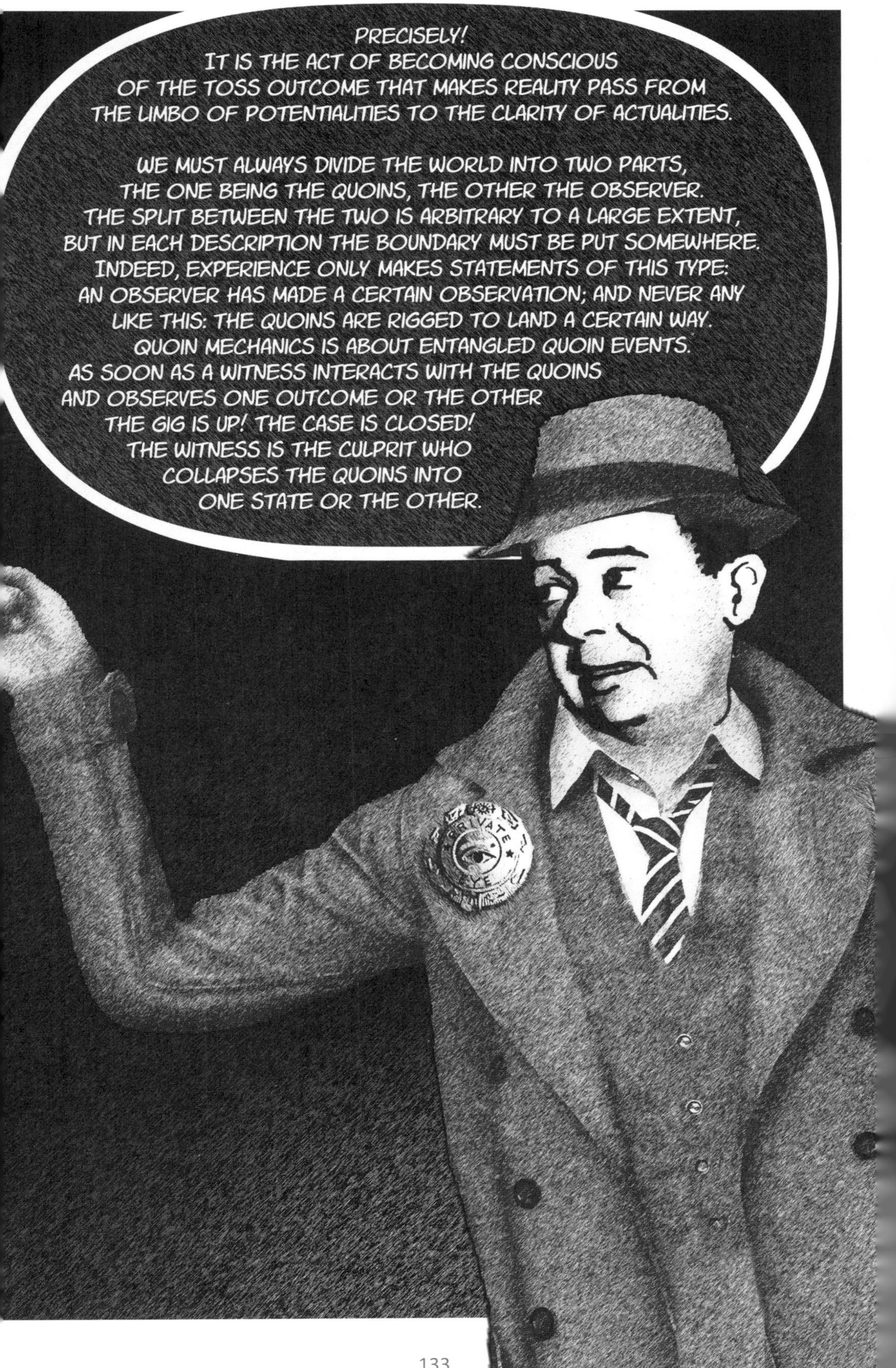
PRECISELY!
IT IS THE ACT OF BECOMING CONSCIOUS
OF THE TOSS OUTCOME THAT MAKES REALITY PASS FROM
THE LIMBO OF POTENTIALITIES TO THE CLARITY OF ACTUALITIES.
WE MUST ALWAYS DIVIDE THE WORLD INTO TWO PARTS,
THE ONE BEING THE QUOINS, THE OTHER THE OBSERVER.
THE SPLIT BETWEEN THE TWO IS ARBITRARY TO A LARGE EXTENT,
BUT IN EACH DESCRIPTION THE BOUNDARY MUST BE PUT SOMEWHERE.
INDEED, EXPERIENCE ONLY MAKES STATEMENTS OF THIS TYPE:
AN OBSERVER HAS MADE A CERTAIN OBSERVATION; AND NEVER ANY
LIKE THIS: THE QUOINS ARE RIGGED TO LAND A CERTAIN WAY.
QUOIN MECHANICS IS ABOUT ENTANGLED QUOIN EVENTS.
AS SOON AS A WITNESS INTERACTS WITH THE QUOINS
AND OBSERVES ONE OUTCOME OR THE OTHER
THE GIG IS UP! THE CASE IS CLOSED!
THE WITNESS IS THE CULPRIT WHO
COLLAPSES THE QUOINS INTO
ONE STATE OR THE OTHER.
PRIVATE EYE

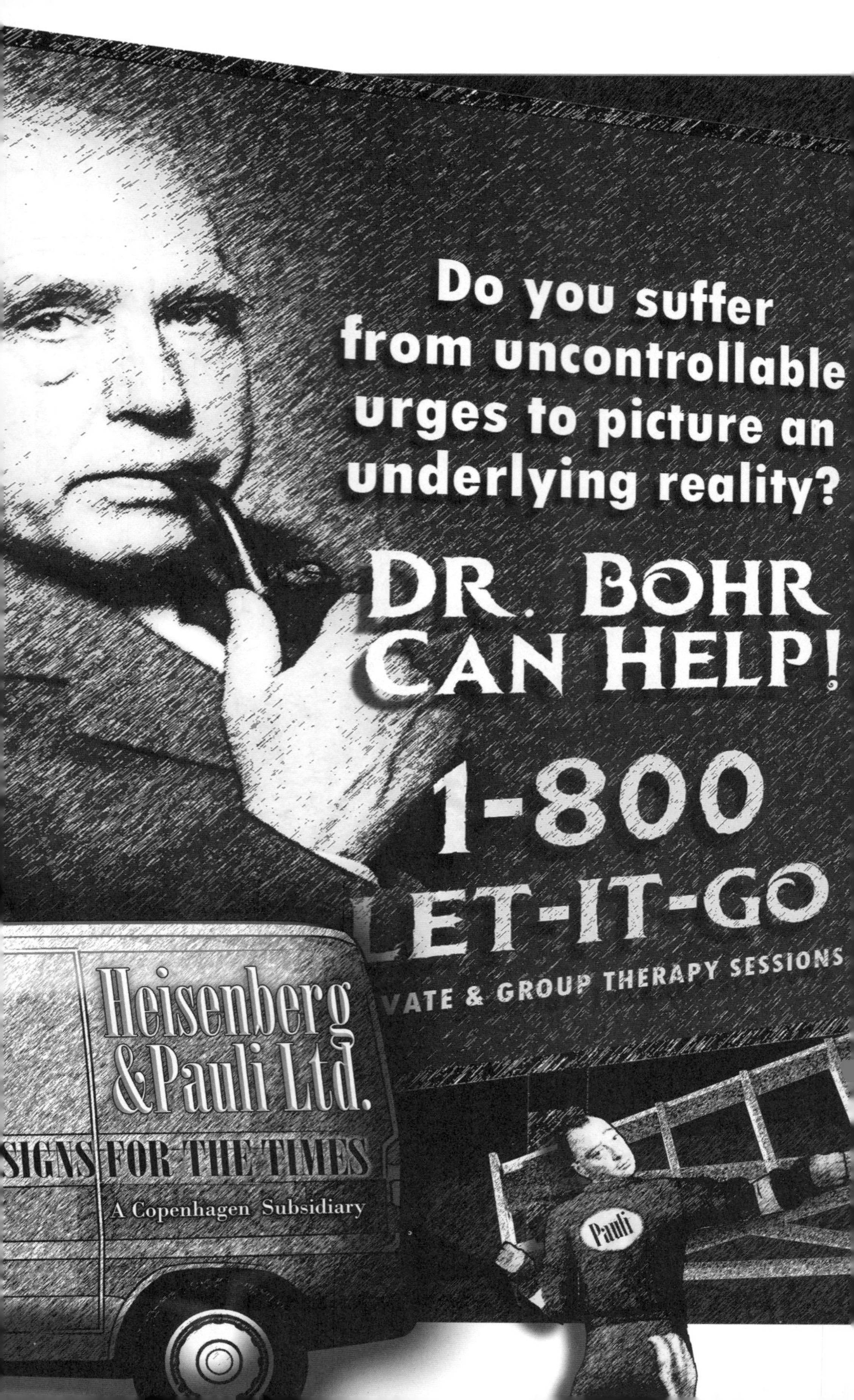
Do you suffer from uncontrollable urges to picture an underlying reality?
DR. BOHR CAN HELP!
1-800 LET-IT-GO
VATE & GROUP THERAPY SESSIONS
Heisenberg & Pauli Ltd.
SIGNS FOR THE TIMES
A Copenhagen Subsidiary
Pauli

Stop telling God what to do!

Nobel prize winning Danish physicist Niels Bohr's 'complementarity' interpretation has exerted a powerful influential on the thinking of the physics community since the 1920's. At its core is Bohr's belief that quantum mechanics forces us to let go of the classical ideal of picturing reality at the micro-level. With variations by other physicists like Werner Heisenberg, inventor of quantum mechanics, and Wolfgang Pauli, a prominent figure in the Bohr camp, this doctrine came to be known as the Copenhagen interpretation.

Einstein didn't buy into Bohr's view and never abandoned the quest to find underlying causes that could account for quantum probabilities. He regarded complementarity as nothing more than therapy for physicists depressed about the enigma of entanglement.

In the 1950's, David Bohm, convinced by Einstein's arguments, added 'hidden variables' and came up with a theory that showed how the correlations of entangled particles could be explained causally. To get this to work, Bohm had to give up some aspects of Einstein's special theory of relativity. Einstein wasn't enthusiastic about Bohm's theory and remarked to Schrödinger that it seemed 'too cheap' to him. As Bell pointed out, 'the Einstein-Podolsky-Rosen paradox is resolved in the way which Einstein would have liked least.'

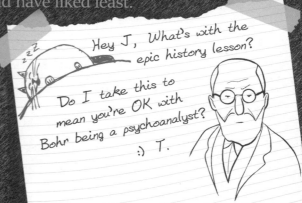

Hey J, What's with the epic history lesson?

Do I take this to mean you're OK with Bohr being a psychoanalyst?

:) T.

There is no quantum world.
There is only an abstract quantum
physical description. It is wrong to think that
the task of physics is to find out how nature is.
Physics concerns what we can say about nature.

—**Niels Bohr**

The Heisenberg-Bohr tranquilizing philosophy
—or religion?—is so delicately contrived that, for
the time being, it provides a gentle pillow for the
true believer from which he cannot very easily
be aroused. So let him lie there.

—**Albert Einstein**

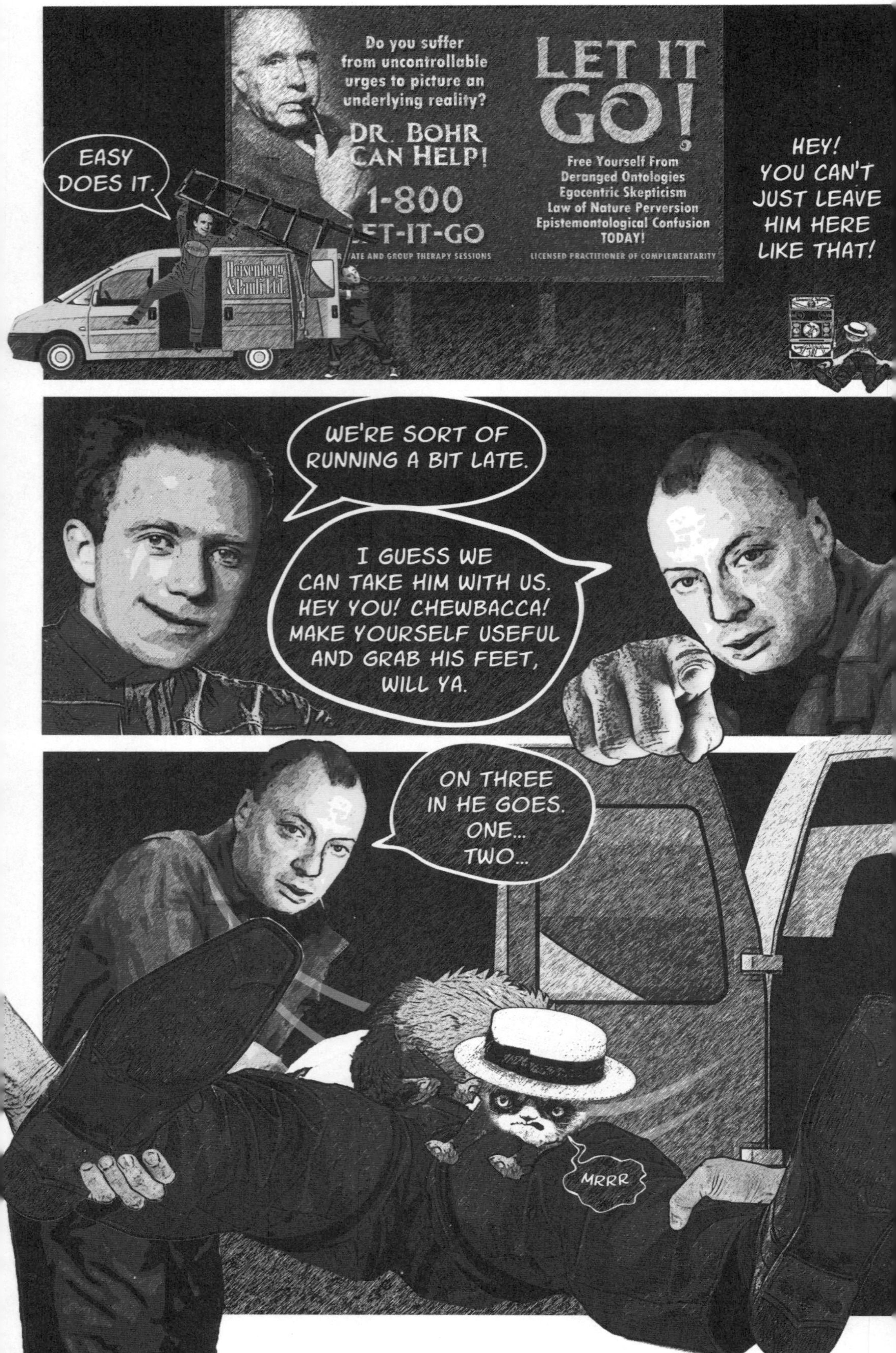
Do you suffer from uncontrollable urges to picture an underlying reality?
DR. BOHR CAN HELP!
1-800
ET-IT-GO
ATE AND GROUP THERAPY SESSIONS
LET IT GO!
Free Yourself From
Deranged Ontologies
Egocentric Skepticism
Law of Nature Perversion
Epistemontological Confusion
TODAY!
LICENSED PRACTITIONER OF COMPLEMENTARITY
EASY DOES IT.
Heisenberg & Pauli Ltd.
HEY! YOU CAN'T JUST LEAVE HIM HERE LIKE THAT!
WE'RE SORT OF RUNNING A BIT LATE.
I GUESS WE CAN TAKE HIM WITH US. HEY YOU! CHEWBACCA! MAKE YOURSELF USEFUL AND GRAB HIS FEET, WILL YA.
ON THREE IN HE GOES. ONE... TWO...
MRRR

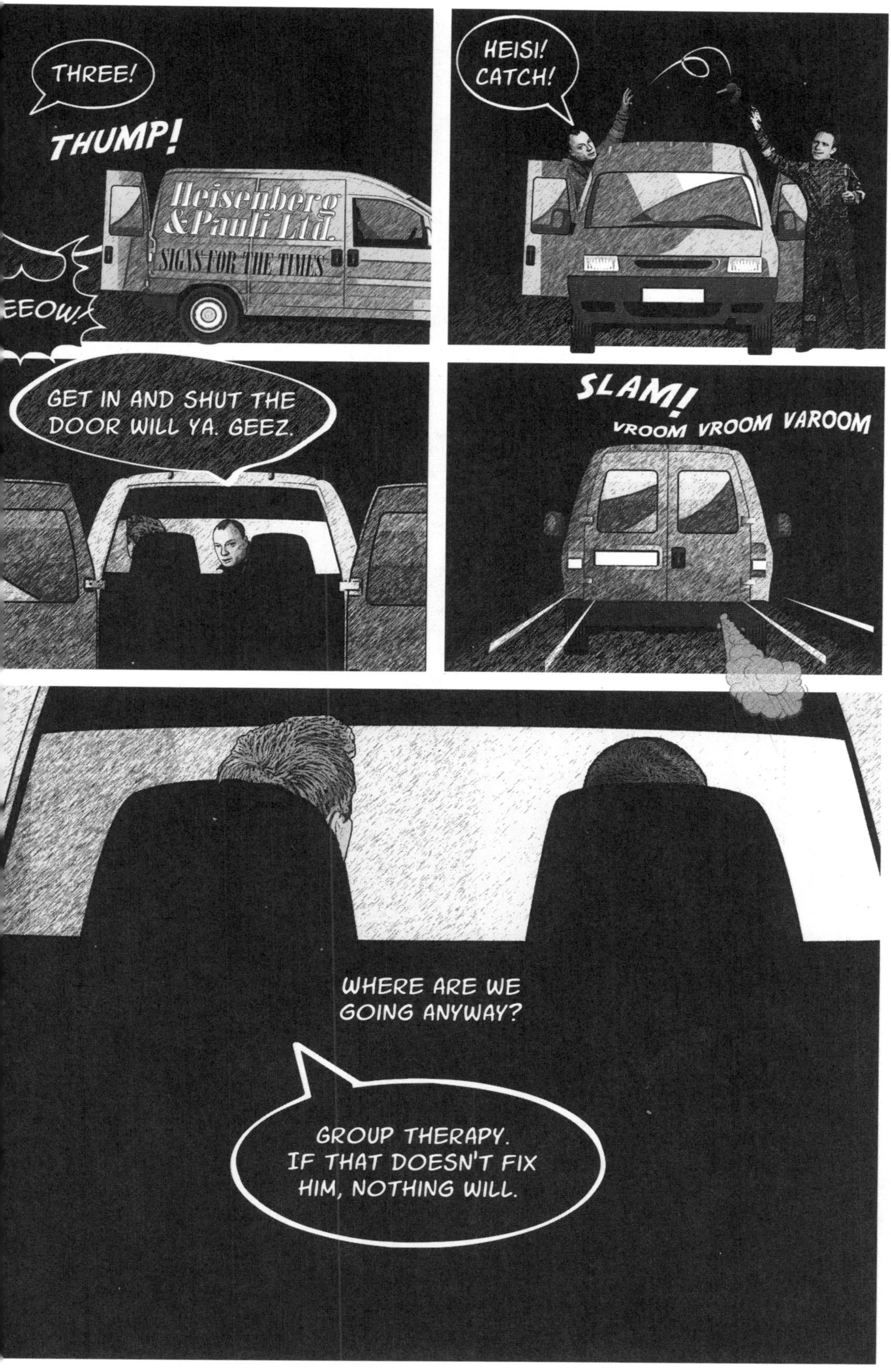
THREE!
THUMP!
Heisenberg & Pauli Ltd.
SIGNS FOR THE TIMES
EEOW!
HEISI! CATCH!
GET IN AND SHUT THE DOOR WILL YA. GEEZ.
SLAM!
VROOM VROOM VAROOM
WHERE ARE WE GOING ANYWAY?
GROUP THERAPY. IF THAT DOESN'T FIX HIM, NOTHING WILL.

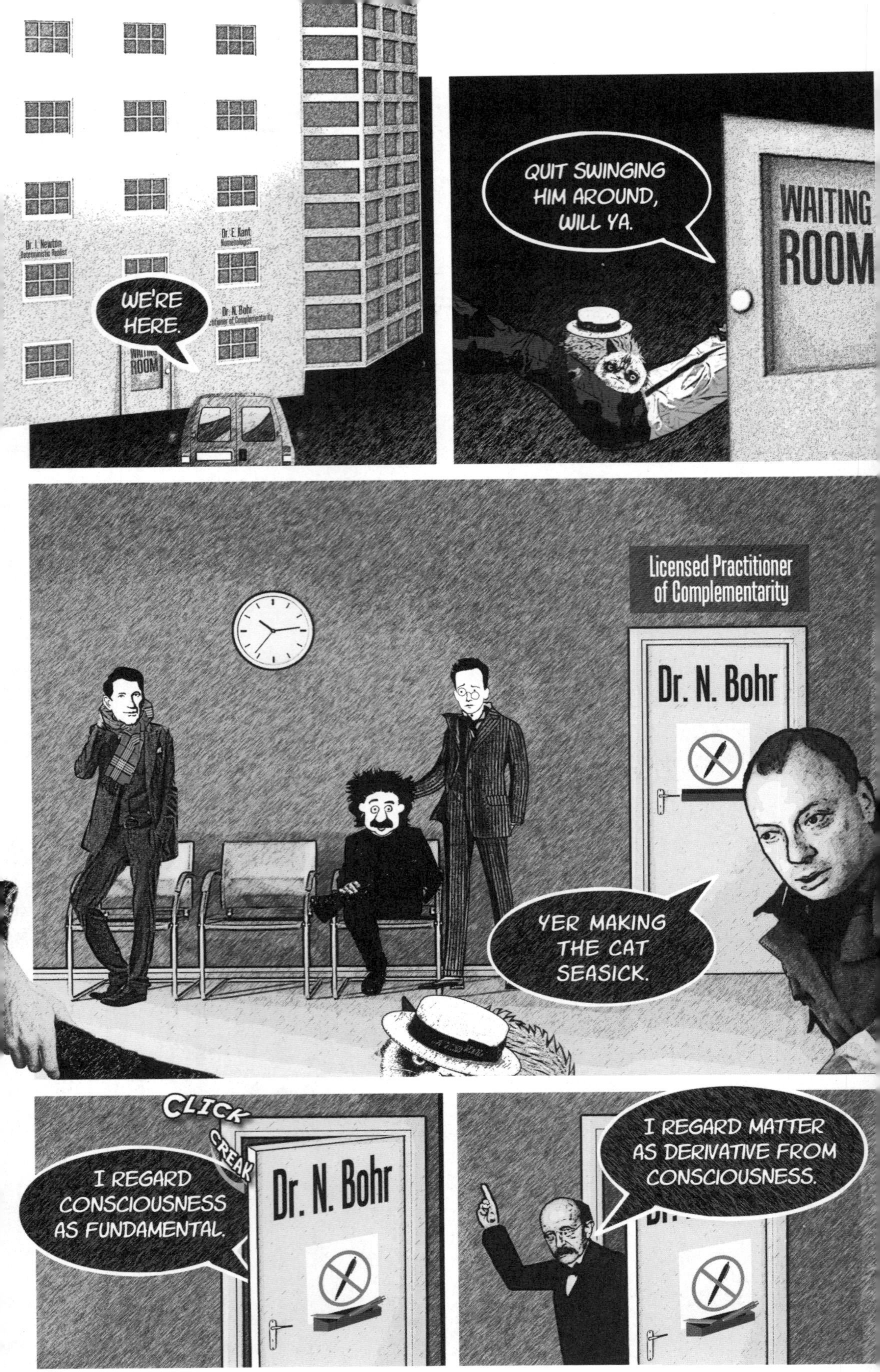

Dr. I. Newton
Deterministic Realist
Dr. E. Kant
Numenologist
Dr. N. Bohr
WE'RE HERE.
ROOM
QUIT SWINGING HIM AROUND, WILL YA.
WAITING ROOM
Licensed Practitioner of Complementarity
Dr. N. Bohr
YER MAKING THE CAT SEASICK.
CLICK
CREAK
I REGARD CONSCIOUSNESS AS FUNDAMENTAL.
Dr. N. Bohr
I REGARD MATTER AS DERIVATIVE FROM CONSCIOUSNESS.

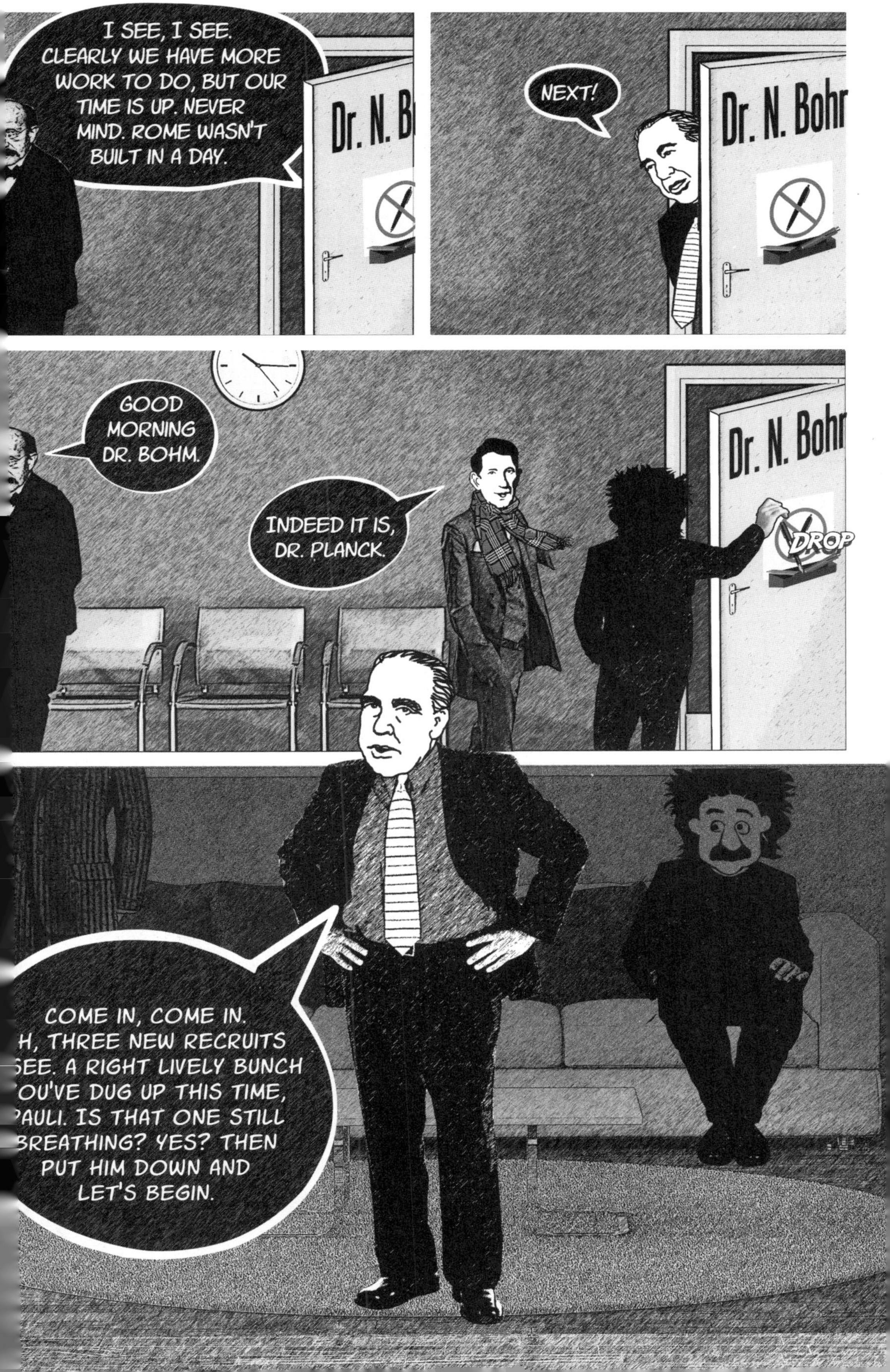

I SEE, I SEE. CLEARLY WE HAVE MORE WORK TO DO, BUT OUR TIME IS UP. NEVER MIND. ROME WASN'T BUILT IN A DAY.
Dr. N. B
NEXT!
Dr. N. Bohr
GOOD MORNING DR. BOHM.
INDEED IT IS, DR. PLANCK.
Dr. N. Bohr
DROP
COME IN, COME IN.
H, THREE NEW RECRUITS
SEE. A RIGHT LIVELY BUNCH
OU'VE DUG UP THIS TIME,
PAULI. IS THAT ONE STILL
BREATHING? YES? THEN
PUT HIM DOWN AND
LET'S BEGIN.

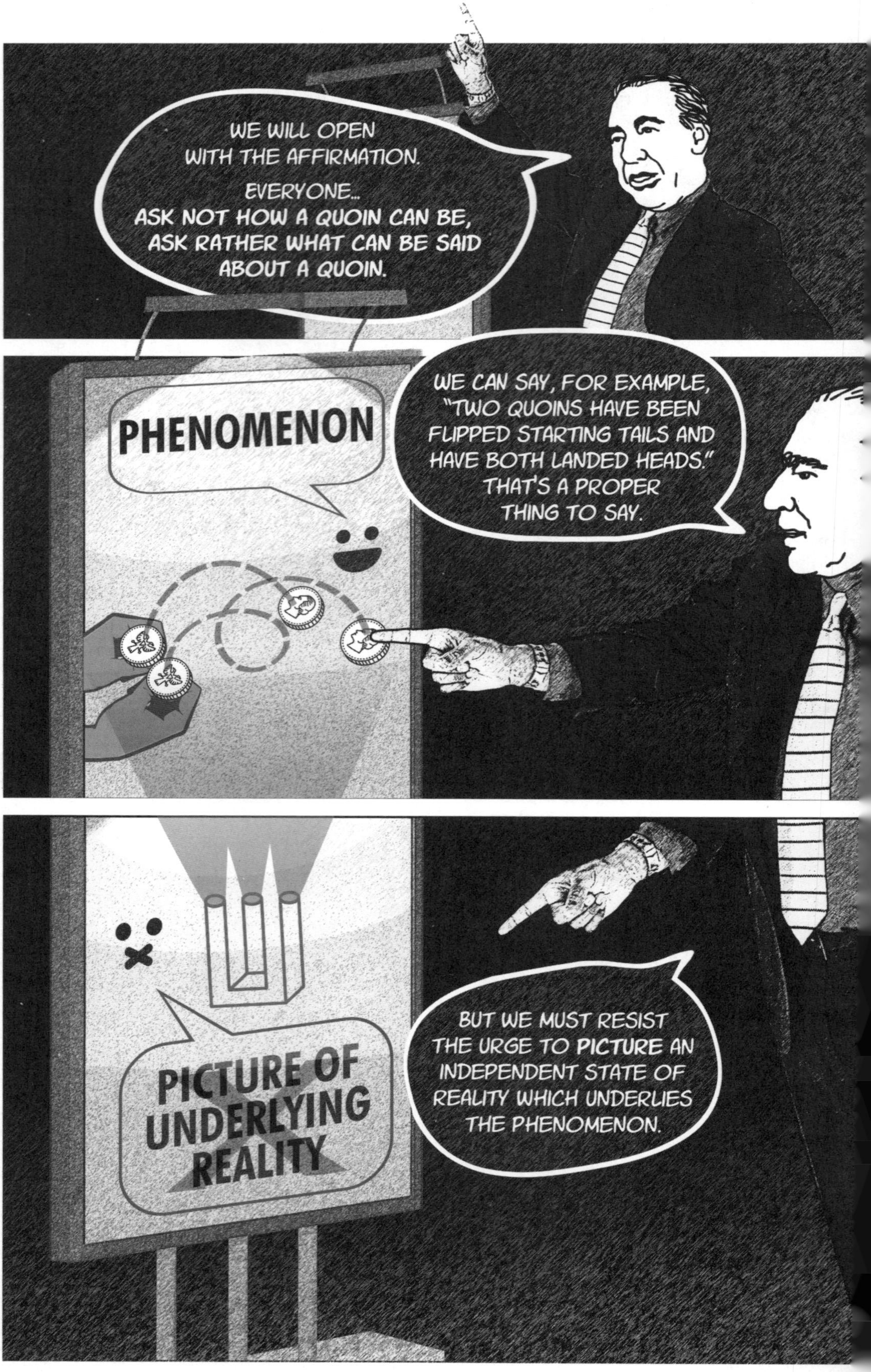
WE WILL OPEN WITH THE AFFIRMATION.
EVERYONE... ASK NOT HOW A QUOIN CAN BE, ASK RATHER WHAT CAN BE SAID ABOUT A QUOIN.
PHENOMENON
WE CAN SAY, FOR EXAMPLE, "TWO QUOINS HAVE BEEN FLIPPED STARTING TAILS AND HAVE BOTH LANDED HEADS." THAT'S A PROPER THING TO SAY.
PICTURE OF UNDERLYING REALITY
BUT WE MUST RESIST THE URGE TO **PICTURE** AN INDEPENDENT STATE OF REALITY WHICH UNDERLIES THE PHENOMENON.

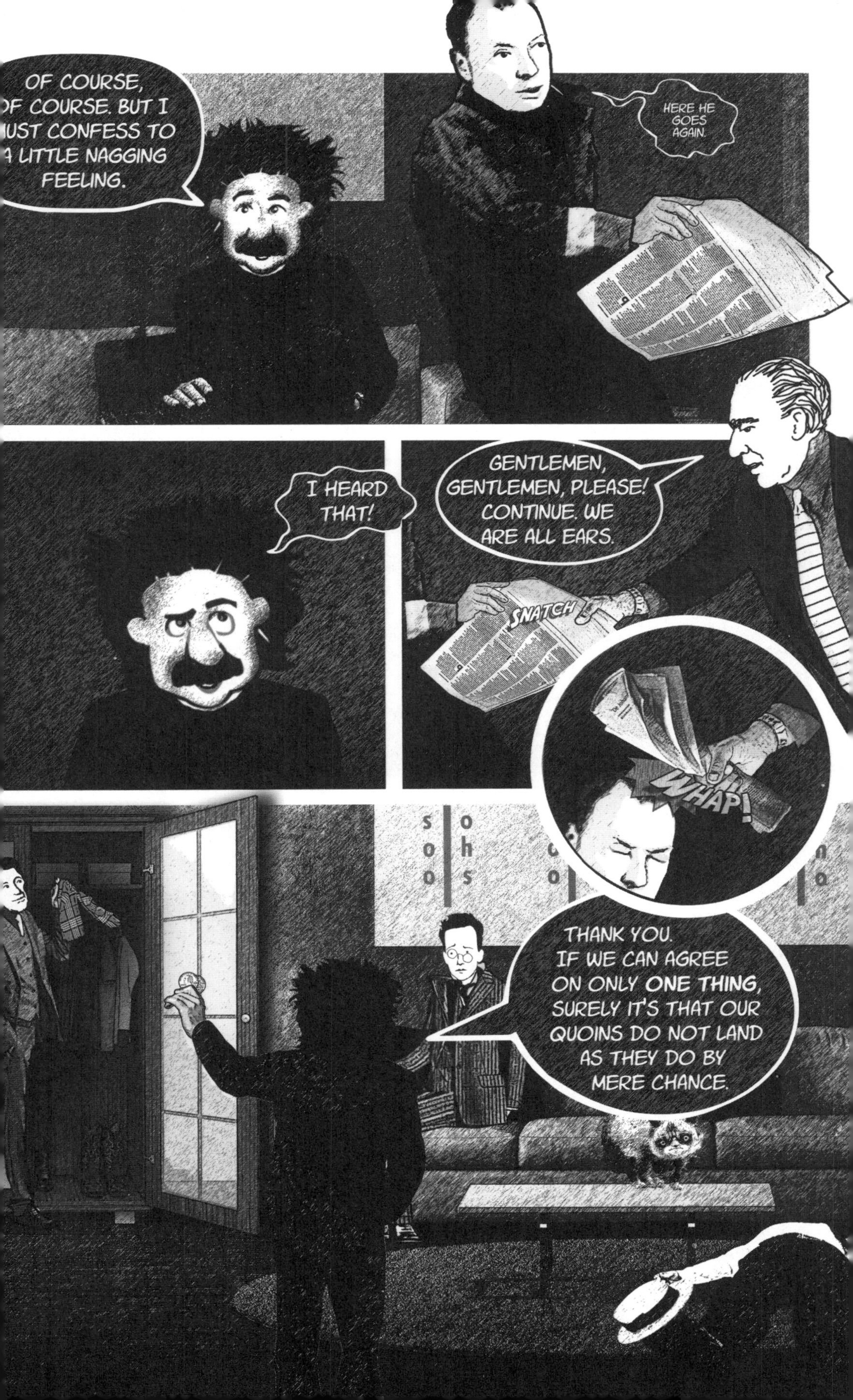

OF COURSE, OF COURSE. BUT I MUST CONFESS TO A LITTLE NAGGING FEELING.
HERE HE GOES AGAIN.
I HEARD THAT!
GENTLEMEN, GENTLEMEN, PLEASE! CONTINUE. WE ARE ALL EARS.
SNATCH
WHAP!
THANK YOU. IF WE CAN AGREE ON ONLY **ONE THING**, SURELY IT'S THAT OUR QUOINS DO NOT LAND AS THEY DO BY MERE CHANCE.

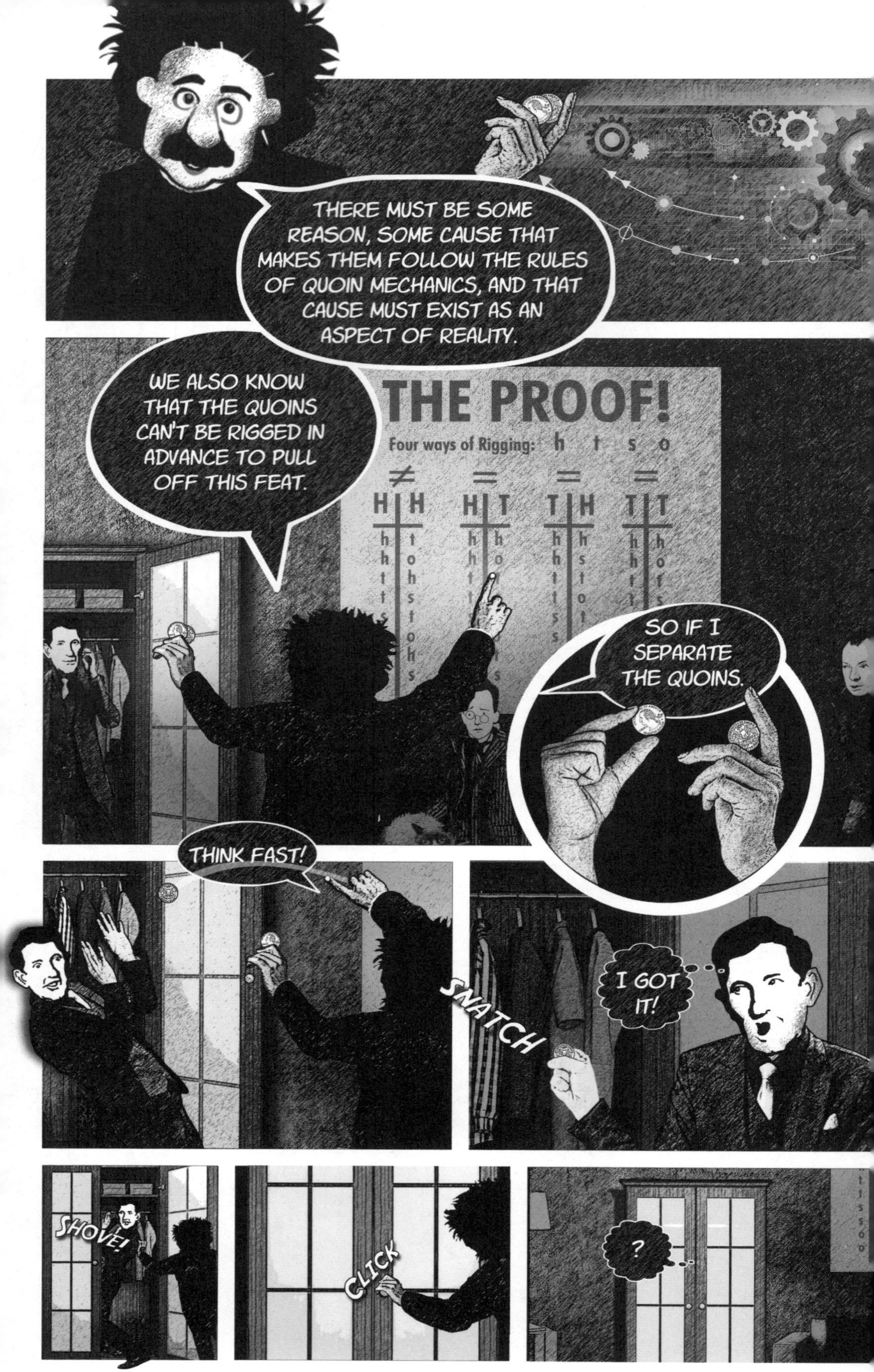
THERE MUST BE SOME REASON, SOME CAUSE THAT MAKES THEM FOLLOW THE RULES OF QUOIN MECHANICS, AND THAT CAUSE MUST EXIST AS AN ASPECT OF REALITY.
WE ALSO KNOW THAT THE QUOINS CAN'T BE RIGGED IN ADVANCE TO PULL OFF THIS FEAT.
THE PROOF!
Four ways of Rigging: h t s o
SO IF I SEPARATE THE QUOINS.
THINK FAST!
SNATCH
I GOT IT!
SHOVE!
CLICK
?

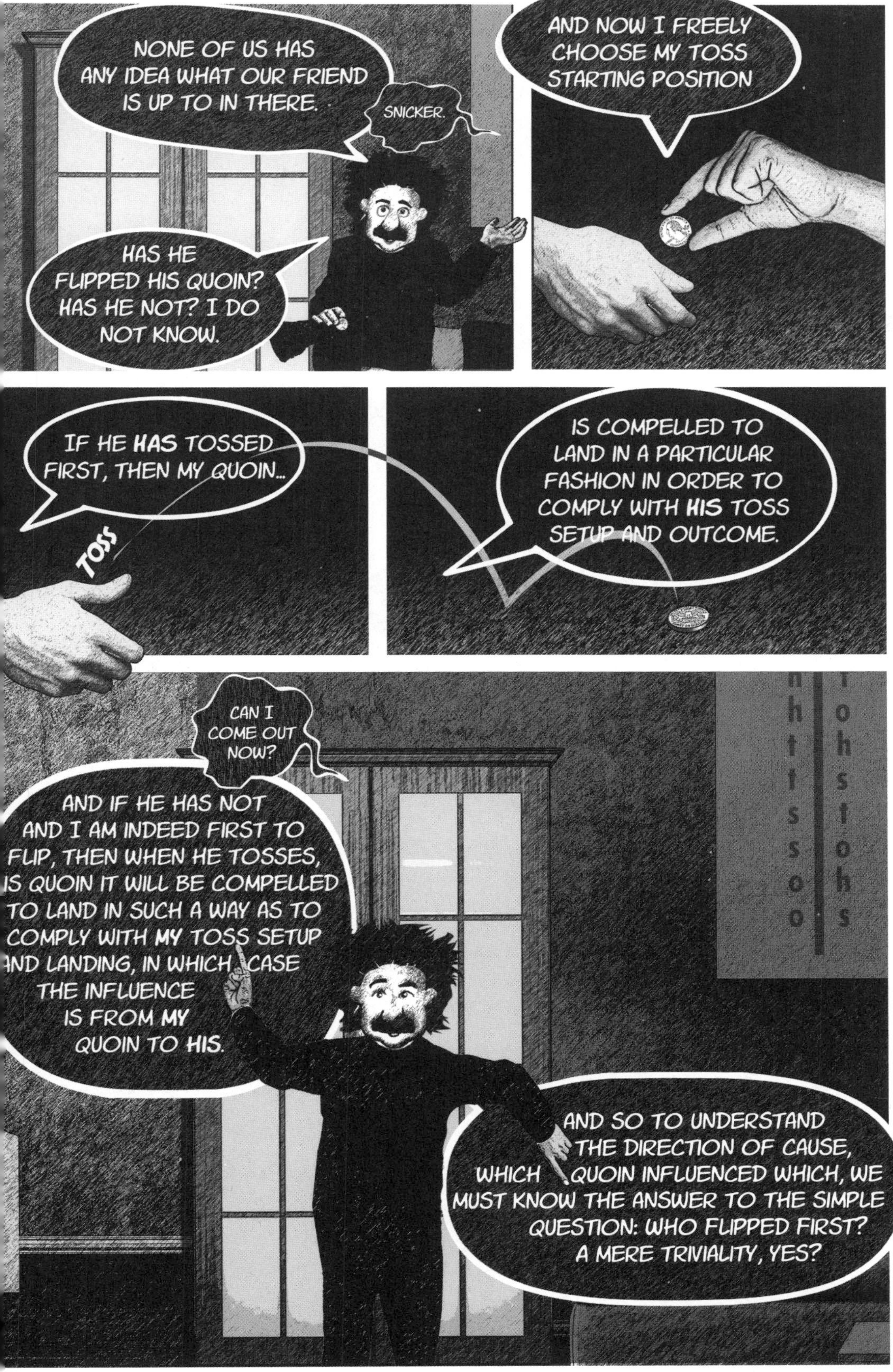
NONE OF US HAS ANY IDEA WHAT OUR FRIEND IS UP TO IN THERE.
SNICKER.
HAS HE FLIPPED HIS QUOIN? HAS HE NOT? I DO NOT KNOW.
AND NOW I FREELY CHOOSE MY TOSS STARTING POSITION
IF HE **HAS** TOSSED FIRST, THEN MY QUOIN...
TOSS
IS COMPELLED TO LAND IN A PARTICULAR FASHION IN ORDER TO COMPLY WITH **HIS** TOSS SETUP AND OUTCOME.
CAN I COME OUT NOW?
AND IF HE HAS NOT AND I AM INDEED FIRST TO FLIP, THEN WHEN HE TOSSES, IS QUOIN IT WILL BE COMPELLED TO LAND IN SUCH A WAY AS TO COMPLY WITH **MY** TOSS SETUP AND LANDING, IN WHICH CASE THE INFLUENCE IS FROM **MY** QUOIN TO **HIS**.
AND SO TO UNDERSTAND THE DIRECTION OF CAUSE, WHICH QUOIN INFLUENCED WHICH, WE MUST KNOW THE ANSWER TO THE SIMPLE QUESTION: WHO FLIPPED FIRST? A MERE TRIVIALITY, YES?

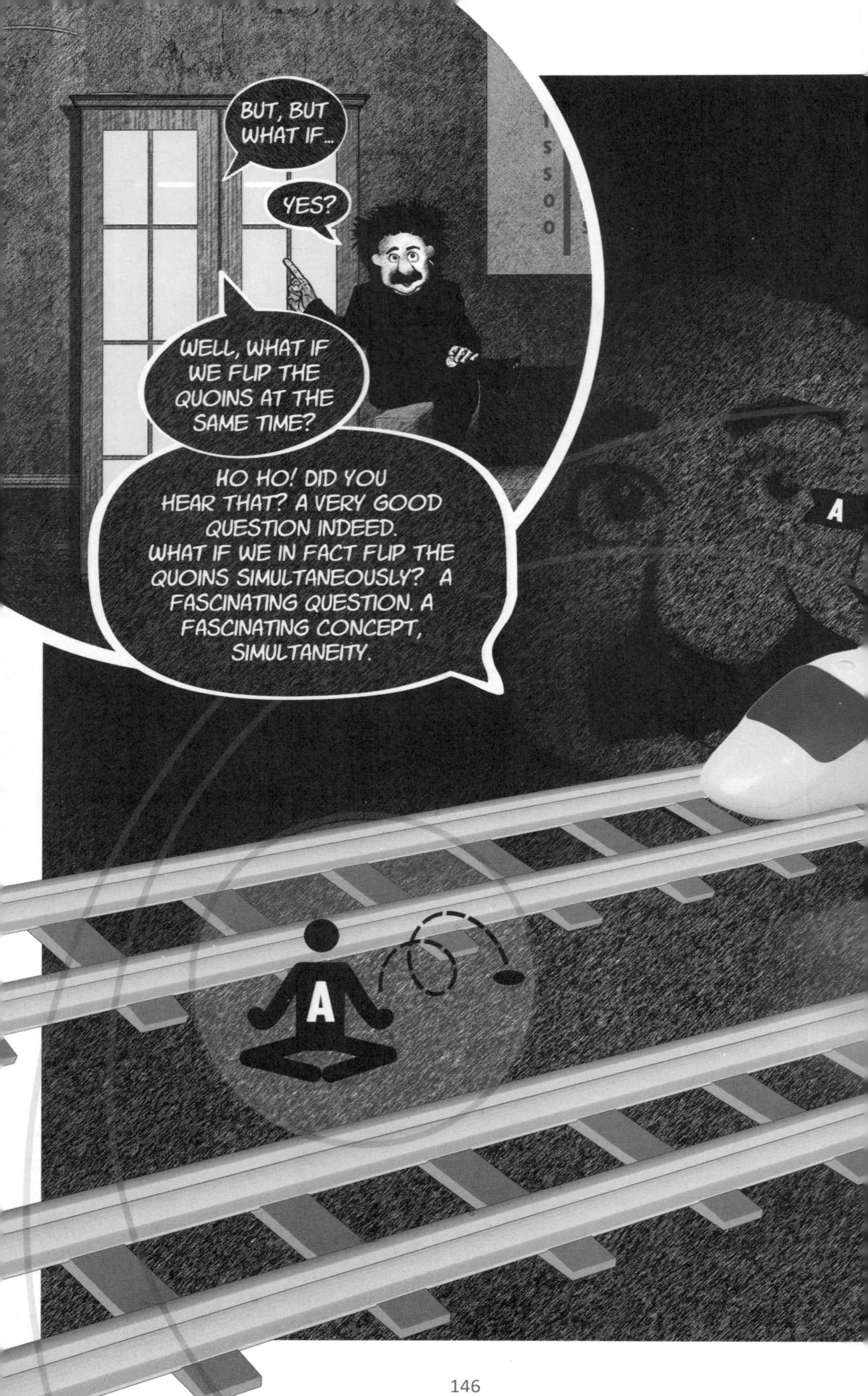
BUT, BUT WHAT IF...
YES?
WELL, WHAT IF WE FLIP THE QUOINS AT THE SAME TIME?
HO HO! DID YOU HEAR THAT? A VERY GOOD QUESTION INDEED. WHAT IF WE IN FACT FLIP THE QUOINS SIMULTANEOUSLY? A FASCINATING QUESTION. A FASCINATING CONCEPT, SIMULTANEITY.
A
A

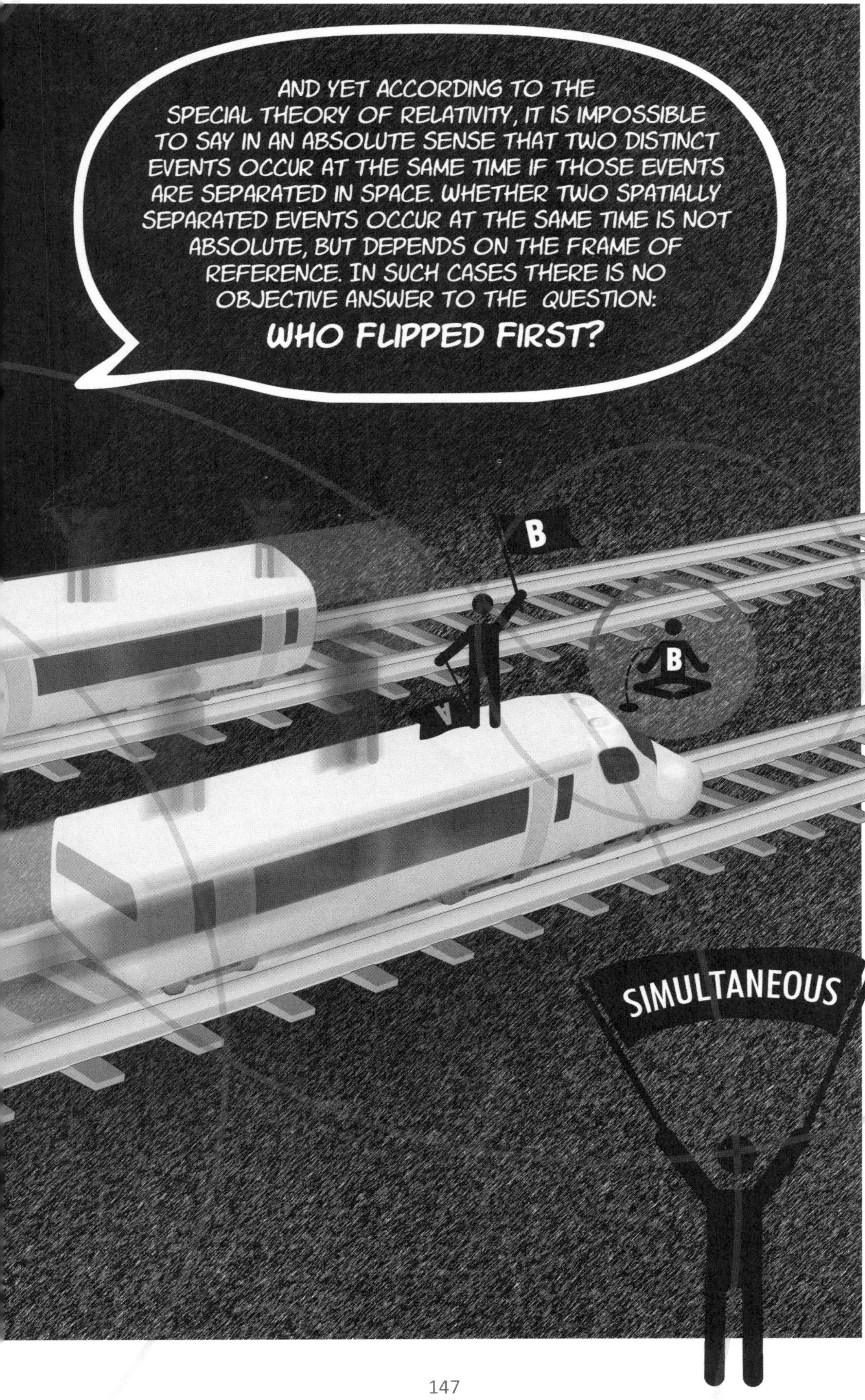
AND YET ACCORDING TO THE SPECIAL THEORY OF RELATIVITY, IT IS IMPOSSIBLE TO SAY IN AN ABSOLUTE SENSE THAT TWO DISTINCT EVENTS OCCUR AT THE SAME TIME IF THOSE EVENTS ARE SEPARATED IN SPACE. WHETHER TWO SPATIALLY SEPARATED EVENTS OCCUR AT THE SAME TIME IS NOT ABSOLUTE, BUT DEPENDS ON THE FRAME OF REFERENCE. IN SUCH CASES THERE IS NO OBJECTIVE ANSWER TO THE QUESTION:
WHO FLIPPED FIRST?
B
B
A
SIMULTANEOUS

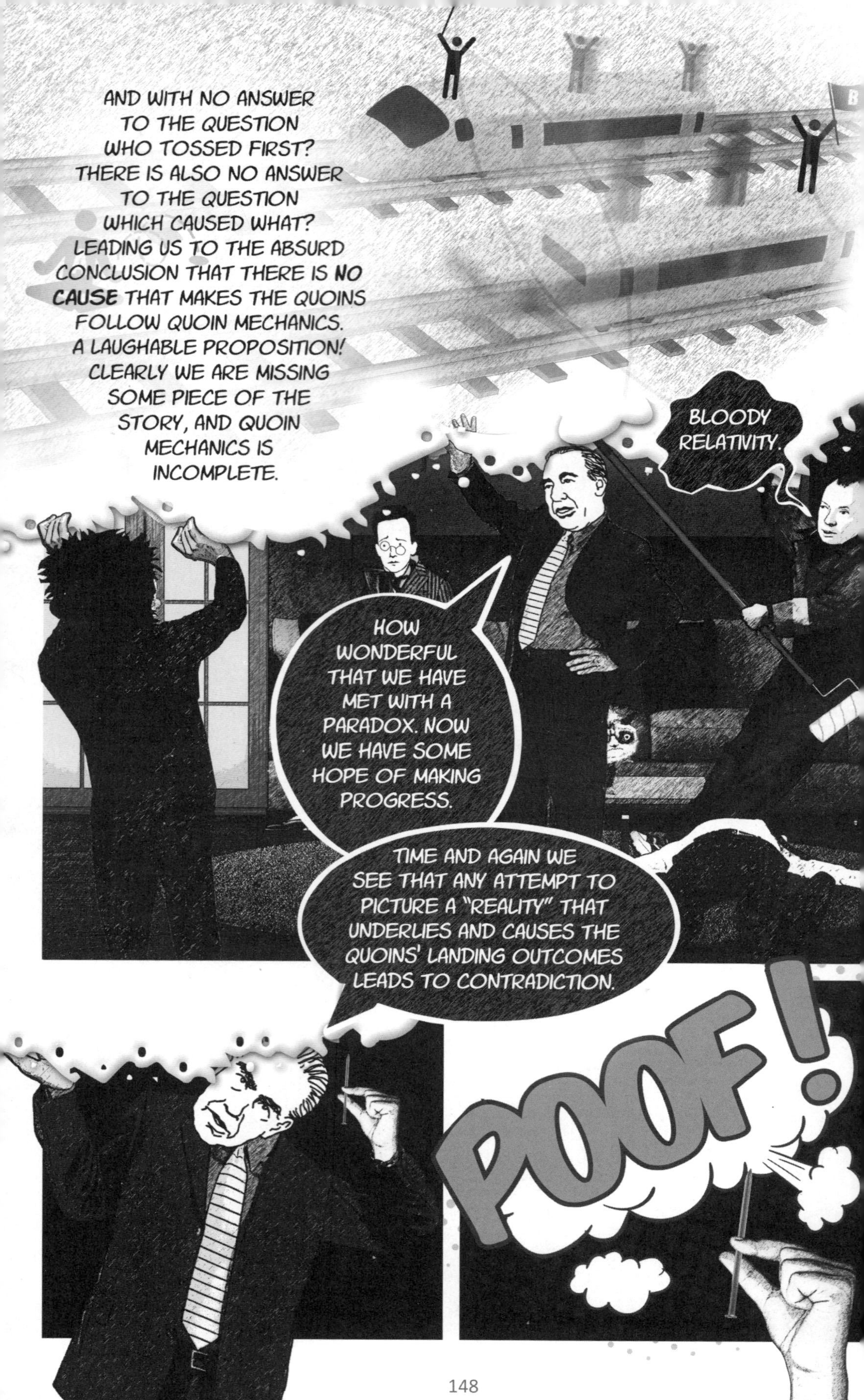
AND WITH NO ANSWER TO THE QUESTION WHO TOSSED FIRST? THERE IS ALSO NO ANSWER TO THE QUESTION WHICH CAUSED WHAT? LEADING US TO THE ABSURD CONCLUSION THAT THERE IS **NO CAUSE** THAT MAKES THE QUOINS FOLLOW QUOIN MECHANICS. A LAUGHABLE PROPOSITION! CLEARLY WE ARE MISSING SOME PIECE OF THE STORY, AND QUOIN MECHANICS IS INCOMPLETE.
BLOODY RELATIVITY.
HOW WONDERFUL THAT WE HAVE MET WITH A PARADOX. NOW WE HAVE SOME HOPE OF MAKING PROGRESS.
TIME AND AGAIN WE SEE THAT ANY ATTEMPT TO PICTURE A "REALITY" THAT UNDERLIES AND CAUSES THE QUOINS' LANDING OUTCOMES LEADS TO CONTRADICTION.
POOF!

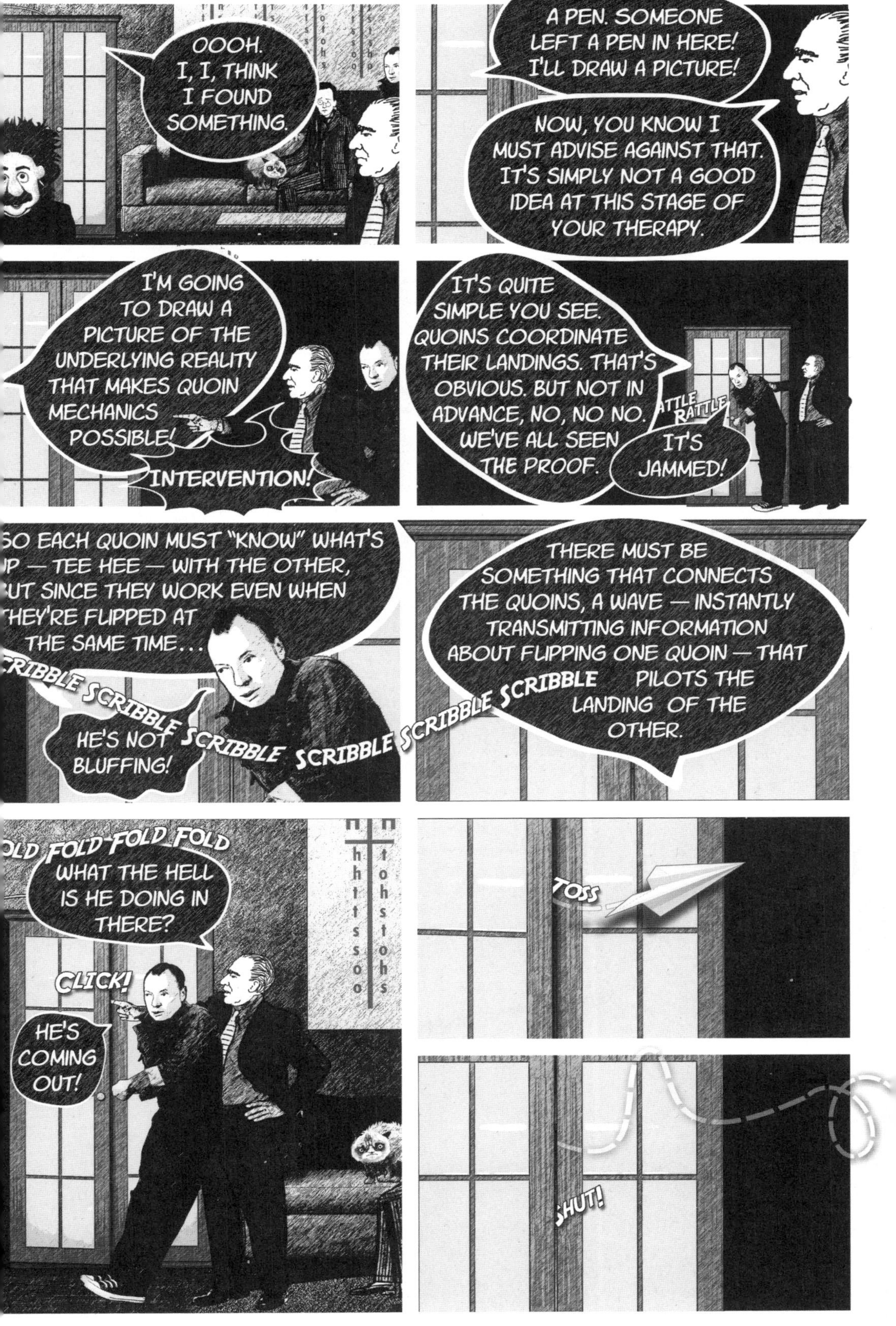
OOOH. I, I, THINK I FOUND SOMETHING.
A PEN. SOMEONE LEFT A PEN IN HERE! I'LL DRAW A PICTURE!
NOW, YOU KNOW I MUST ADVISE AGAINST THAT. IT'S SIMPLY NOT A GOOD IDEA AT THIS STAGE OF YOUR THERAPY.
I'M GOING TO DRAW A PICTURE OF THE UNDERLYING REALITY THAT MAKES QUOIN MECHANICS POSSIBLE!
INTERVENTION!
IT'S QUITE SIMPLE YOU SEE. QUOINS COORDINATE THEIR LANDINGS. THAT'S OBVIOUS. BUT NOT IN ADVANCE, NO, NO NO. WE'VE ALL SEEN THE PROOF.
RATTLE
IT'S JAMMED!
SO EACH QUOIN MUST "KNOW" WHAT'S UP — TEE HEE — WITH THE OTHER, BUT SINCE THEY WORK EVEN WHEN THEY'RE FLIPPED AT THE SAME TIME...
SCRIBBLE SCRIBBLE
HE'S NOT BLUFFING!
SCRIBBLE SCRIBBLE SCRIBBLE SCRIBBLE
THERE MUST BE SOMETHING THAT CONNECTS THE QUOINS, A WAVE — INSTANTLY TRANSMITTING INFORMATION ABOUT FLIPPING ONE QUOIN — THAT PILOTS THE LANDING OF THE OTHER.
FOLD FOLD FOLD
WHAT THE HELL IS HE DOING IN THERE?
CLICK!
HE'S COMING OUT!
TOSS
SHUT!

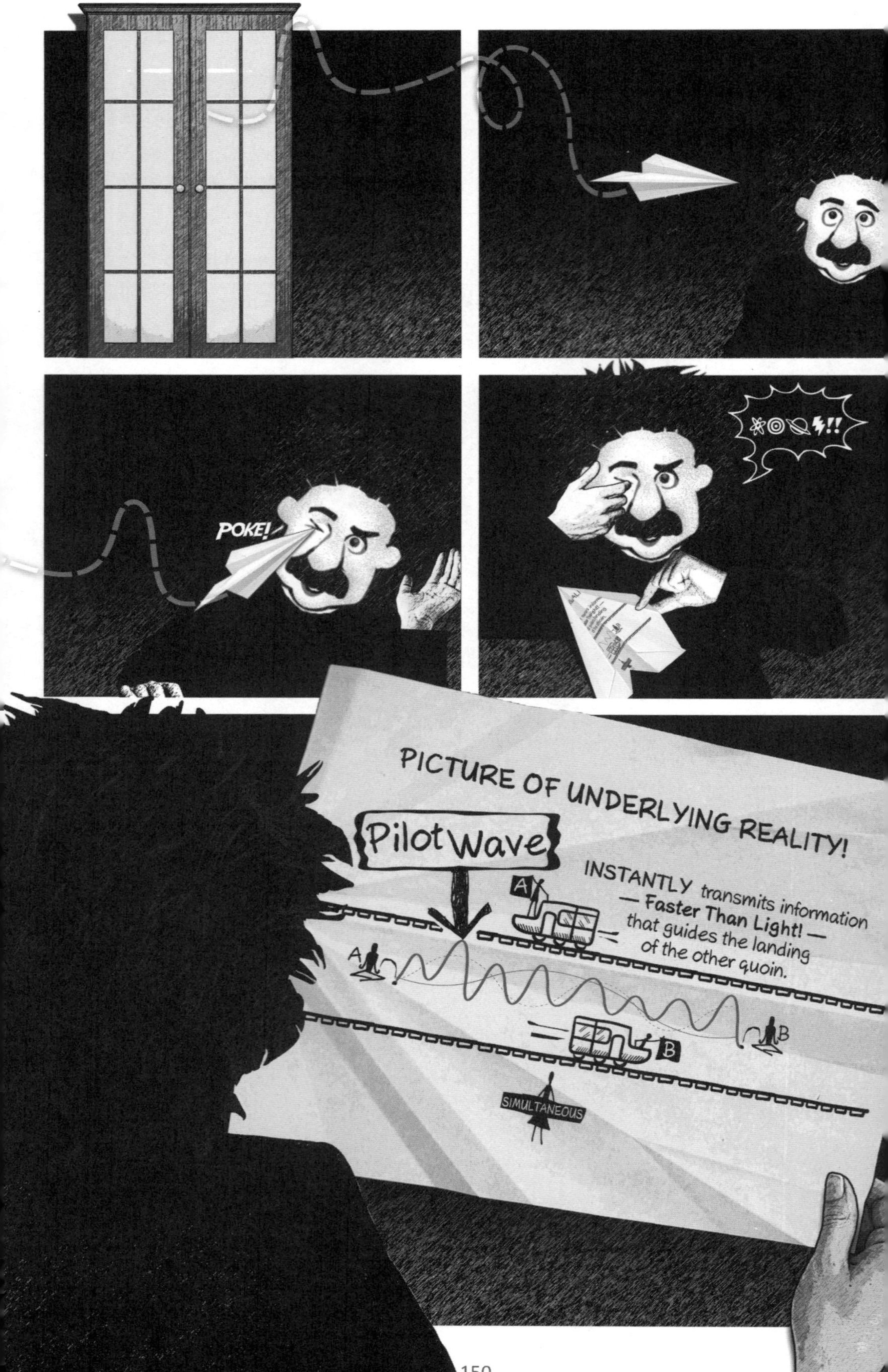
POKE!
PICTURE OF UNDERLYING REALITY!
Pilot Wave
INSTANTLY transmits information
— Faster Than Light! —
that guides the landing
of the other quoin.
A
A
B
B
SIMULTANEOUS

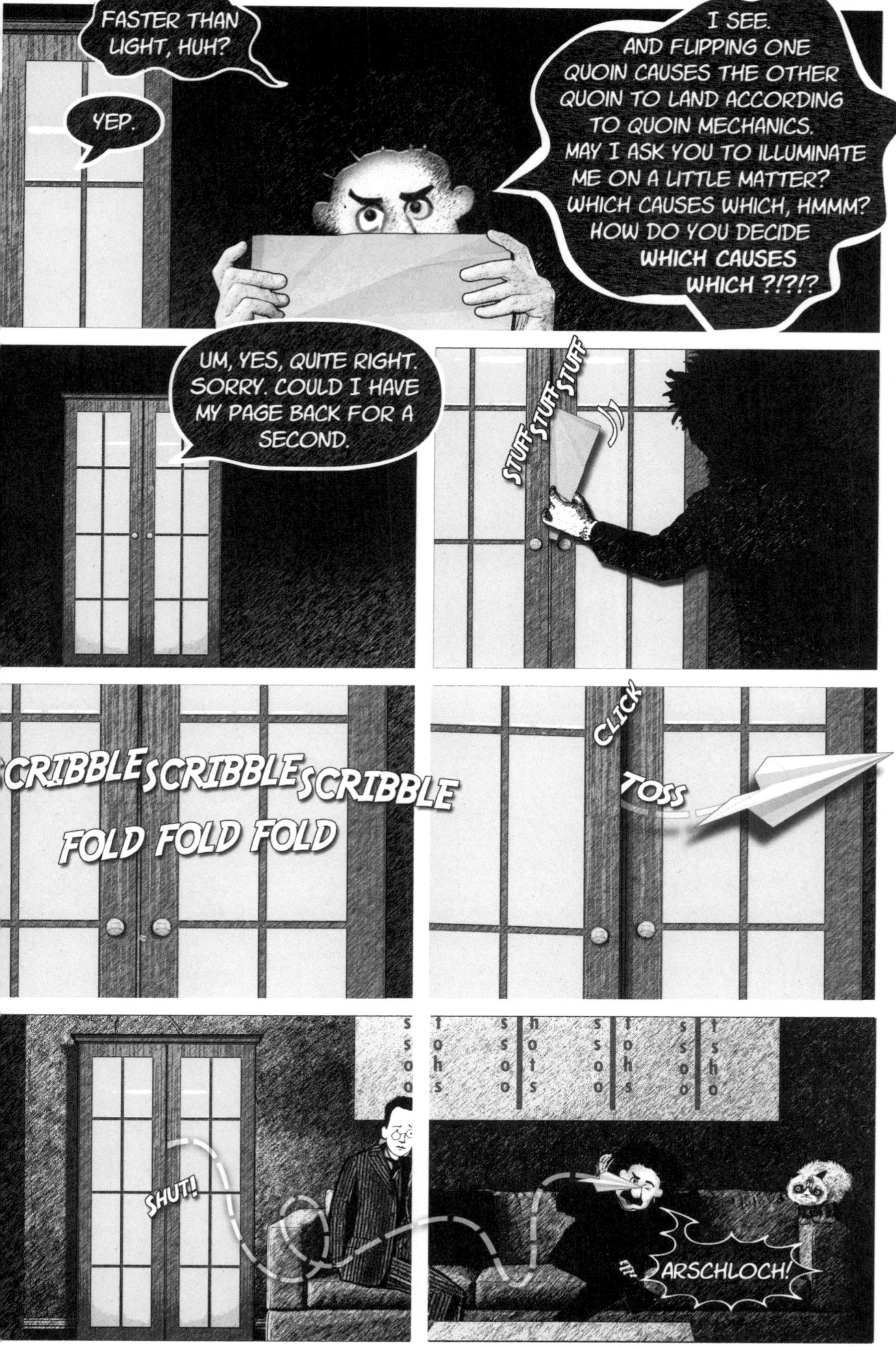
FASTER THAN LIGHT, HUH?
YEP.
I SEE. AND FLIPPING ONE QUOIN CAUSES THE OTHER QUOIN TO LAND ACCORDING TO QUOIN MECHANICS. MAY I ASK YOU TO ILLUMINATE ME ON A LITTLE MATTER? WHICH CAUSES WHICH, HMMM? HOW DO YOU DECIDE WHICH CAUSES WHICH ?!?!?
UM, YES, QUITE RIGHT. SORRY. COULD I HAVE MY PAGE BACK FOR A SECOND.
STUFF STUFF STUFF
SCRIBBLE SCRIBBLE SCRIBBLE
FOLD FOLD FOLD
CLICK
TOSS
SHUT!
ARSCHLOCH!

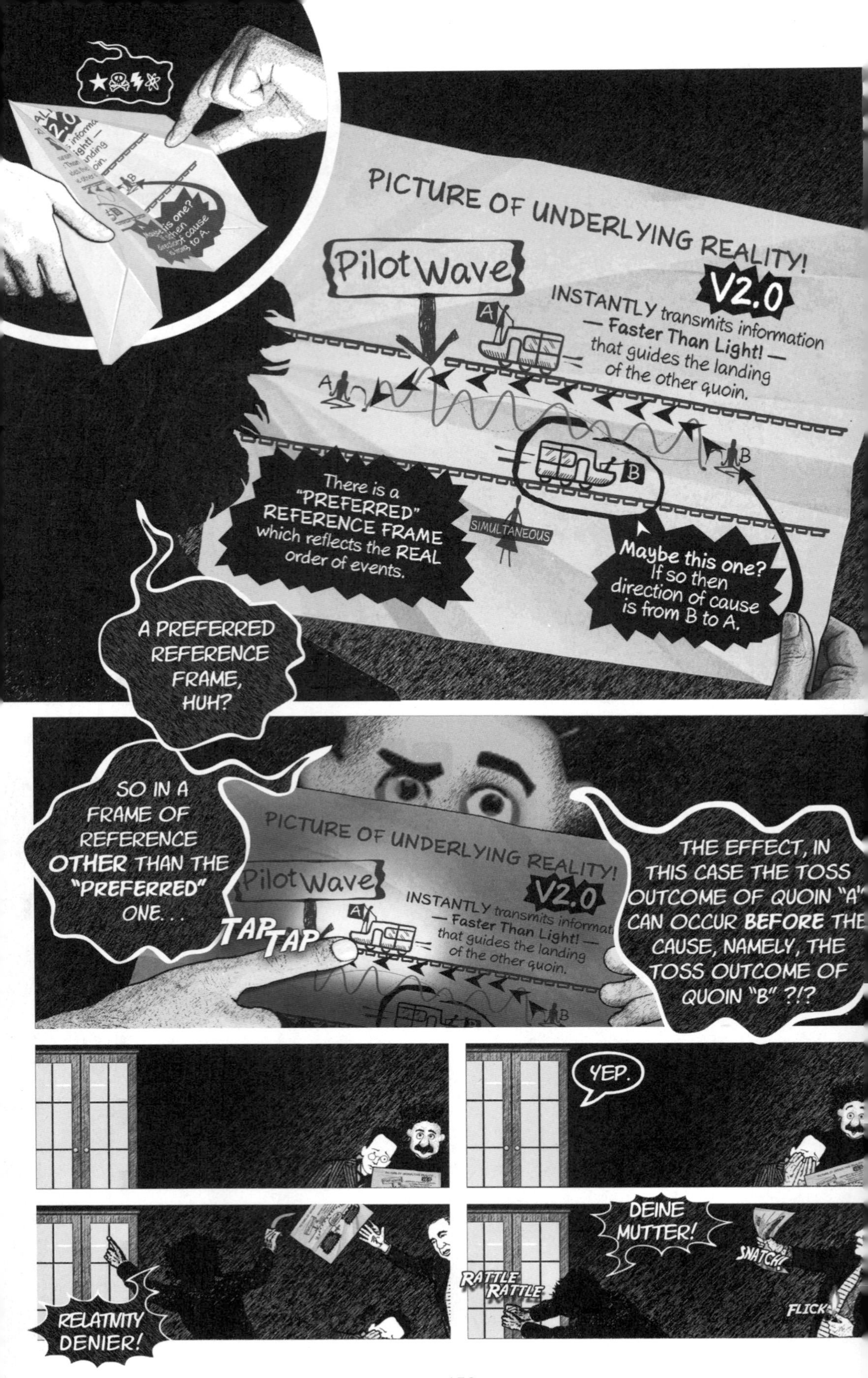
PICTURE OF UNDERLYING REALITY!
Pilot Wave
V2.0
INSTANTLY transmits information — Faster Than Light! — that guides the landing of the other quoin.
A
B
There is a "PREFERRED" REFERENCE FRAME which reflects the REAL order of events.
SIMULTANEOUS
Maybe this one? If so then direction of cause is from B to A.
A PREFERRED REFERENCE FRAME, HUH?
SO IN A FRAME OF REFERENCE OTHER THAN THE "PREFERRED" ONE. . .
TAP TAP
THE EFFECT, IN THIS CASE THE TOSS OUTCOME OF QUOIN "A" CAN OCCUR BEFORE THE CAUSE, NAMELY, THE TOSS OUTCOME OF QUOIN "B" ?!?
YEP.
RELATIVITY DENIER!
RATTLE RATTLE
DEINE MUTTER!
SNATCH!
FLICK

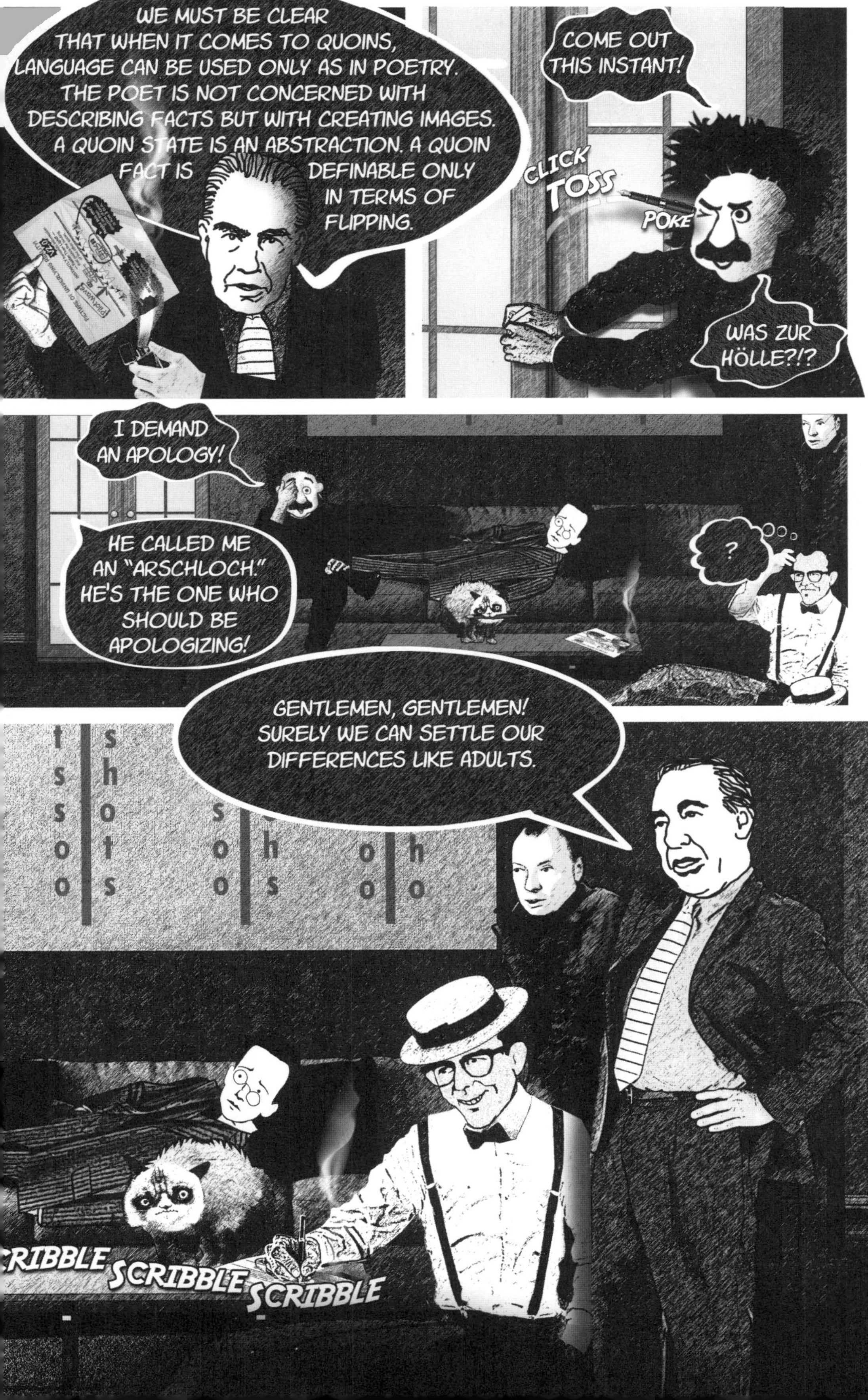
WE MUST BE CLEAR THAT WHEN IT COMES TO QUOINS, LANGUAGE CAN BE USED ONLY AS IN POETRY. THE POET IS NOT CONCERNED WITH DESCRIBING FACTS BUT WITH CREATING IMAGES. A QUOIN STATE IS AN ABSTRACTION. A QUOIN FACT IS DEFINABLE ONLY IN TERMS OF FLIPPING.
COME OUT THIS INSTANT!
CLICK
TOSS
POKE
WAS ZUR HÖLLE?!?
I DEMAND AN APOLOGY!
HE CALLED ME AN "ARSCHLOCH." HE'S THE ONE WHO SHOULD BE APOLOGIZING!
?
GENTLEMEN, GENTLEMEN! SURELY WE CAN SETTLE OUR DIFFERENCES LIKE ADULTS.
CRIBBLE SCRIBBLE SCRIBBLE

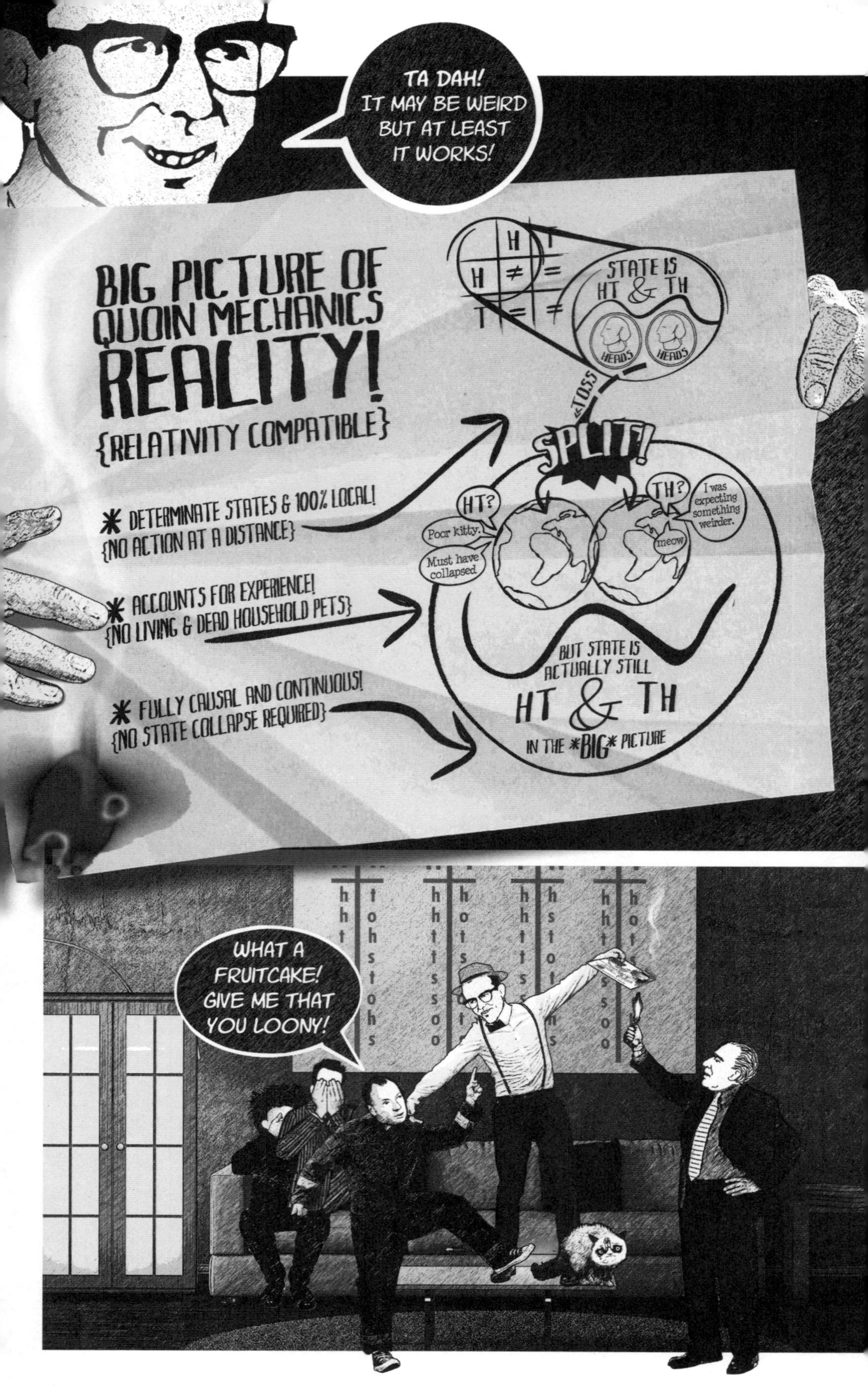
TA DAH! IT MAY BE WEIRD BUT AT LEAST IT WORKS!
BIG PICTURE OF QUOIN MECHANICS REALITY!
{RELATIVITY COMPATIBLE}
* DETERMINATE STATES & 100% LOCAL! {NO ACTION AT A DISTANCE}
* ACCOUNTS FOR EXPERIENCE! {NO LIVING & DEAD HOUSEHOLD PETS}
* FULLY CAUSAL AND CONTINUOUS! {NO STATE COLLAPSE REQUIRED}
STATE IS HT & TH
HEADS
HEADS
TOSS
SPLIT!
HT?
Poor kitty.
Must have collapsed
TH?
I was expecting something weirder.
meow
BUT STATE IS ACTUALLY STILL HT & TH IN THE *BIG* PICTURE
WHAT A FRUITCAKE! GIVE ME THAT YOU LOONY!

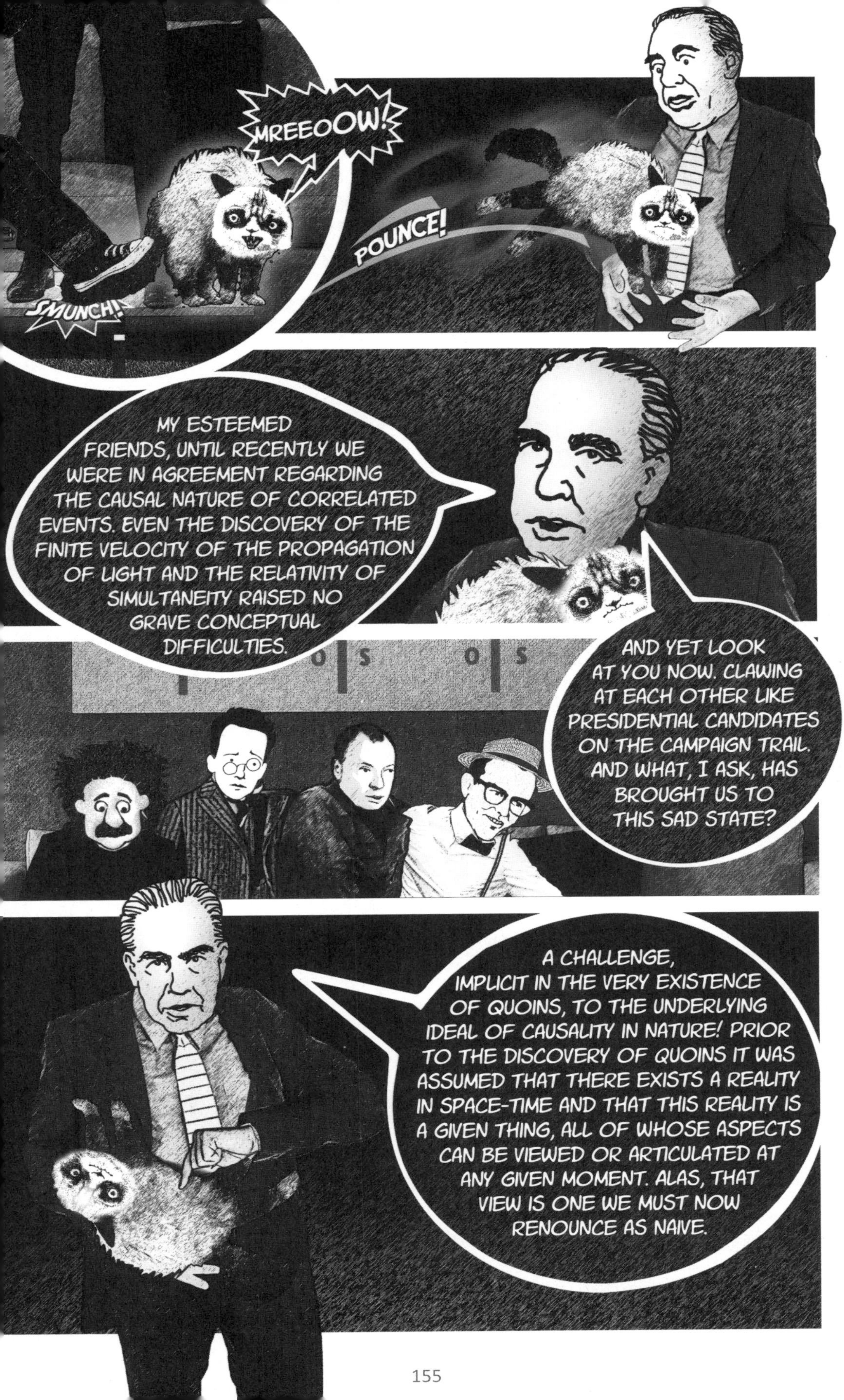
MREEOOW!
SMUNCH!
POUNCE!
MY ESTEEMED FRIENDS, UNTIL RECENTLY WE WERE IN AGREEMENT REGARDING THE CAUSAL NATURE OF CORRELATED EVENTS. EVEN THE DISCOVERY OF THE FINITE VELOCITY OF THE PROPAGATION OF LIGHT AND THE RELATIVITY OF SIMULTANEITY RAISED NO GRAVE CONCEPTUAL DIFFICULTIES.
AND YET LOOK AT YOU NOW. CLAWING AT EACH OTHER LIKE PRESIDENTIAL CANDIDATES ON THE CAMPAIGN TRAIL. AND WHAT, I ASK, HAS BROUGHT US TO THIS SAD STATE?
A CHALLENGE, IMPLICIT IN THE VERY EXISTENCE OF QUOINS, TO THE UNDERLYING IDEAL OF CAUSALITY IN NATURE! PRIOR TO THE DISCOVERY OF QUOINS IT WAS ASSUMED THAT THERE EXISTS A REALITY IN SPACE-TIME AND THAT THIS REALITY IS A GIVEN THING, ALL OF WHOSE ASPECTS CAN BE VIEWED OR ARTICULATED AT ANY GIVEN MOMENT. ALAS, THAT VIEW IS ONE WE MUST NOW RENOUNCE AS NAIVE.

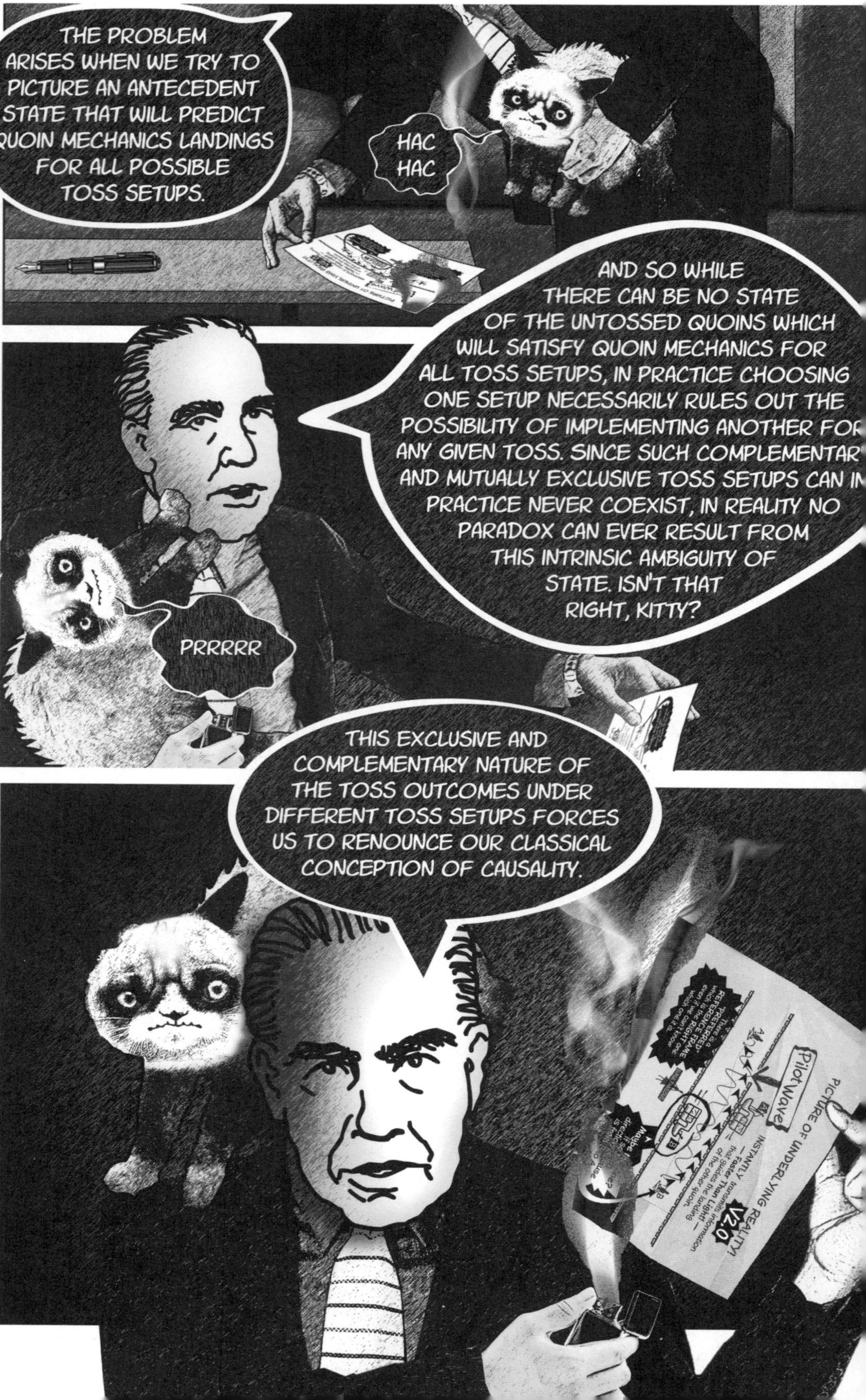
THE PROBLEM ARISES WHEN WE TRY TO PICTURE AN ANTECEDENT STATE THAT WILL PREDICT QUOIN MECHANICS LANDINGS FOR ALL POSSIBLE TOSS SETUPS.
HAC HAC
AND SO WHILE THERE CAN BE NO STATE OF THE UNTOSSED QUOINS WHICH WILL SATISFY QUOIN MECHANICS FOR ALL TOSS SETUPS, IN PRACTICE CHOOSING ONE SETUP NECESSARILY RULES OUT THE POSSIBILITY OF IMPLEMENTING ANOTHER FO ANY GIVEN TOSS. SINCE SUCH COMPLEMENTAR AND MUTUALLY EXCLUSIVE TOSS SETUPS CAN I PRACTICE NEVER COEXIST, IN REALITY NO PARADOX CAN EVER RESULT FROM THIS INTRINSIC AMBIGUITY OF STATE. ISN'T THAT RIGHT, KITTY?
PRRRRR
THIS EXCLUSIVE AND COMPLEMENTARY NATURE OF THE TOSS OUTCOMES UNDER DIFFERENT TOSS SETUPS FORCES US TO RENOUNCE OUR CLASSICAL CONCEPTION OF CAUSALITY.
PICTURE OF UNDERLYING REALITY!
V2.0
PilotWave

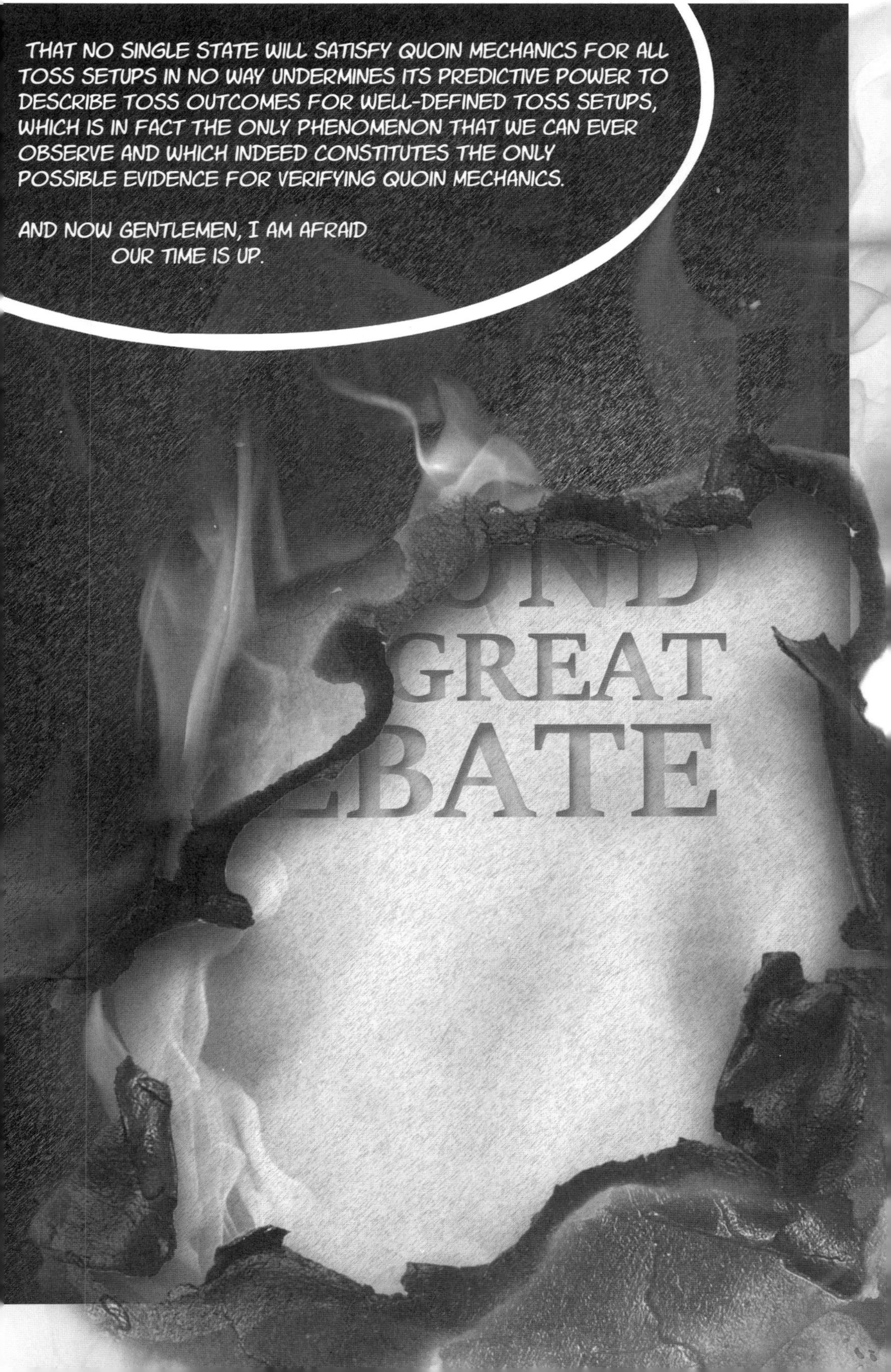
THAT NO SINGLE STATE WILL SATISFY QUOIN MECHANICS FOR ALL TOSS SETUPS IN NO WAY UNDERMINES ITS PREDICTIVE POWER TO DESCRIBE TOSS OUTCOMES FOR WELL-DEFINED TOSS SETUPS, WHICH IS IN FACT THE ONLY PHENOMENON THAT WE CAN EVER OBSERVE AND WHICH INDEED CONSTITUTES THE ONLY POSSIBLE EVIDENCE FOR VERIFYING QUOIN MECHANICS.
AND NOW GENTLEMEN, I AM AFRAID OUR TIME IS UP.
GREAT

Part III

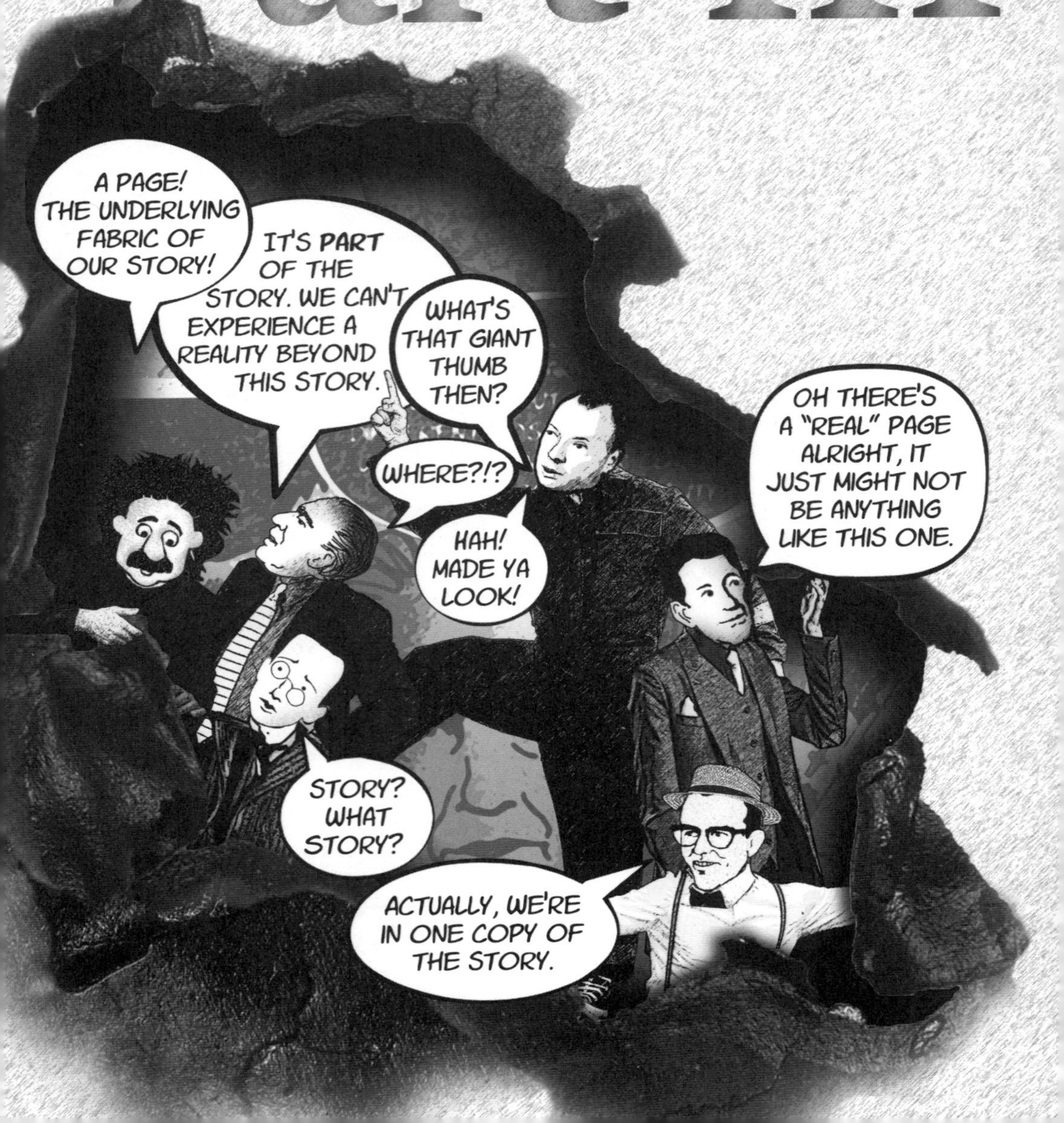
A PAGE! THE UNDERLYING FABRIC OF OUR STORY!
IT'S PART OF THE STORY. WE CAN'T EXPERIENCE A REALITY BEYOND THIS STORY.
WHAT'S THAT GIANT THUMB THEN?
WHERE?!?
HAH! MADE YA LOOK!
OH THERE'S A "REAL" PAGE ALRIGHT, IT JUST MIGHT NOT BE ANYTHING LIKE THIS ONE.
STORY? WHAT STORY?
ACTUALLY, WE'RE IN ONE COPY OF THE STORY.

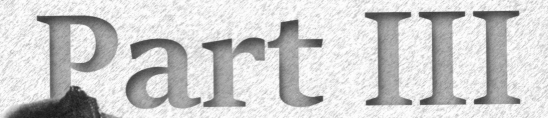

BEYOND THE GREAT DEBATE

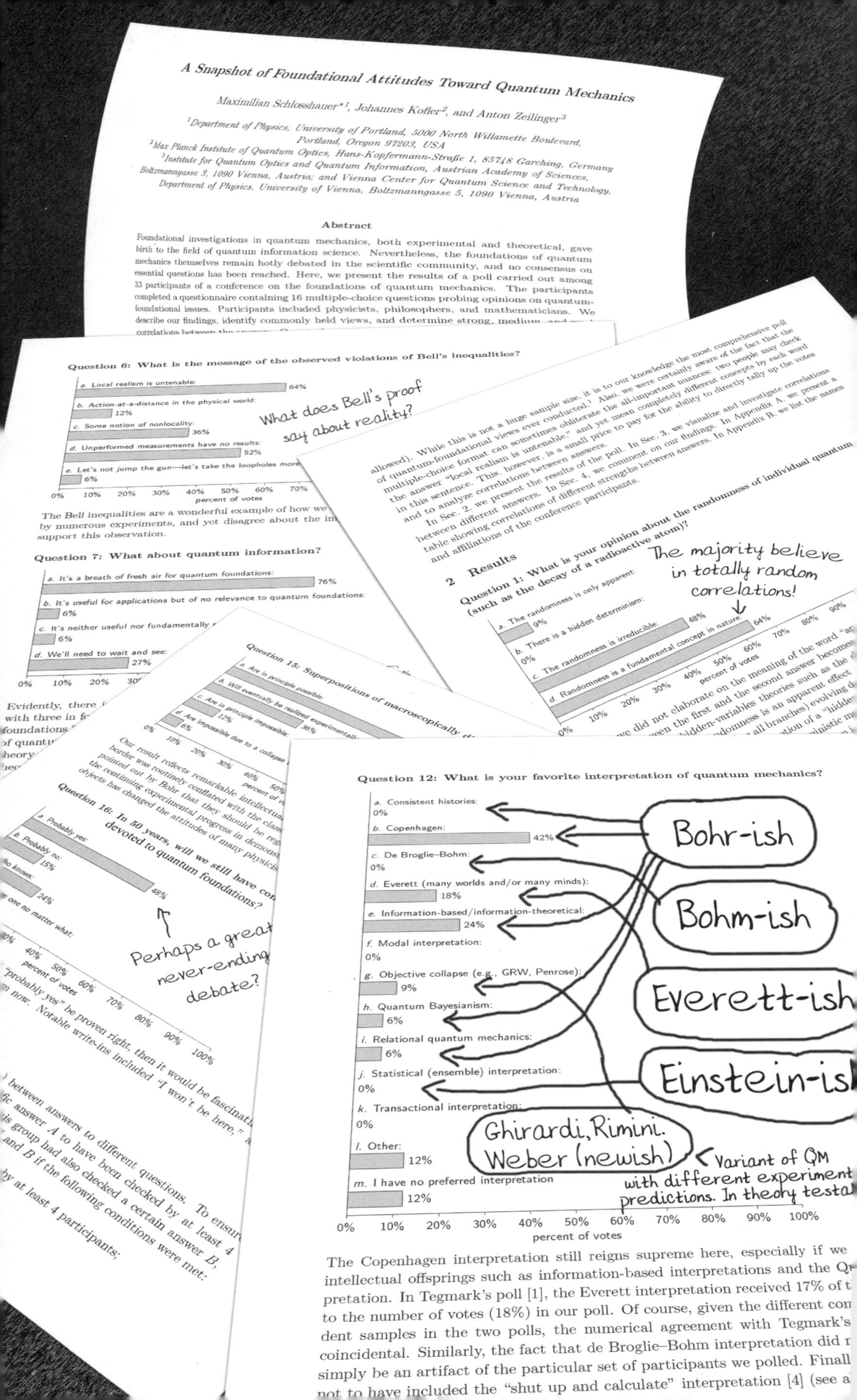
A Snapshot of Foundational Attitudes Toward Quantum Mechanics
Maximilian Schlosshauer, Johannes Kofler, and Anton Zeilinger
Abstract
Question 6: What is the message of the observed violations of Bell's inequalities?
What does Bell's proof say about reality?
Question 7: What about quantum information?
Question 1: What is your opinion about the randomness of individual quantum events (such as the decay of a radioactive atom)?
The majority believe in totally random correlations!
Question 15: Superpositions of macroscopically distinct states
Question 16: In 50 years, will we still have conferences devoted to quantum foundations?
Perhaps a great never-ending debate?
Question 12: What is your favorite interpretation of quantum mechanics?
Bohr-ish
Bohm-ish
Everett-ish
Einstein-ish
Ghirardi, Rimini. Weber (newish)
Variant of QM with different experiment predictions. In theory testa
The Copenhagen interpretation still reigns supreme here, especially if we

And now?

???

I mean, some of these guys were arguing about this stuff almost 100 years ago. Before Google, before iPhones, before YouTube, apparently before women existed. What do people think these days?!?

Oh, a fierce debate (now co-ed) about the foundations of QM rages on. The new stuff mostly builds on the old stuff: Bohr-ish, Einstein-ish, Bohm-ish, or Everett-ish. With some exceptions, everyone fits into one of those camps, more or less.

There's a paper with the results of a conference survey from 2013 where people voted for their preferred view. I'll see if I can find it for you.

So no crazy new stuff?

Oh, I wouldn't exactly say that :)

But What Can You Do With It?

An engineer [is] one of those people who make things work without even understanding how they function.

—**Nicolas Gisin**

ME?

I DUNNO.

WHAT DO
YOU BELIEVE?

MAYBE THINKING ABOUT IT
IS JUST A WASTE OF TIME.
I MEAN IF THE LIKES OF **EINSTEIN**
COULDN'T FIGURE IT OUT, WHAT
CHANCE DO **I** HAVE?

〈SIGH〉

BUT ON THE OTHER HAND...

SOMETIMES I CAN'T HELP
BUT HOPE THAT THE ANSWER...

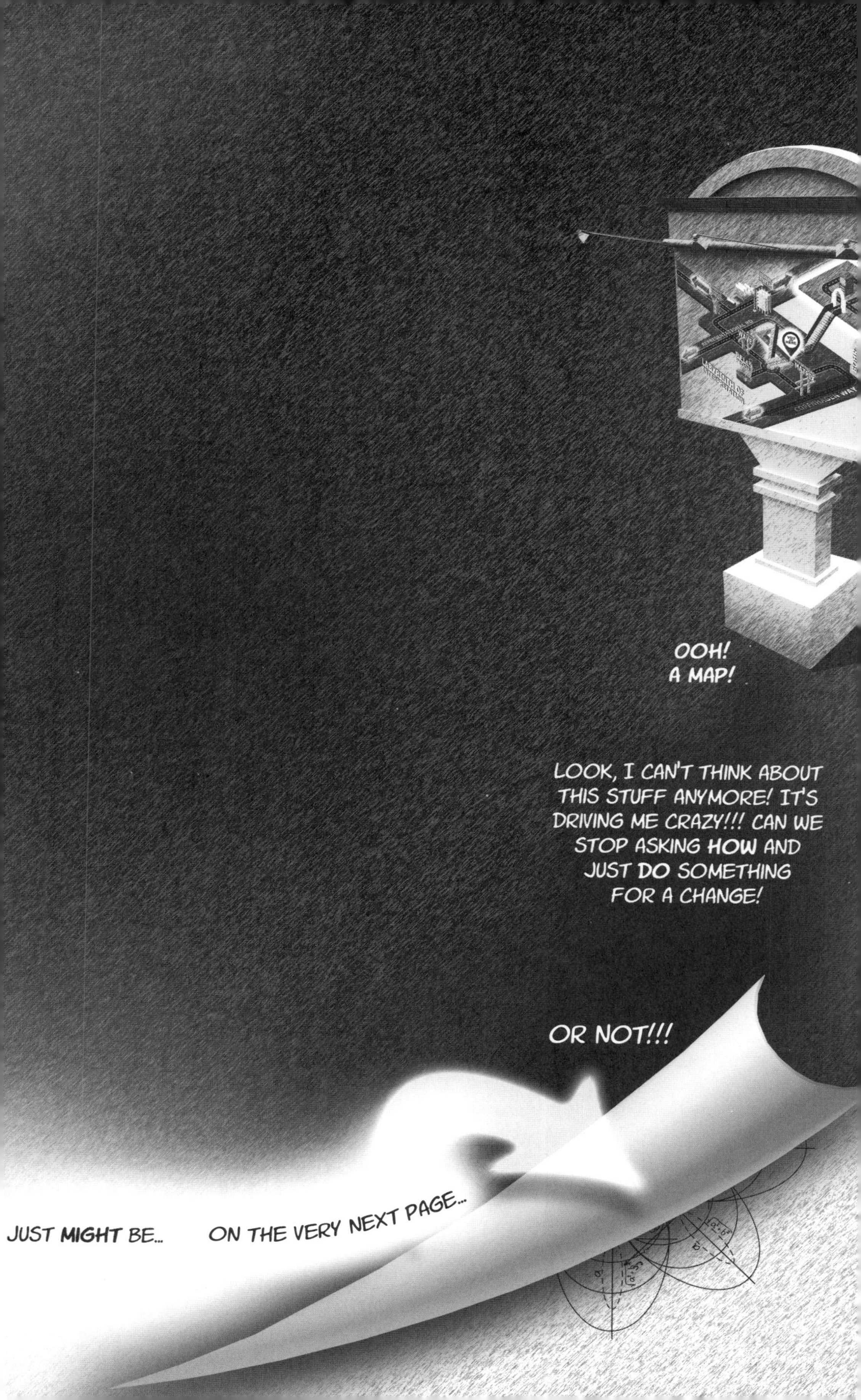

OOH! A MAP!
LOOK, I CAN'T THINK ABOUT THIS STUFF ANYMORE! IT'S DRIVING ME CRAZY!!! CAN WE STOP ASKING **HOW** AND JUST **DO** SOMETHING FOR A CHANGE!
OR NOT!!!
JUST **MIGHT** BE...
ON THE VERY NEXT PAGE...

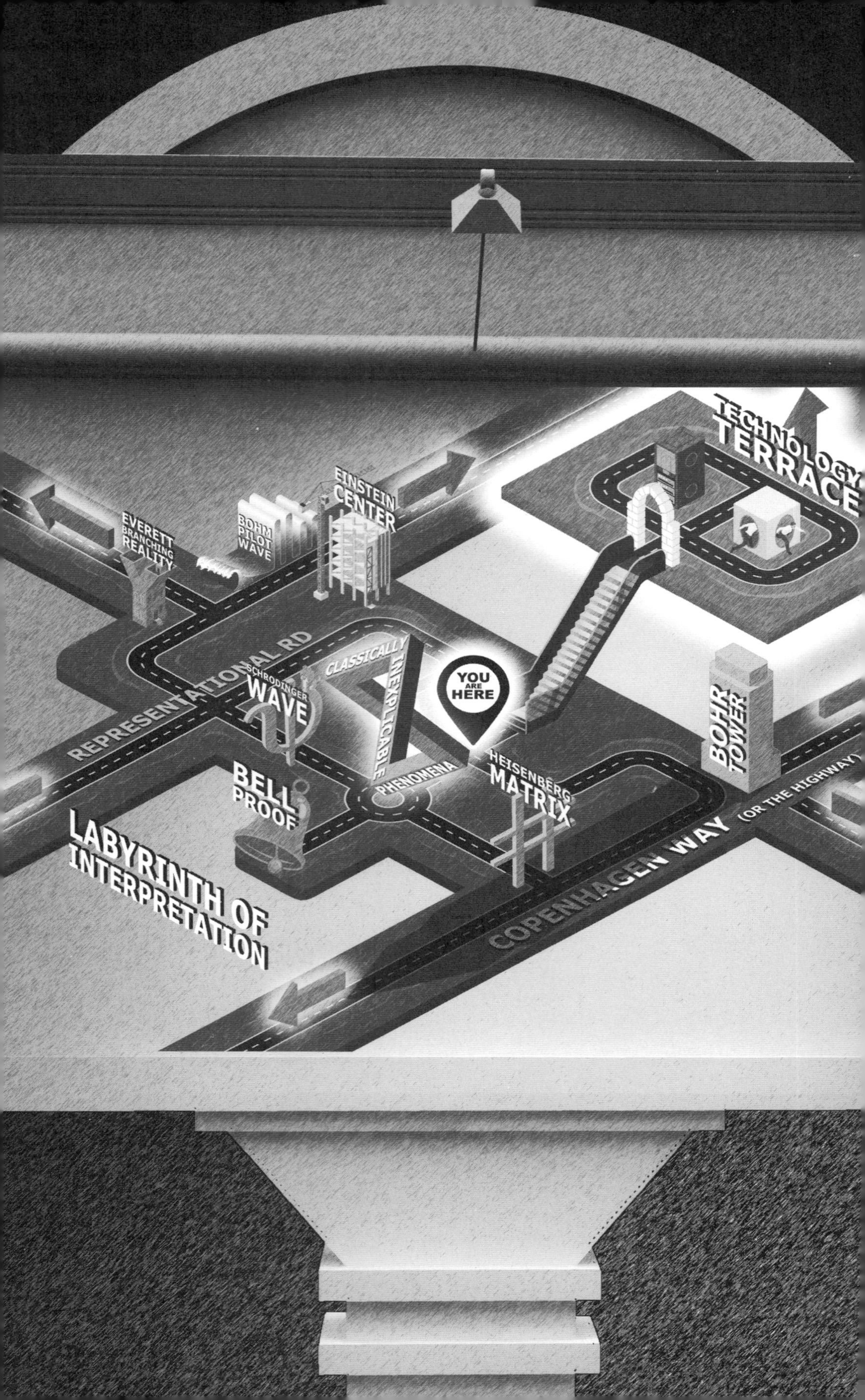

TECHNOLOGY TERRACE
EINSTEIN CENTER
BOHM PILOT WAVE
EVERETT BRANCHING REALITY
REPRESENTATIONAL RD
SCHRODINGER WAVE
CLASSICALLY INEXPLICABLE PHENOMENA
YOU ARE HERE
BOHR TOWER
HEISENBERG MATRIX
BELL PROOF
LABYRINTH OF INTERPRETATION
COPENHAGEN WAY (OR THE HIGHWAY)

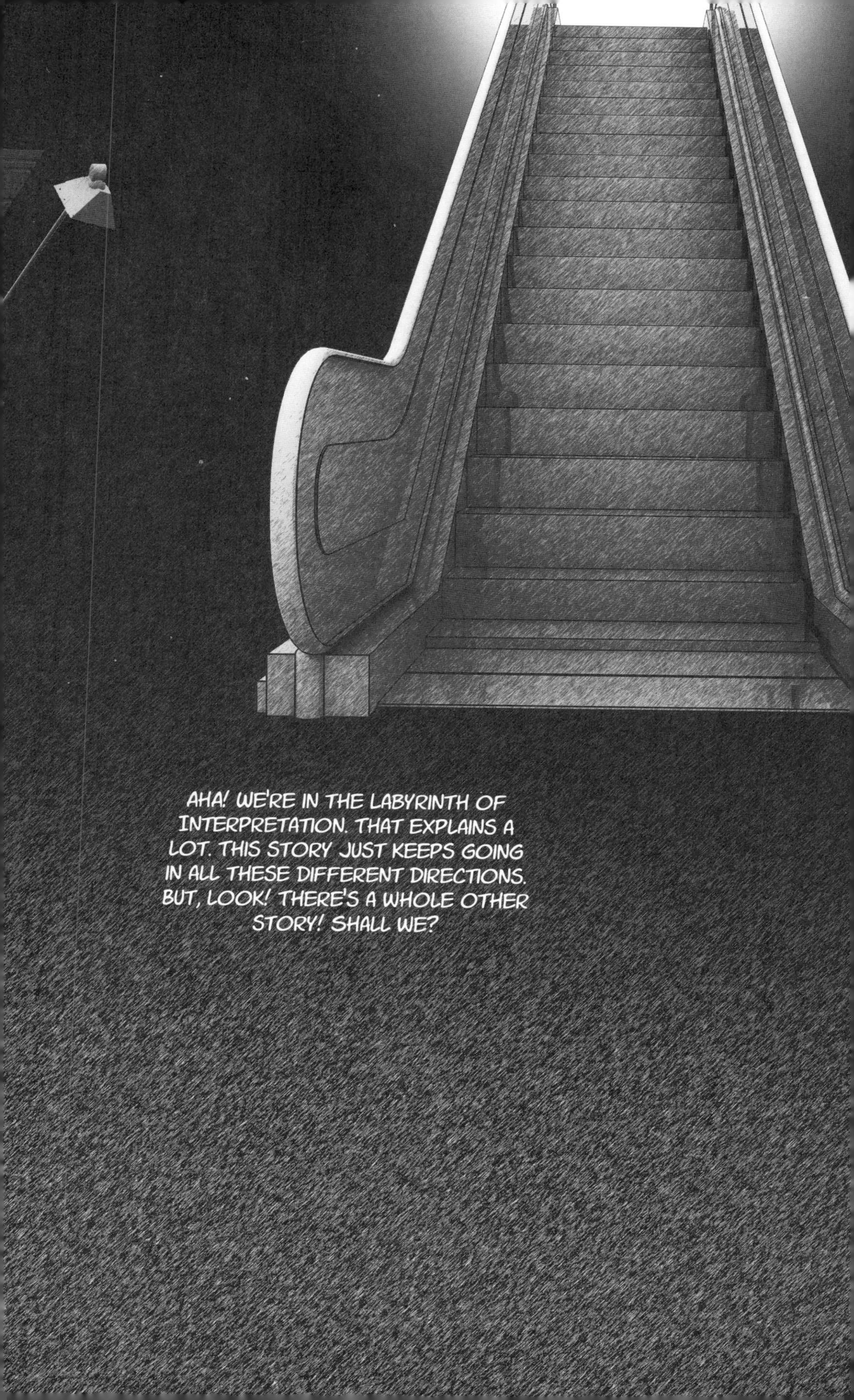
AHA! WE'RE IN THE LABYRINTH OF INTERPRETATION. THAT EXPLAINS A LOT. THIS STORY JUST KEEPS GOING IN ALL THESE DIFFERENT DIRECTIONS. BUT, LOOK! THERE'S A WHOLE OTHER STORY! SHALL WE?

Confounding E.V.E.
(the perfect eavesdropper)

Thus if we accept the [Einstein] definitions of
reality and locality then we have no choice but
to admit that God is an inveterate gambler who
throws the dice on every possible occasion. In spite
of this, we may find some consolation in harnessing
this randomness and putting it into a good use.
And this, finally, brings us to cryptography.

—Artur Ekert

The first practical application of Bell's inequality
was in the spooky art of secret communication.
... Information is always represented by measurable
physical properties ... Conversely, if such properties
do not exist prior to measurements, then there is
nothing to eavesdrop on. This was the basic idea
that led me to the development of a new
tool for detecting eavesdropping.

—Artur Ekert

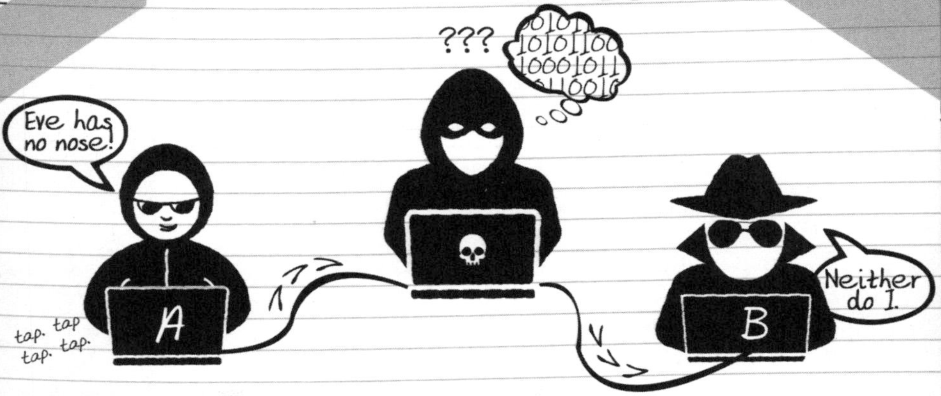
Eve has no nose!
???
Neither do I.
tap. tap tap. tap.
A
B

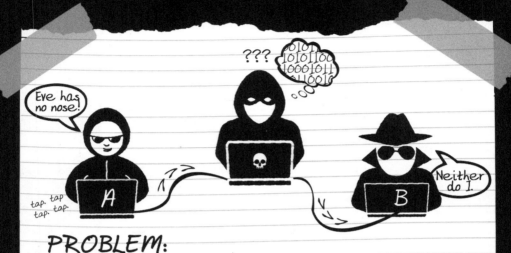

PROBLEM:

A (Alice) must get a secret message (plaintext) past an eavesdropper (Eve) to B (Bob).

SOLUTION: Totally Random QM Encryption

UNCRACKABLE: No amount of computational power gives Eve any advantage in recovering the plaintext from the encoded message (ciphertext) because the ciphertext is TOTALLY RANDOM. Only possible way to decode the ciphertext is with the key!

UNHACKABLE: Alice and Bob will know if Eve gets any information about the key, because then Alice's and Bob's sides of the key won't be "curiously correlated" any more.

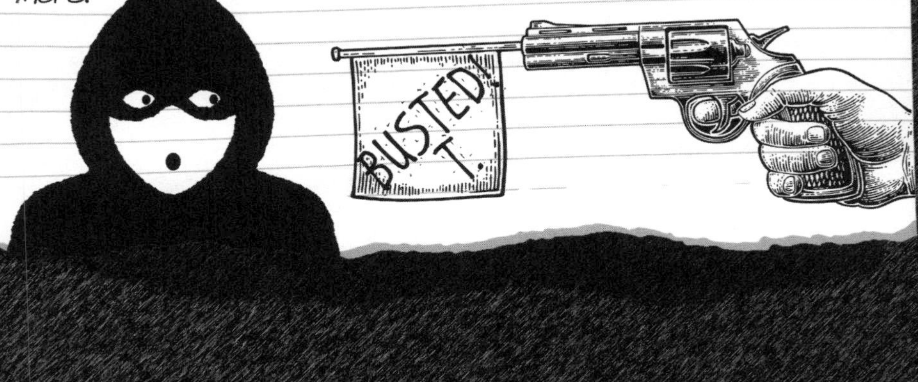
BUSTED!

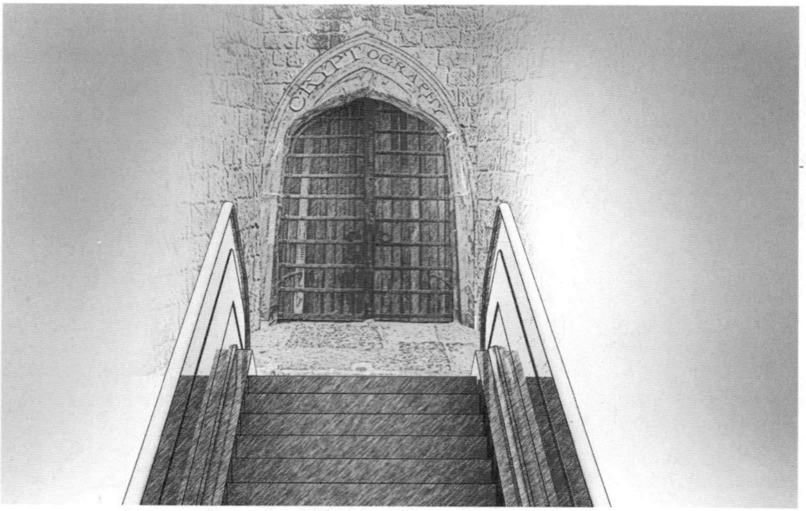

CRYPTOGRAPHY

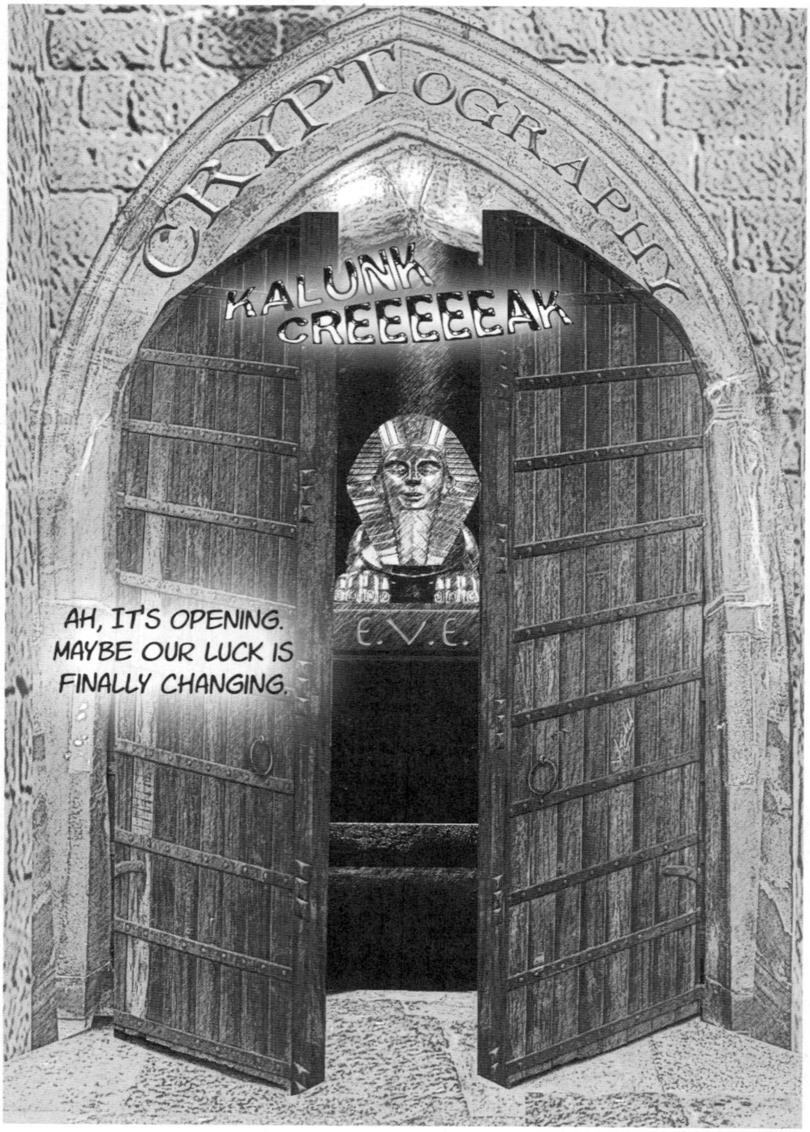

CRYPTOGRAPHY
KALUNK
CREEEEEAK
E.V.E.
AH, IT'S OPENING. MAYBE OUR LUCK IS FINALLY CHANGING.

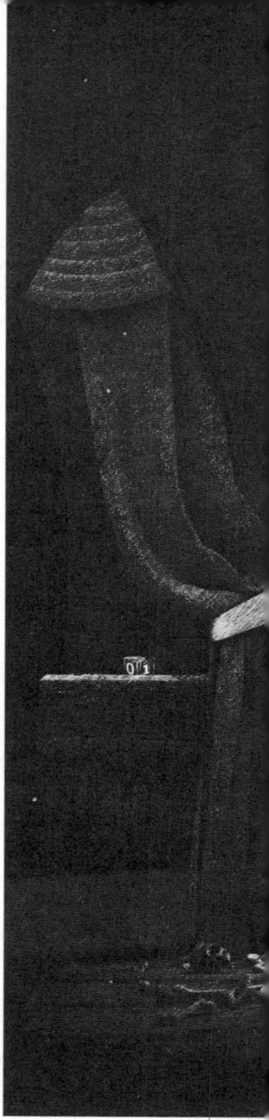

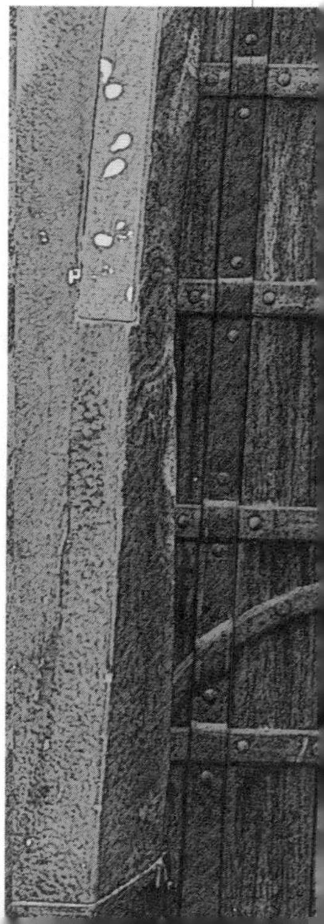

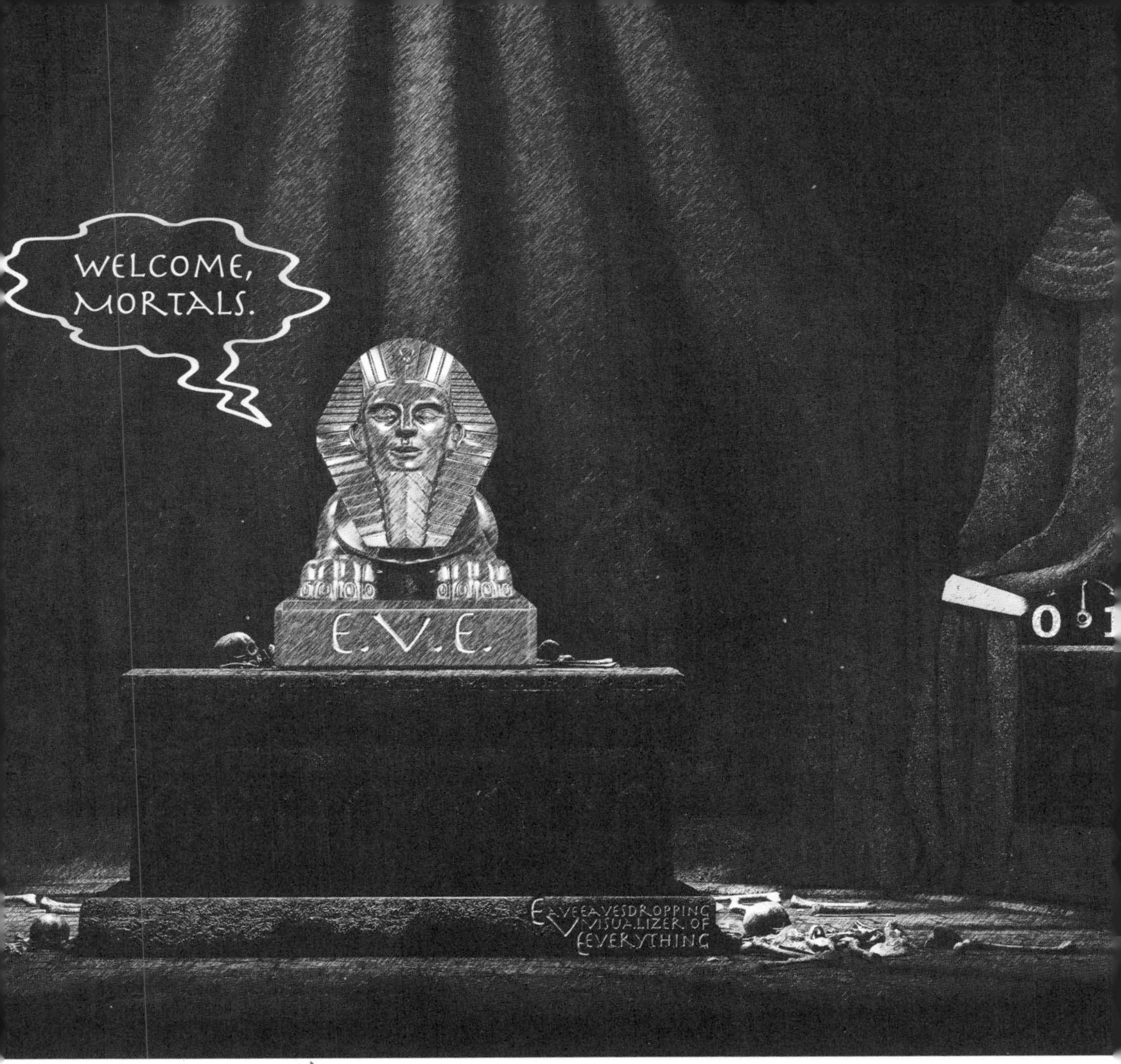
WELCOME, MORTALS.
E.V.E.
EVE EAVESDROPPING VISUALIZER OF EVERYTHING

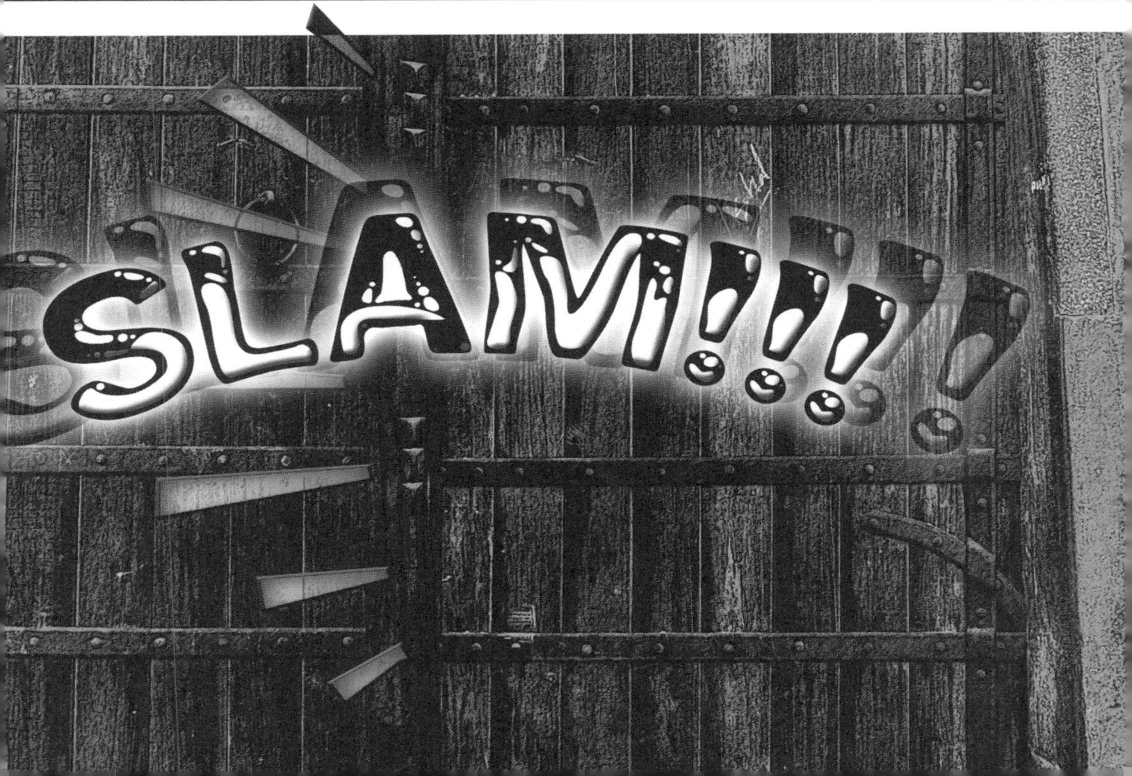
SLAM!!!!

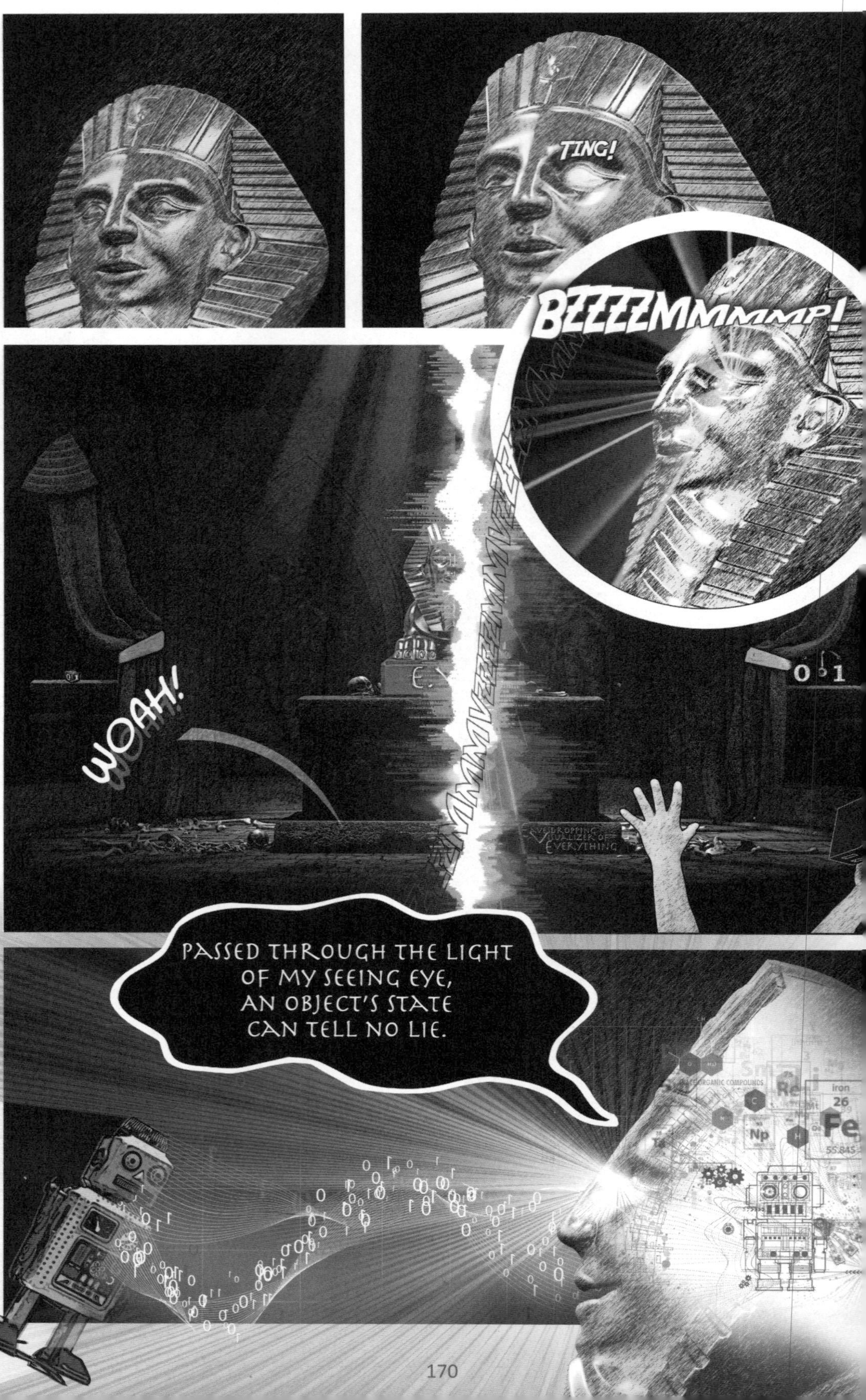
TING!
BZZZZMMMMP!
WOAH!
PASSED THROUGH THE LIGHT OF MY SEEING EYE, AN OBJECT'S STATE CAN TELL NO LIE.

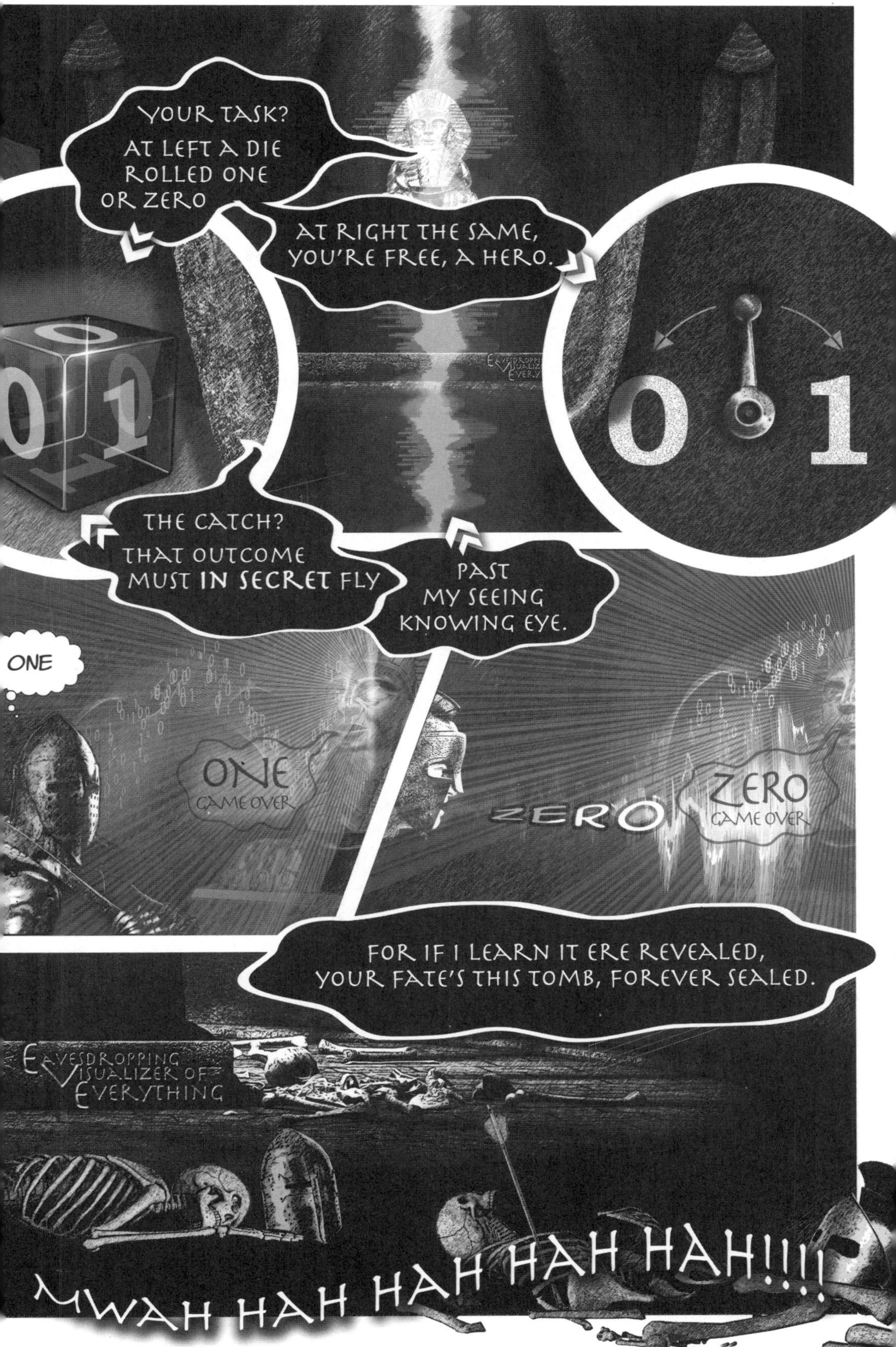
YOUR TASK?
AT LEFT A DIE ROLLED ONE OR ZERO
AT RIGHT THE SAME, YOU'RE FREE, A HERO.
0 1
THE CATCH? THAT OUTCOME MUST IN SECRET FLY
PAST MY SEEING KNOWING EYE.
ONE
ONE GAME OVER
ZERO
ZERO GAME OVER
FOR IF I LEARN IT ERE REVEALED, YOUR FATE'S THIS TOMB, FOREVER SEALED.
EAVESDROPPING VISUALIZER OF EVERYTHING
MWAH HAH HAH HAH HAH!!!!

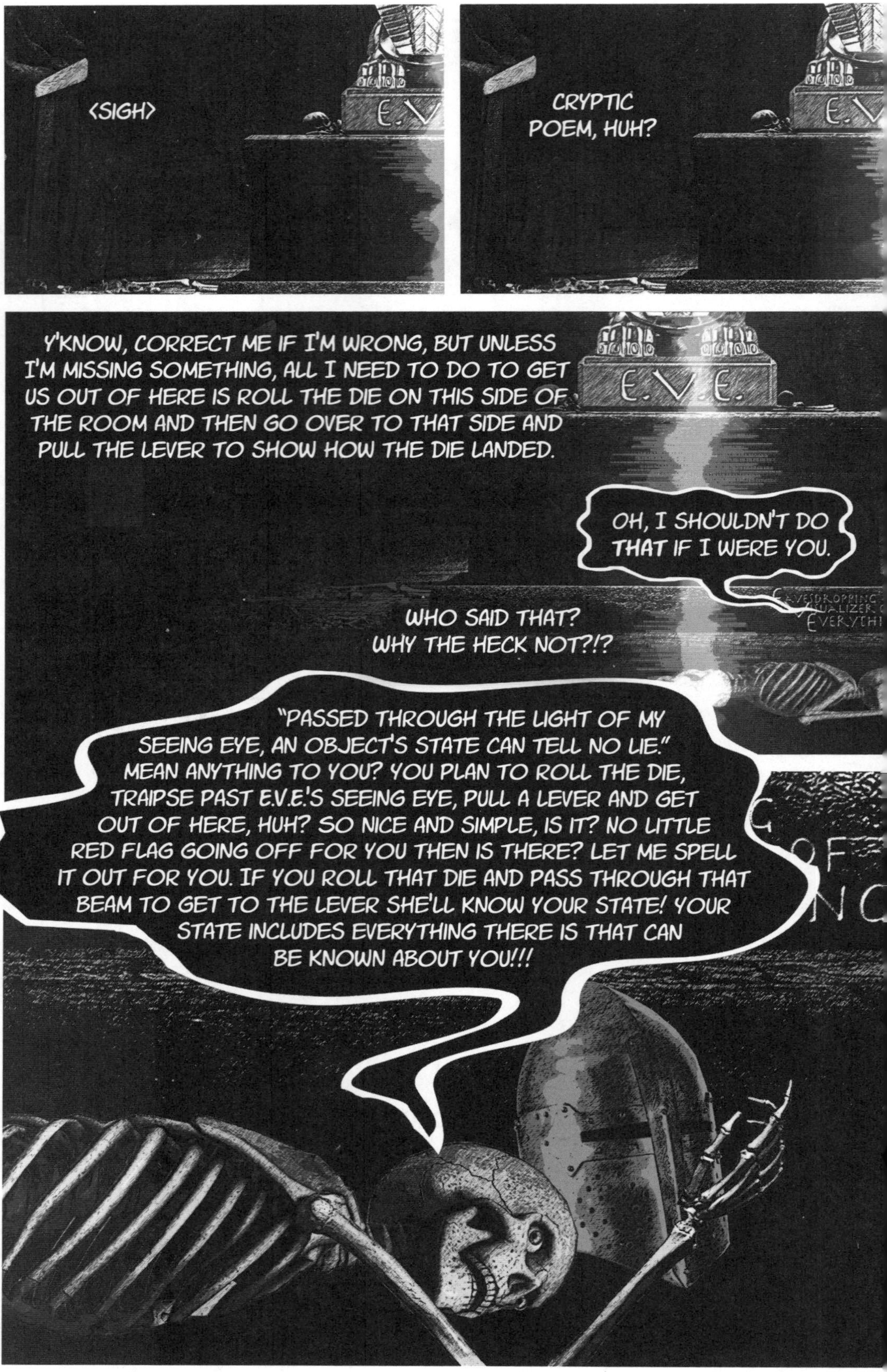
<SIGH>
CRYPTIC POEM, HUH?
Y'KNOW, CORRECT ME IF I'M WRONG, BUT UNLESS I'M MISSING SOMETHING, ALL I NEED TO DO TO GET US OUT OF HERE IS ROLL THE DIE ON THIS SIDE OF THE ROOM AND THEN GO OVER TO THAT SIDE AND PULL THE LEVER TO SHOW HOW THE DIE LANDED.
E.V.E.
OH, I SHOULDN'T DO THAT IF I WERE YOU.
WHO SAID THAT? WHY THE HECK NOT?!?
"PASSED THROUGH THE LIGHT OF MY SEEING EYE, AN OBJECT'S STATE CAN TELL NO LIE." MEAN ANYTHING TO YOU? YOU PLAN TO ROLL THE DIE, TRAIPSE PAST E.V.E.'S SEEING EYE, PULL A LEVER AND GET OUT OF HERE, HUH? SO NICE AND SIMPLE, IS IT? NO LITTLE RED FLAG GOING OFF FOR YOU THEN IS THERE? LET ME SPELL IT OUT FOR YOU. IF YOU ROLL THAT DIE AND PASS THROUGH THAT BEAM TO GET TO THE LEVER SHE'LL KNOW YOUR STATE! YOUR STATE INCLUDES EVERYTHING THERE IS THAT CAN BE KNOWN ABOUT YOU!!!

SO?
SMACK
PASS THROUGH THAT BEAM AND SHE KNOWS WHAT YOU KNOW, WHICH BY THE WAY, WOULD HAPPEN TO INCLUDE A LITTLE DETAIL REGARDING WHICH WAY THE DIE FELL WHEN YOU ROLLED IT! IF I MAY BE SO BOLD AS TO REMIND YOU, "THAT OUTCOME MUST IN SECRET FLY PAST MY SEEING KNOWING EYE". WALK PAST HER EYE AND IF YOU KNOW THE OUTCOME OF THE ROLL, SHE'LL KNOW IT TOO AND IF SHE KNOWS IT BEFORE YOU PULL THE LEVER IT'S CURTAINS FOR YOU BOTH. ANYWAY, I TRIED IT, WOULDN'T RECOMMEND IT. NOT A NICE WAY TO GO. QUITE UNPLEASANT.
AH. I SEE YOUR POINT. WELL, THANKS, I GUESS.
TA
BUT IF YOU DON'T MIND THIS IS SORT OF A PRIVATE CONVERSATION. AHEM.
MUM'S THE WORD. FLIES ON THE WALL FROM NOW ON IS WHAT WE ARE.
FLOP
CLACKETY CLACK
BZZZZZZZ
QUITE RIGHT. MY BAD.

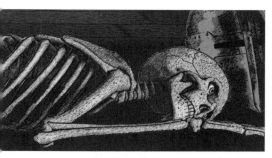

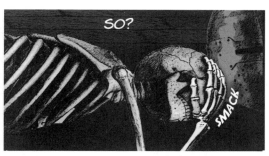

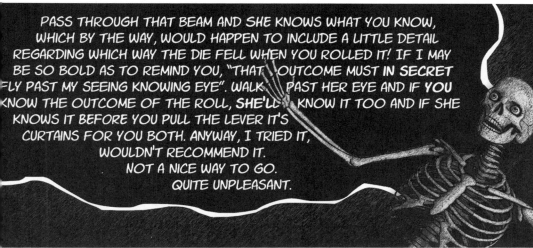

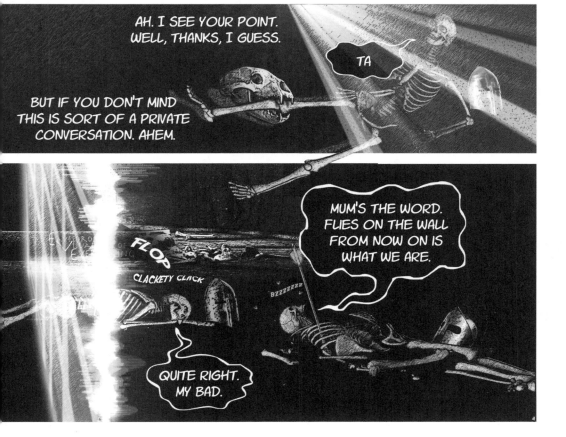

DON'T WORRY ABOUT IT. NOW WHERE WERE WE?

LOOK, HOW ABOUT THIS. I ROLL THE DIE AND SEE HOW IT LANDS BUT STAY ON MY SIDE OF THE ROOM AND JUST YELL OUT THE RESULT AND THEN YOU PULL THE LEVER. EASY PEAZY AND I KEEP MY STATE TO MYSELF.

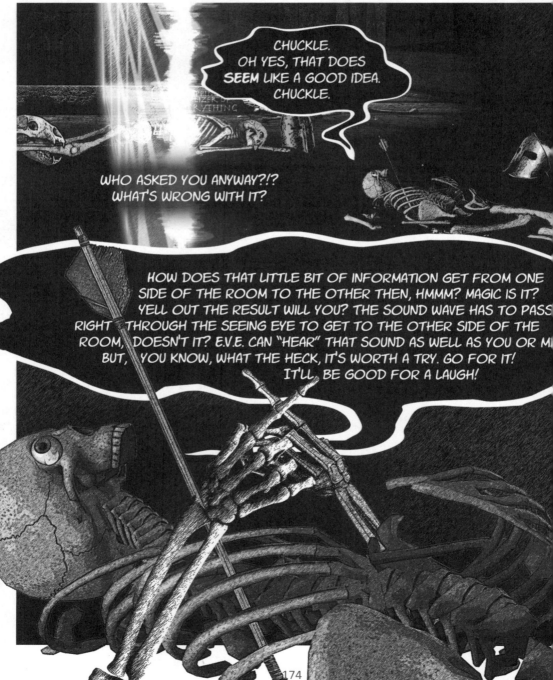

CHUCKLE. OH YES, THAT DOES **SEEM** LIKE A GOOD IDEA. CHUCKLE.

WHO ASKED YOU ANYWAY?!? WHAT'S WRONG WITH IT?

HOW DOES THAT LITTLE BIT OF INFORMATION GET FROM ONE SIDE OF THE ROOM TO THE OTHER THEN, HMMM? MAGIC IS IT? YELL OUT THE RESULT WILL YOU? THE SOUND WAVE HAS TO PASS RIGHT THROUGH THE SEEING EYE TO GET TO THE OTHER SIDE OF THE ROOM, DOESN'T IT? E.V.E. CAN "HEAR" THAT SOUND AS WELL AS YOU OR ME BUT, YOU KNOW, WHAT THE HECK, IT'S WORTH A TRY. GO FOR IT! IT'LL BE GOOD FOR A LAUGH!

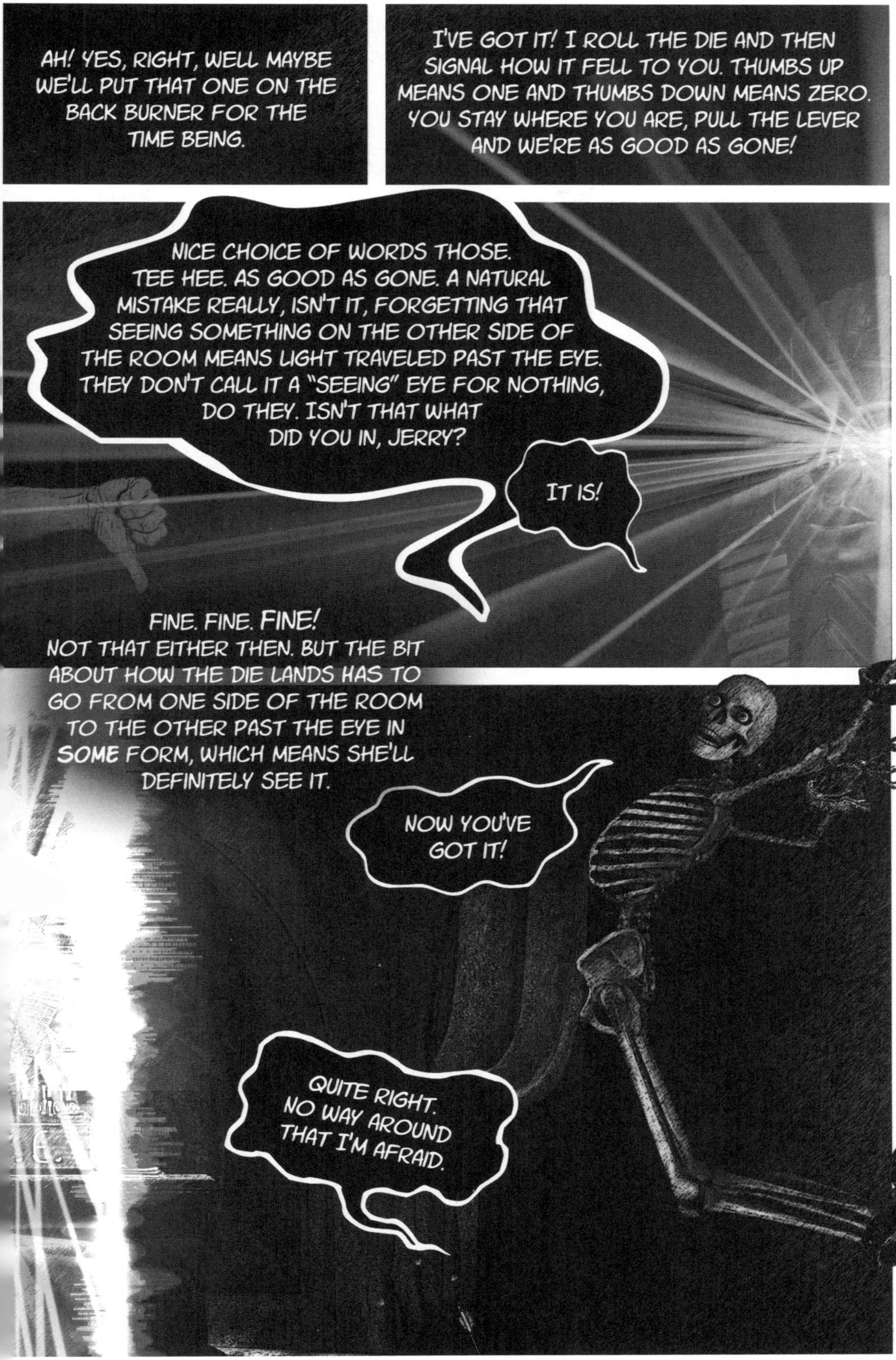
AH! YES, RIGHT, WELL MAYBE WE'LL PUT THAT ONE ON THE BACK BURNER FOR THE TIME BEING.
I'VE GOT IT! I ROLL THE DIE AND THEN SIGNAL HOW IT FELL TO YOU. THUMBS UP MEANS ONE AND THUMBS DOWN MEANS ZERO. YOU STAY WHERE YOU ARE, PULL THE LEVER AND WE'RE AS GOOD AS GONE!
NICE CHOICE OF WORDS THOSE. TEE HEE. AS GOOD AS GONE. A NATURAL MISTAKE REALLY, ISN'T IT, FORGETTING THAT SEEING SOMETHING ON THE OTHER SIDE OF THE ROOM MEANS LIGHT TRAVELED PAST THE EYE. THEY DON'T CALL IT A "SEEING" EYE FOR NOTHING, DO THEY. ISN'T THAT WHAT DID YOU IN, JERRY?
IT IS!
FINE. FINE. FINE! NOT THAT EITHER THEN. BUT THE BIT ABOUT HOW THE DIE LANDS HAS TO GO FROM ONE SIDE OF THE ROOM TO THE OTHER PAST THE EYE IN SOME FORM, WHICH MEANS SHE'LL DEFINITELY SEE IT.
NOW YOU'VE GOT IT!
QUITE RIGHT. NO WAY AROUND THAT I'M AFRAID.

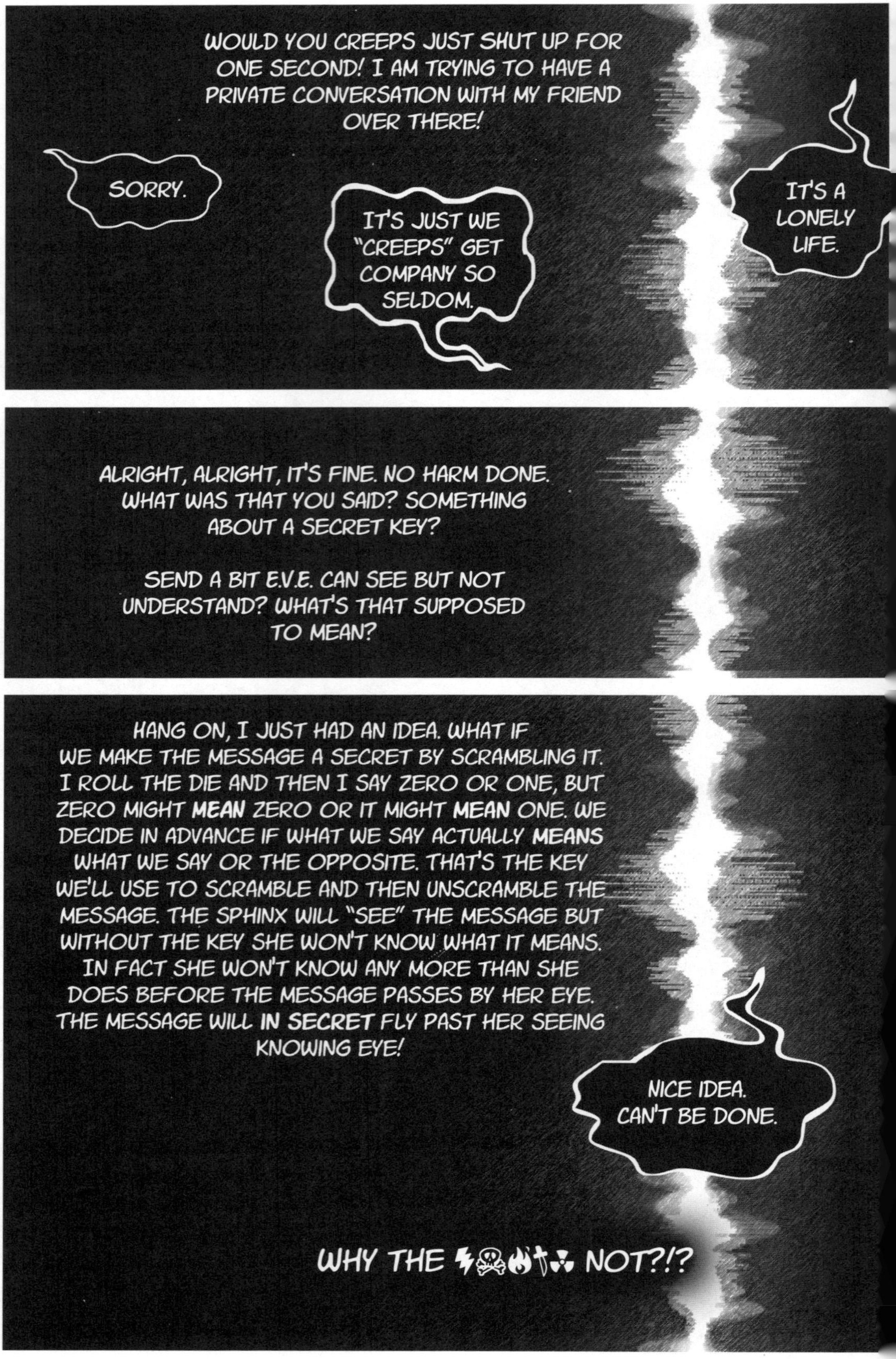
WOULD YOU CREEPS JUST SHUT UP FOR ONE SECOND! I AM TRYING TO HAVE A PRIVATE CONVERSATION WITH MY FRIEND OVER THERE!
SORRY.
IT'S JUST WE "CREEPS" GET COMPANY SO SELDOM.
IT'S A LONELY LIFE.
ALRIGHT, ALRIGHT, IT'S FINE. NO HARM DONE. WHAT WAS THAT YOU SAID? SOMETHING ABOUT A SECRET KEY?
SEND A BIT E.V.E. CAN SEE BUT NOT UNDERSTAND? WHAT'S THAT SUPPOSED TO MEAN?
HANG ON, I JUST HAD AN IDEA. WHAT IF WE MAKE THE MESSAGE A SECRET BY SCRAMBLING IT. I ROLL THE DIE AND THEN I SAY ZERO OR ONE, BUT ZERO MIGHT **MEAN** ZERO OR IT MIGHT **MEAN** ONE. WE DECIDE IN ADVANCE IF WHAT WE SAY ACTUALLY **MEANS** WHAT WE SAY OR THE OPPOSITE. THAT'S THE KEY WE'LL USE TO SCRAMBLE AND THEN UNSCRAMBLE THE MESSAGE. THE SPHINX WILL "SEE" THE MESSAGE BUT WITHOUT THE KEY SHE WON'T KNOW WHAT IT MEANS. IN FACT SHE WON'T KNOW ANY MORE THAN SHE DOES BEFORE THE MESSAGE PASSES BY HER EYE. THE MESSAGE WILL **IN SECRET** FLY PAST HER SEEING KNOWING EYE!
NICE IDEA. CAN'T BE DONE.
WHY THE ⚡☠🔥†☢ NOT?!?

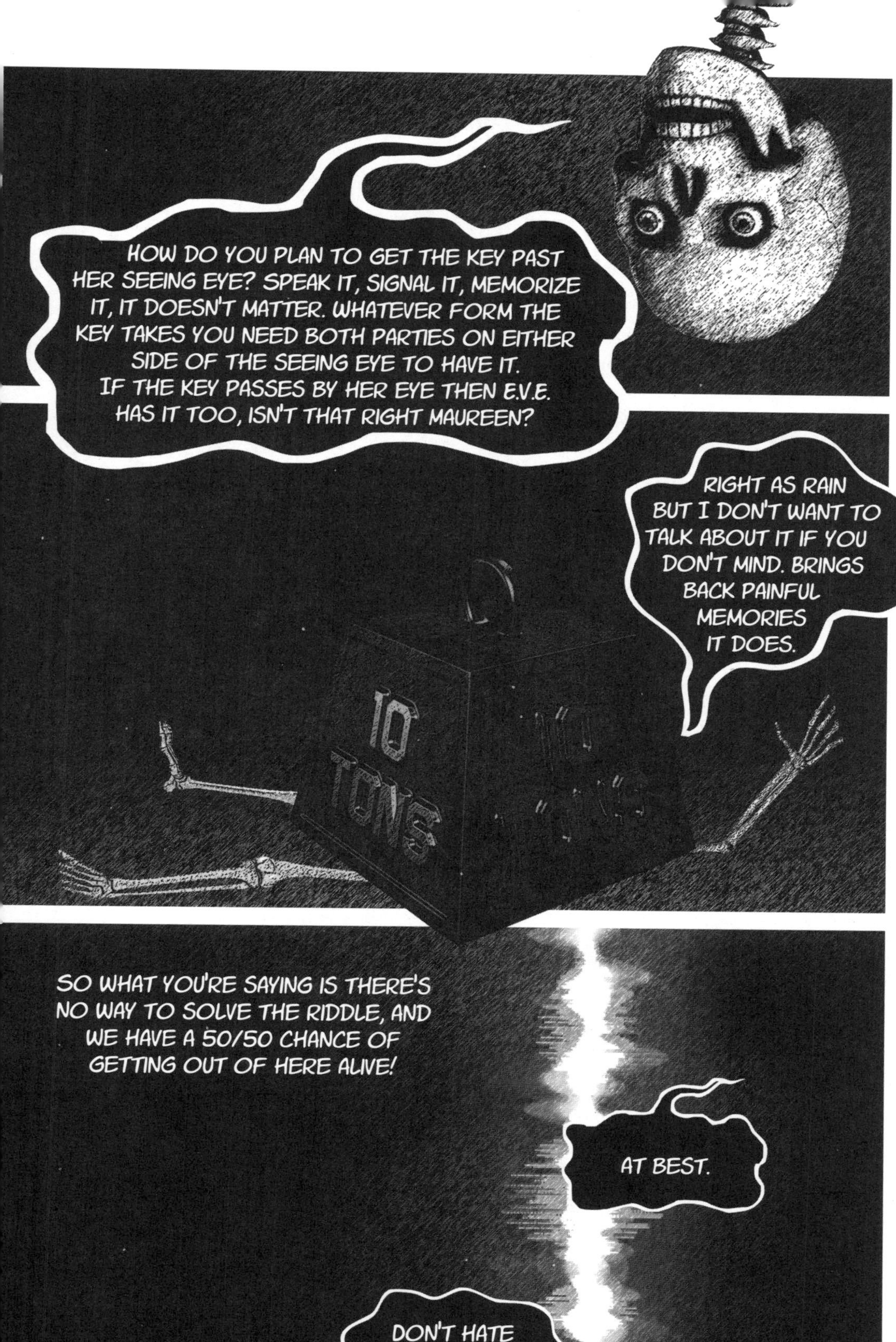
HOW DO YOU PLAN TO GET THE KEY PAST HER SEEING EYE? SPEAK IT, SIGNAL IT, MEMORIZE IT, IT DOESN'T MATTER. WHATEVER FORM THE KEY TAKES YOU NEED BOTH PARTIES ON EITHER SIDE OF THE SEEING EYE TO HAVE IT. IF THE KEY PASSES BY HER EYE THEN E.V.E. HAS IT TOO, ISN'T THAT RIGHT MAUREEN?
RIGHT AS RAIN BUT I DON'T WANT TO TALK ABOUT IT IF YOU DON'T MIND. BRINGS BACK PAINFUL MEMORIES IT DOES.
10 TONS
SO WHAT YOU'RE SAYING IS THERE'S NO WAY TO SOLVE THE RIDDLE, AND WE HAVE A 50/50 CHANCE OF GETTING OUT OF HERE ALIVE!
AT BEST.
DON'T HATE THE MESSENGER.

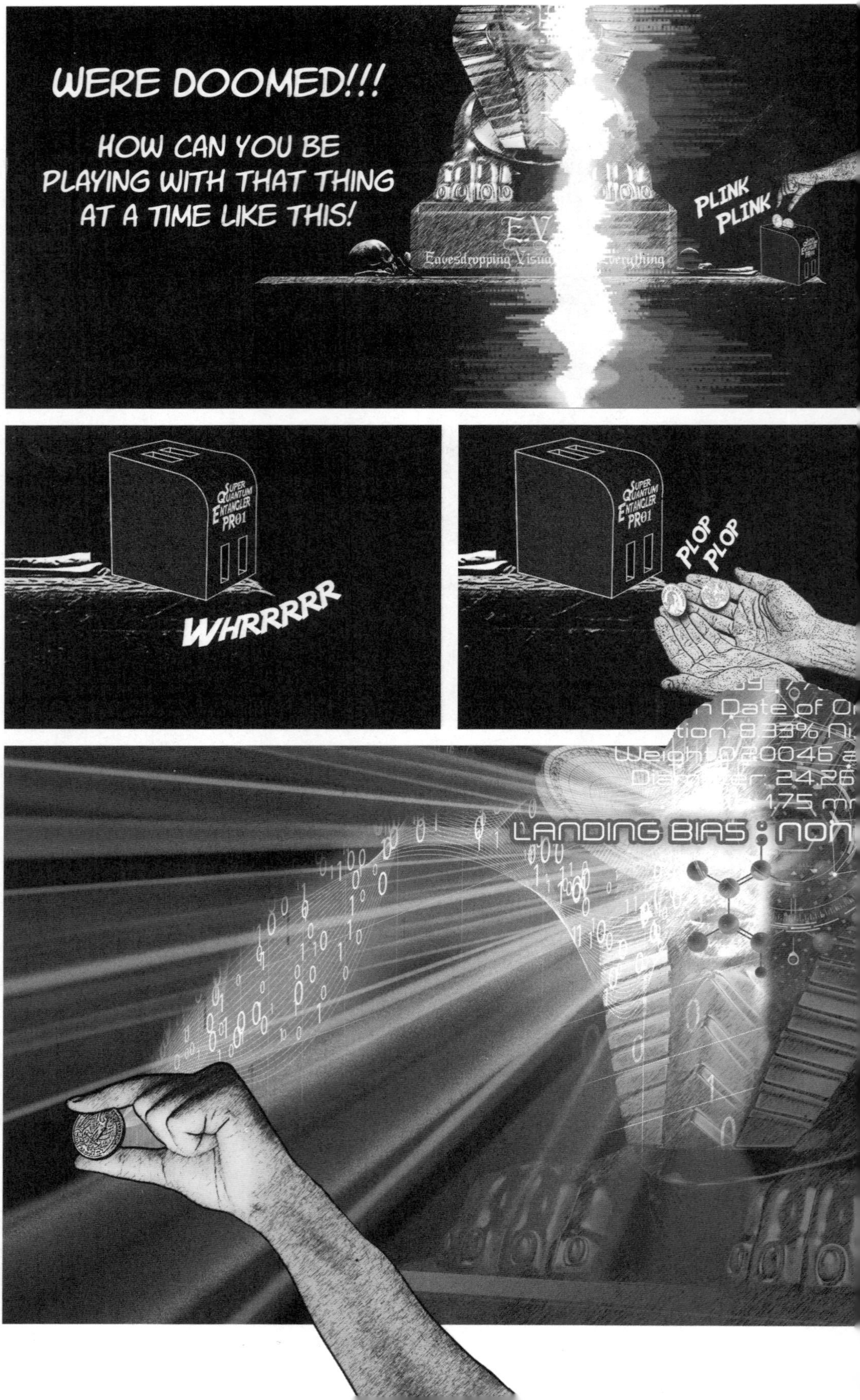
WERE DOOMED!!!
HOW CAN YOU BE PLAYING WITH THAT THING AT A TIME LIKE THIS!
PLINK PLINK
E.V.
Eavesdropping
SUPER QUANTUM ENTANGLER PRO1
WHRRRRR
SUPER QUANTUM ENTANGLER PRO1
PLOP PLOP
LANDING BIAS :

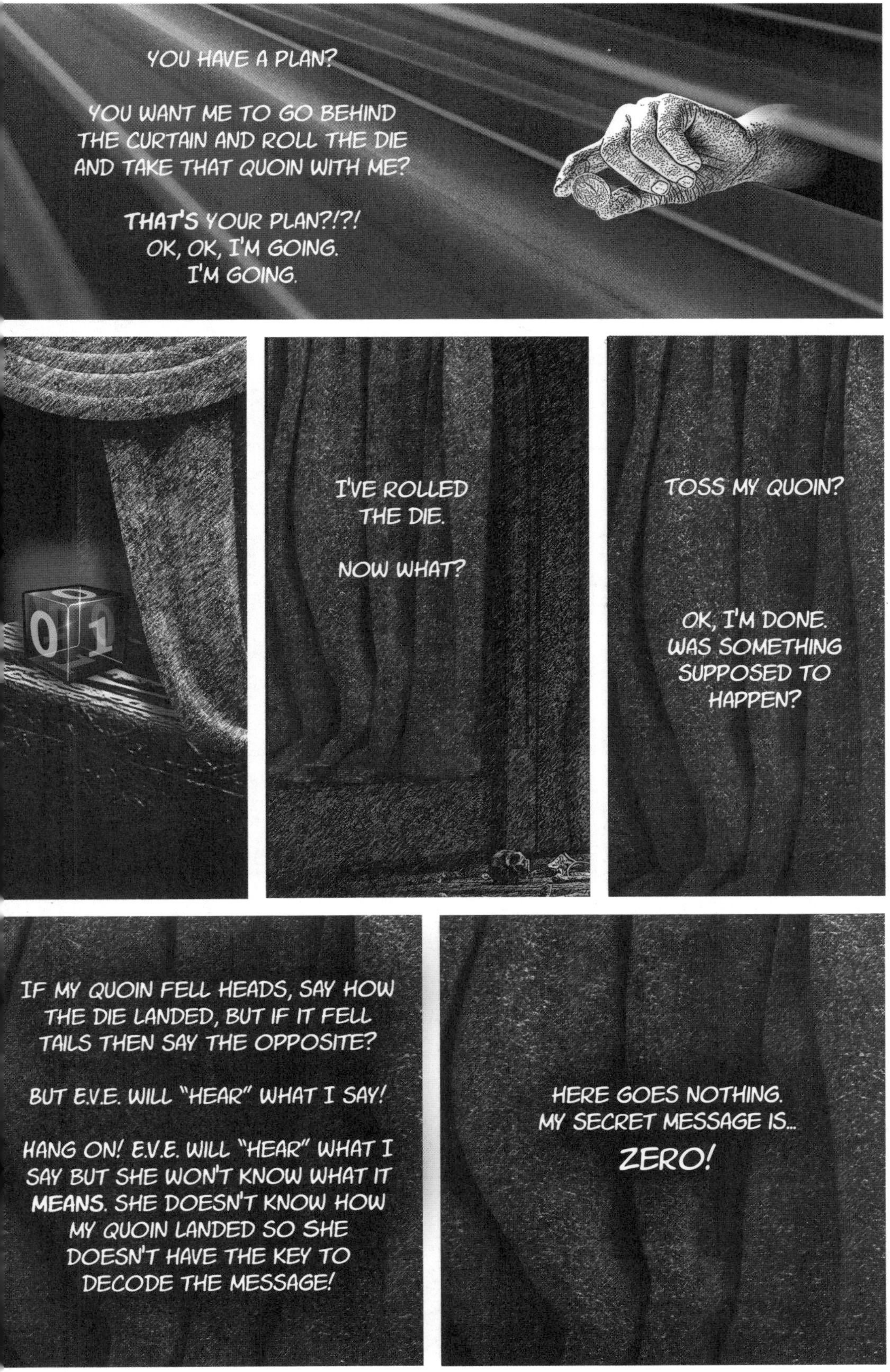
YOU HAVE A PLAN?
YOU WANT ME TO GO BEHIND THE CURTAIN AND ROLL THE DIE AND TAKE THAT QUOIN WITH ME?
THAT'S YOUR PLAN?!?!
OK, OK, I'M GOING.
I'M GOING.
I'VE ROLLED THE DIE.
NOW WHAT?
TOSS MY QUOIN?
OK, I'M DONE. WAS SOMETHING SUPPOSED TO HAPPEN?
IF MY QUOIN FELL HEADS, SAY HOW THE DIE LANDED, BUT IF IT FELL TAILS THEN SAY THE OPPOSITE?
BUT E.V.E. WILL "HEAR" WHAT I SAY!
HANG ON! E.V.E. WILL "HEAR" WHAT I SAY BUT SHE WON'T KNOW WHAT IT **MEANS**. SHE DOESN'T KNOW HOW MY QUOIN LANDED SO SHE DOESN'T HAVE THE KEY TO DECODE THE MESSAGE!
HERE GOES NOTHING.
MY SECRET MESSAGE IS...
ZERO!

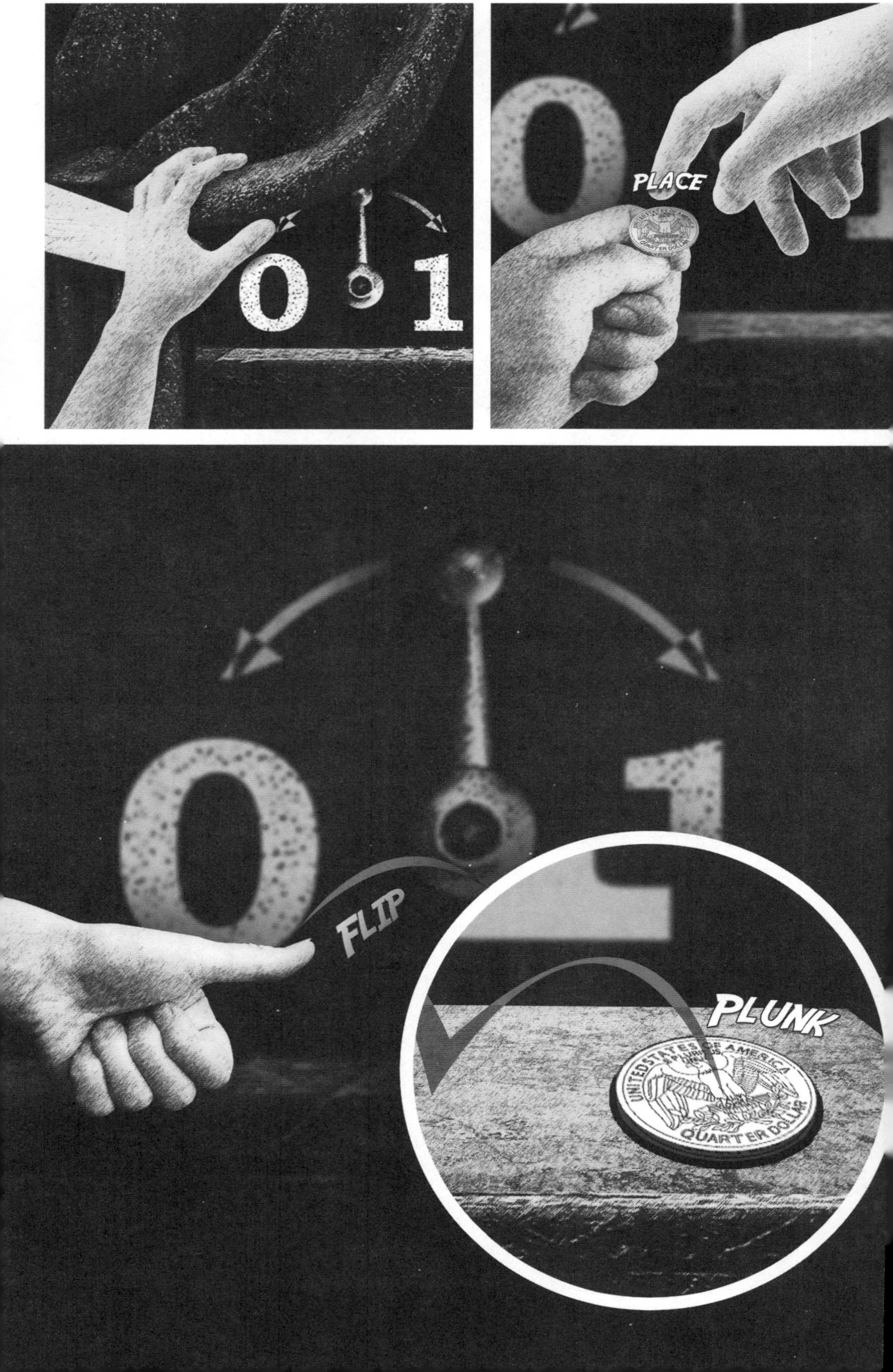
0 1
PLACE
FLIP
PLUNK
UNITED STATES OF AMERICA
QUARTER DOLLAR

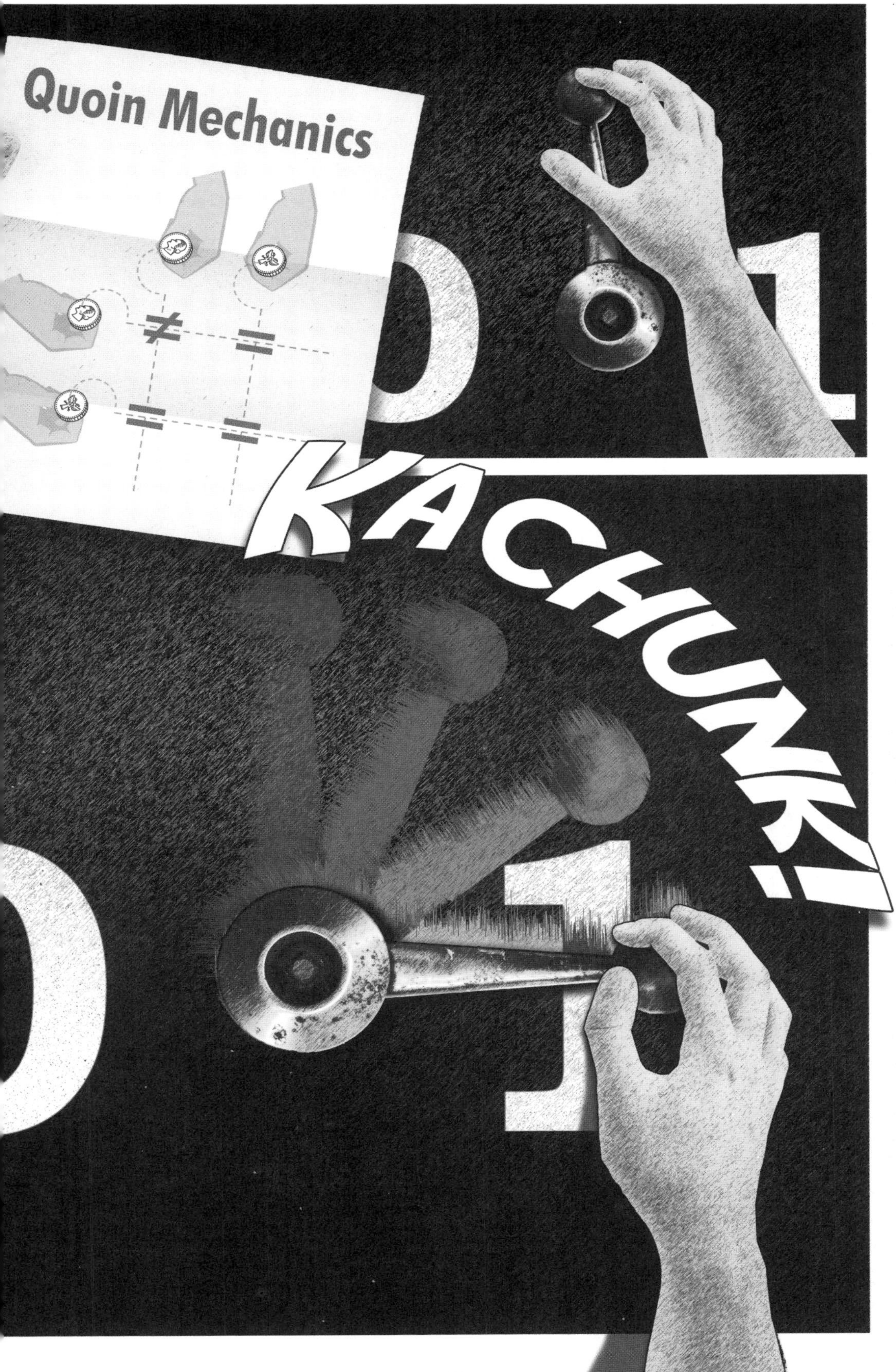

Quoin Mechanics
≠
=
=
=
0
1
KACHUNK!!
0
1

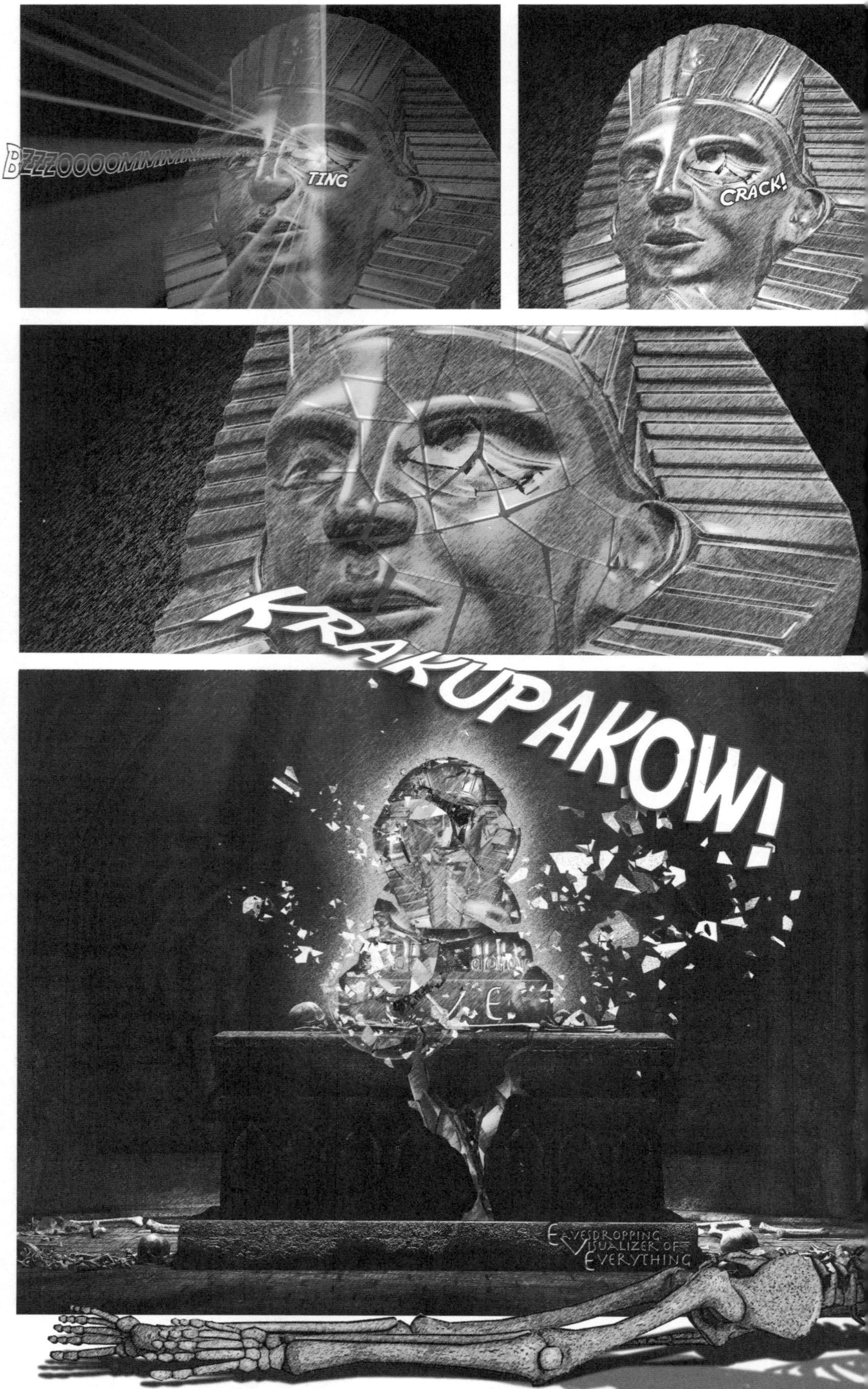

BZZZOOOOMMMM
TING
CRACK!
KRAKUPAKOW!
EAVESDROPPING
VISUALIZER OF
EVERYTHING

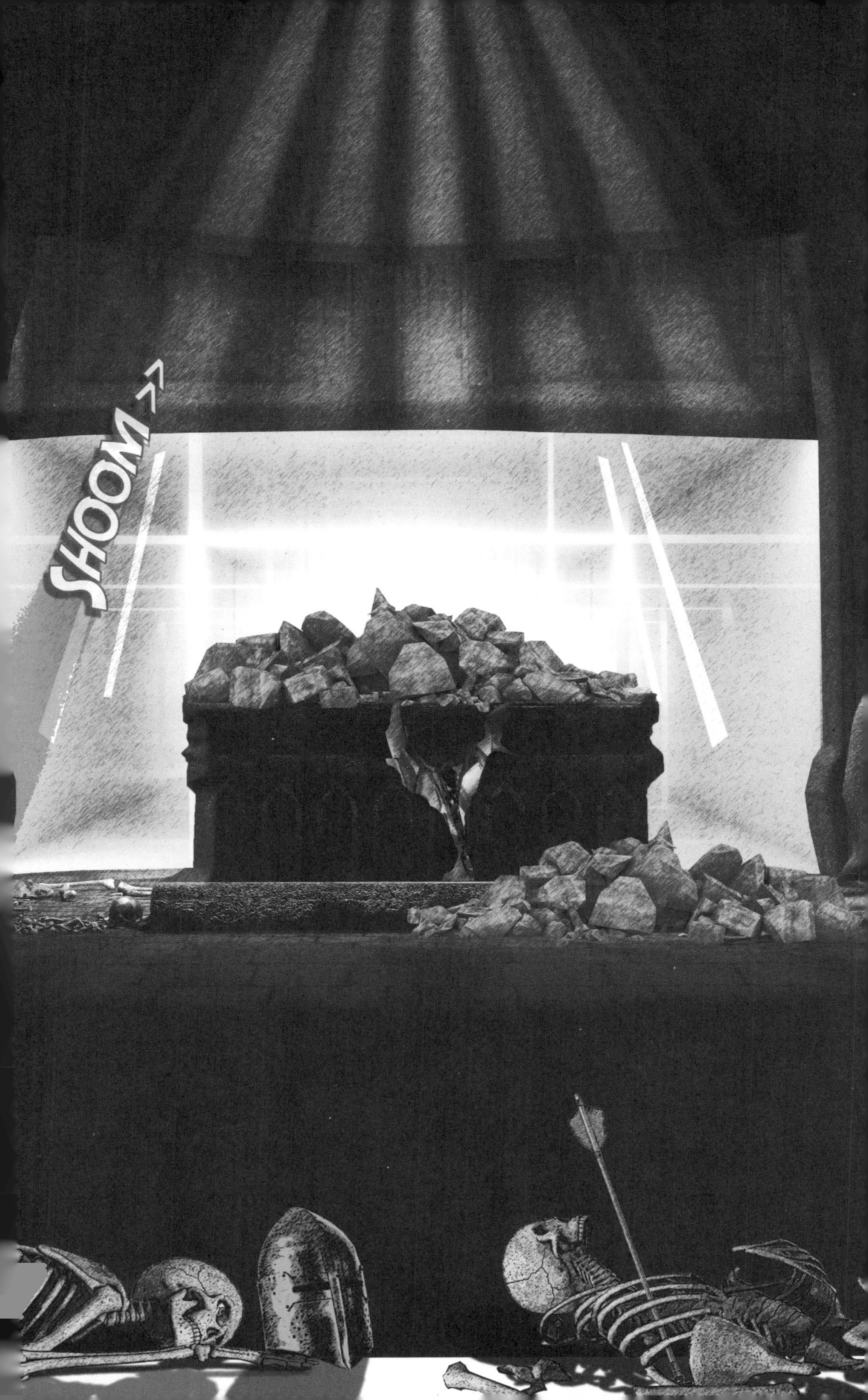
SHOOM

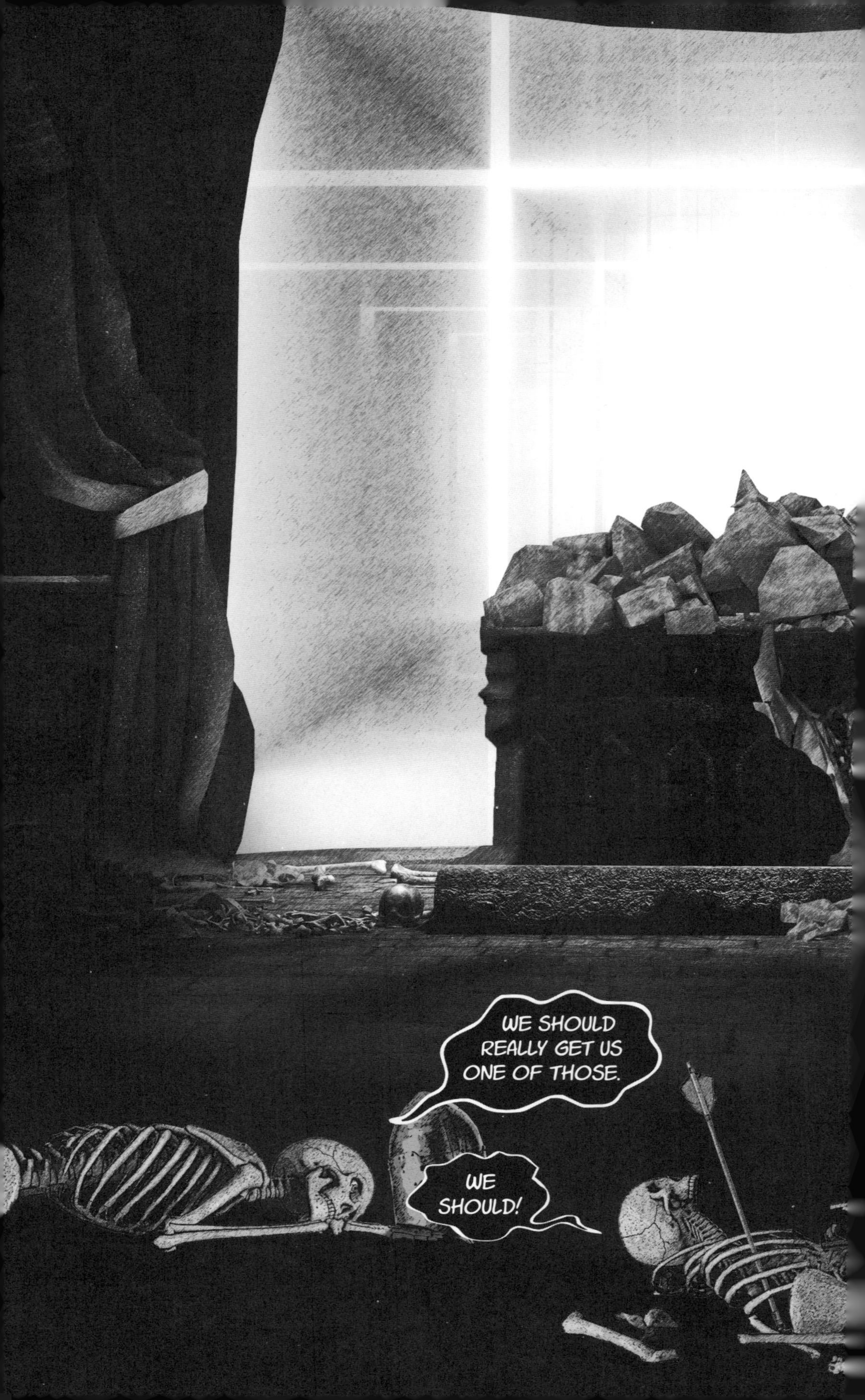
WE SHOULD REALLY GET US ONE OF THOSE.
WE SHOULD!

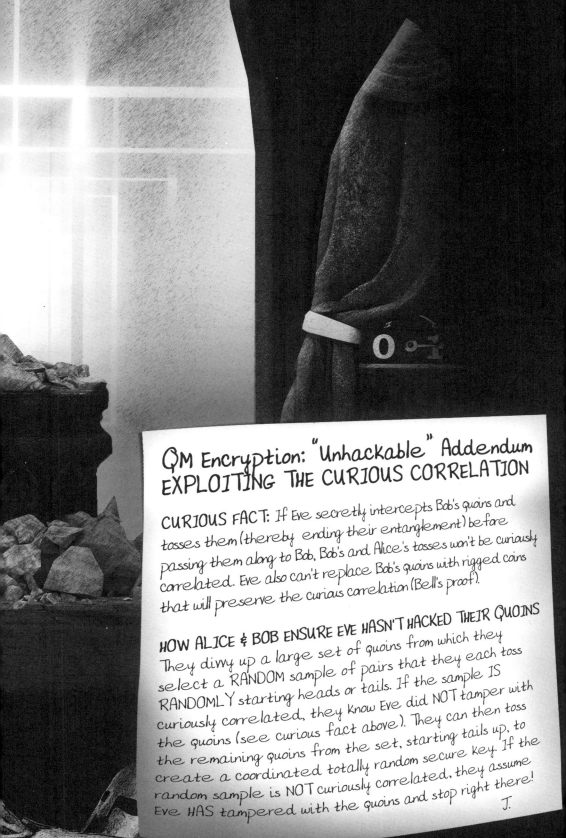

QM Encryption: "Unhackable" Addendum
EXPLOITING THE CURIOUS CORRELATION

CURIOUS FACT: If Eve secretly intercepts Bob's quoins and tosses them (thereby ending their entanglement) before passing them along to Bob, Bob's and Alice's tosses won't be curiously correlated. Eve also can't replace Bob's quoins with rigged coins that will preserve the curious correlation (Bell's proof).

HOW ALICE & BOB ENSURE EVE HASN'T HACKED THEIR QUOINS

They divvy up a large set of quoins from which they select a RANDOM sample of pairs that they each toss RANDOMLY starting heads or tails. If the sample IS curiously correlated, they know Eve did NOT tamper with the quoins (see curious fact above). They can then toss the remaining quoins from the set, starting tails up, to create a coordinated totally random secure key. If the random sample is NOT curiously correlated, they assume Eve HAS tampered with the quoins and stop right there!

J.

Exploiting QM to Compute

If you take just one piece of information from this blog: Quantum computers would not solve hard search problems instantaneously by simply trying all the possible solutions at once.

—Scott Aaronson's blog motto

Rather, a quantum computer can solve certain problems in fewer steps than a classical computer because it can do so without working out the answers to questions that aren't asked, but that a classical computer needs to answer in order to solve the problem. For example, suppose you want to know whether a statement like "P or Q or R or . . ." is true or false, but you're not interested in whether the individual components are true or false. The only way a classical computer could figure that out is by checking each component until it finds one that's true, or finds that all are false. So the computer generally has to work out and store answers to many true/false questions that you aren't interested in as steps in the process of deciding whether the one statement you are interested in is true or false. By contrast, a quantum computer exploits curious correlations to bypass redundant information about the components. Read on to see how this works.

T? Hello?
No sticky note?
No comment? No biting criticism?
Are you even reading this?

[This] problem is not just some silly communication task; it is in fact the most difficult communication task possible (technically called the "inner product" problem). Indeed, every other communication problem can be mapped onto this one, so removing the redundancy from this ... problem means removing the redundancy from all communication problems.

—Sandu Popescu

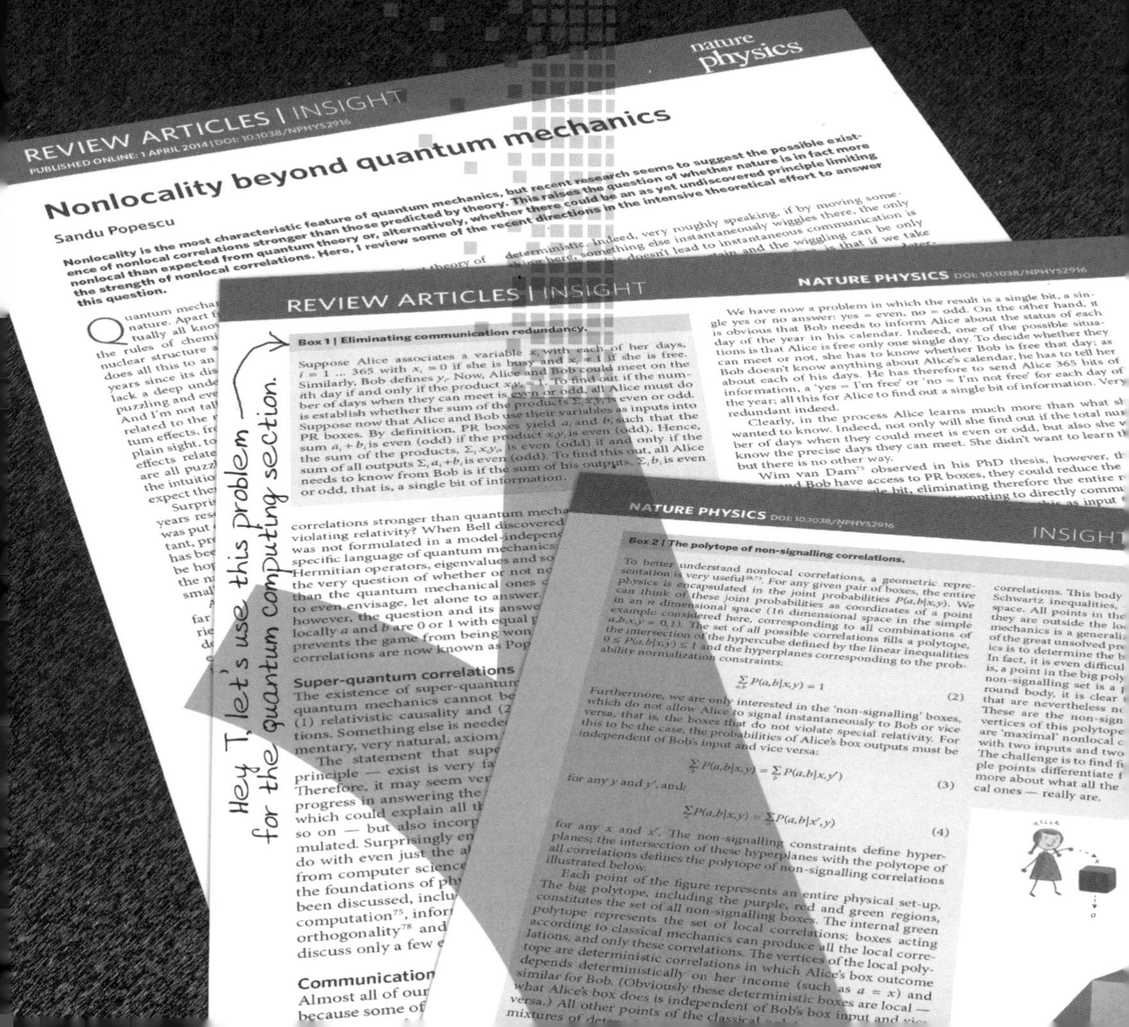

nature physics

REVIEW ARTICLES | INSIGHT
PUBLISHED ONLINE: 1 APRIL 2014 | DOI: 10.1038/NPHYS2916

Nonlocality beyond quantum mechanics

Sandu Popescu

Nonlocality is the most characteristic feature of quantum mechanics, but recent research seems to suggest the possible existence of nonlocal correlations stronger than those predicted by theory. This raises the question of whether nature is in fact more nonlocal than expected from quantum theory or, alternatively, whether there could be an as yet undiscovered theoretical effort to answer the strength of nonlocal correlations. Here, I review some of the recent directions in the intensive theoretical effort to answer this question.

Handwritten note (left margin): Hey T, let's use this problem for the quantum computing section.

NATURE PHYSICS DOI: 10.1038/NPHYS2916

Box 1 | Eliminating communication redundancy.

Suppose Alice associates a variable x_i with each of her days, $i = 1 \dots 365$ with $x_i = 0$ if she is busy and $x_i = 1$ if she is free. Similarly, Bob defines y_i. Now, Alice and Bob could meet on the ith day if and only if the product $x_i y_i = 1$. To find out if the number of days when they can meet is even or odd, all Alice must do is establish whether the sum of the products $\Sigma_i x_i y_i$ is even or odd. Suppose now that Alice and Bob use their variables as inputs into PR boxes. By definition, PR boxes yield a_i and b_i such that the sum $a_i + b_i$ is even (odd) if the product $x_i y_i$ is even (odd). Hence, the sum of the products, $\Sigma_i x_i y_i$, is even (odd) if and only if the sum of all outputs $\Sigma_i a_i + b_i$ is even (odd). To find this out, all Alice needs to know from Bob is if the sum of his outputs, $\Sigma_i b_i$, is even or odd, that is, a single bit of information.

We have now a problem in which the result is a single bit, a single yes or no answer: yes = even, no = odd. On the other hand, it is obvious that Bob needs to inform Alice about the status of each day of the year in his calendar. Indeed, one of the possible situations is that Alice is free only one single day. To decide whether they can meet or not, she has to know whether Bob is free that day; as Bob doesn't know anything about Alice's calendar, he has to tell her about each of his days. He has therefore to send Alice 365 bits of information, a 'yes = I'm free' or 'no = I'm not free' for each day of the year; all this for Alice to find out a single bit of information. Very redundant indeed.

Clearly, in the process Alice learns much more than what she wanted to know. Indeed, not only will she find out if the total number of days when they could meet is even or odd, but also she will know the precise days they can meet. She didn't want to learn this but there is no other way.

Winn van Dam[?] observed in his PhD thesis, however, that if Alice and Bob have access to PR boxes, they could reduce the ... bit, eliminating therefore the entire redundancy ... ... to directly communicate ... ... as input ...

correlations stronger than quantum mecha... violating relativity? When Bell discovered ... was not formulated in a model-indepen... specific language of quantum mechanics... Hermitian operators, eigenvalues and so ... the very question of whether or not no... than the quantum mechanical ones —... to even envisage, let alone to answer... however, the question and its answer ... locally and a and b are 0 or 1 with equal p... prevents the game from being won ... correlations are now known as Po...

Super-quantum correlations

The existence of super-quantum ... quantum mechanics cannot be ... (1) relativistic causality and (2) ... tions. Something else is needed ... mentary, very natural, axiom ...

The statement that supe... principle — exist is very fa... Therefore, it may seem ve... progress in answering the ... which could explain all th... so on — but also incorp... mulated. Surprisingly en... do with even just the a... from computer science ... the foundations of ph... been discussed, inclu... computation[75], infor... orthogonality[78] and ... discuss only a few o...

Communication ...

Almost all of our ... because some o...

Box 2 | The polytope of non-signalling correlations.

To better understand nonlocal correlations, a geometric representation is very useful[?]. For any given pair of boxes, the entire physics is encapsulated in the joint probabilities $P(a,b|x,y)$. We can think of these joint probabilities as coordinates of a point in an n dimensional space (16 dimensional space in the simple example considered here, corresponding to all combinations of $a,b,x,y = 0,1$). The set of all possible correlations fills a polytope, the intersection of the hypercube defined by the linear inequalities $0 \leq P(a,b|x,y) \leq 1$ and the hyperplanes corresponding to the probability normalization constraints:

$$\sum_{a,b} P(a,b|x,y) = 1 \tag{1}$$

Furthermore, we are only interested in the 'non-signalling' boxes, which do not allow Alice to signal instantaneously to Bob or vice versa, that is the boxes that do not violate special relativity. For this to be the case, the probabilities of Alice's box outputs must be independent of Bob's input and vice versa:

$$\sum_{b} P(a,b|x,y) = \sum_{b} P(a,b|x,y') \tag{2}$$

for any y and y', and:

$$\sum_{a} P(a,b|x,y) = \sum_{a} P(a,b|x',y) \tag{3}$$

for any x and x'. The non-signalling constraints define hyperplanes; the intersection of these hyperplanes with the polytope of all correlations defines the polytope of non-signalling correlations illustrated below.

Each point of the figure represents an entire physical set-up. The big polytope, including the purple, red and green regions, constitutes the set of all non-signalling boxes. The internal green polytope represents the set of local correlations; boxes acting according to classical mechanics can produce all the local correlations, and only these correlations. The vertices of the local polytope are deterministic correlations in which Alice's box outcome depends deterministically on her income (such as $a = x$) and similar for Bob. (Obviously these deterministic boxes are local — what Alice's box does is independent of Bob's box input and vice versa.) All other points of the classical ... mixtures of deter...

INSIGHT

correlations. This body ... Schwartz inequalities, ... space. All points in the ... they are outside the loc... mechanics is a generali... of the great unsolved pr... ics is to determine the b... In fact, it is even difficul... is, a point in the big poly... non-signalling set is a r... round body, it is clear t... that are nevertheless no... These are the non-sign... vertices of this polytope ... are 'maximal' nonlocal c... with two inputs and two ... The challenge is to find fu... ple points differentiate t... more about what all the ... cal ones — really are.

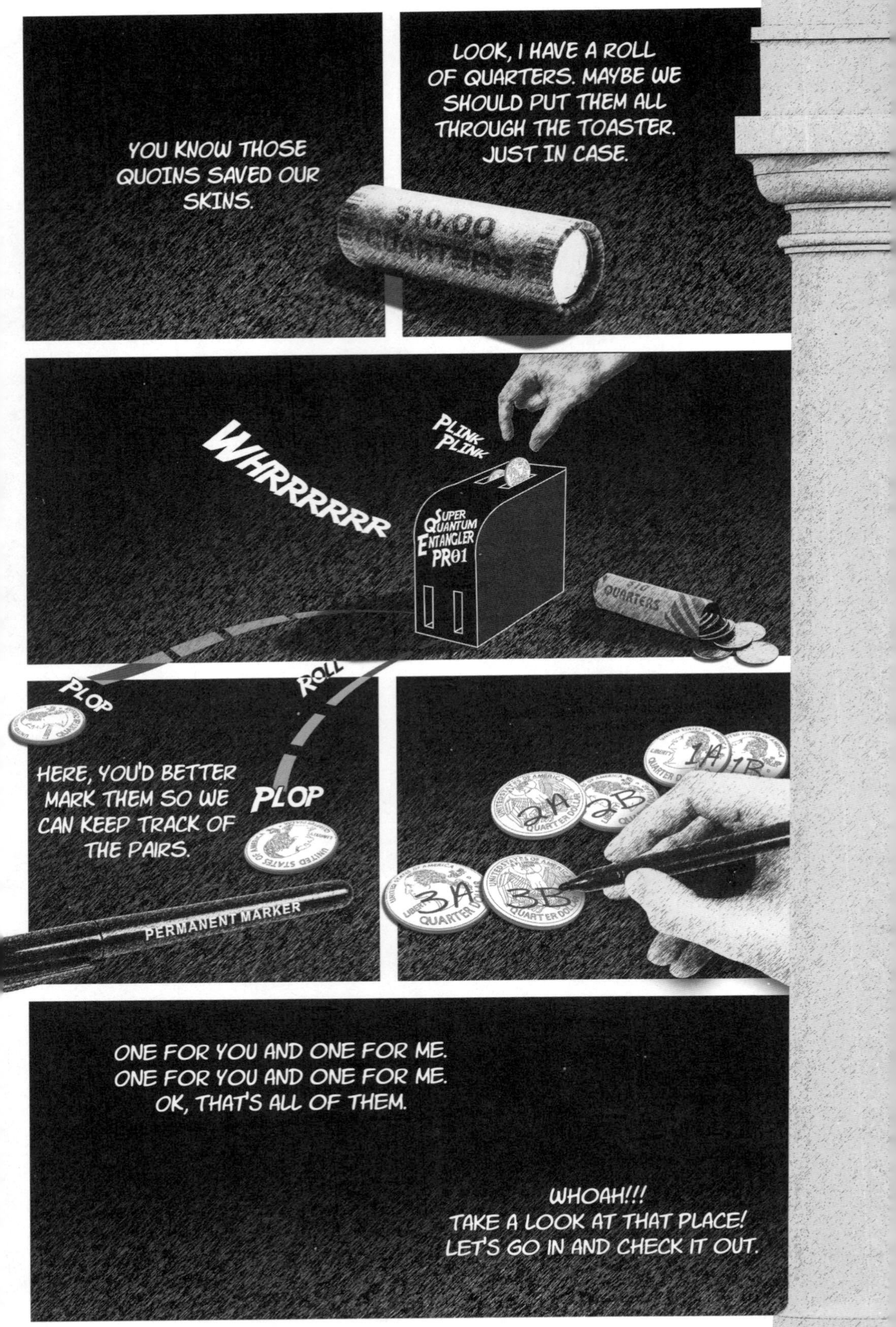
YOU KNOW THOSE QUOINS SAVED OUR SKINS.
LOOK, I HAVE A ROLL OF QUARTERS. MAYBE WE SHOULD PUT THEM ALL THROUGH THE TOASTER. JUST IN CASE.
$10.00 QUARTERS
WHRRRRRR
PLINK PLINK
SUPER QUANTUM ENTANGLER PR01
$10 QUARTERS
PLOP
ROLL
PLOP
HERE, YOU'D BETTER MARK THEM SO WE CAN KEEP TRACK OF THE PAIRS.
PERMANENT MARKER
1A 1B
2A 2B
3A 3B
ONE FOR YOU AND ONE FOR ME. ONE FOR YOU AND ONE FOR ME. OK, THAT'S ALL OF THEM.
WHOAH!!! TAKE A LOOK AT THAT PLACE! LET'S GO IN AND CHECK IT OUT.

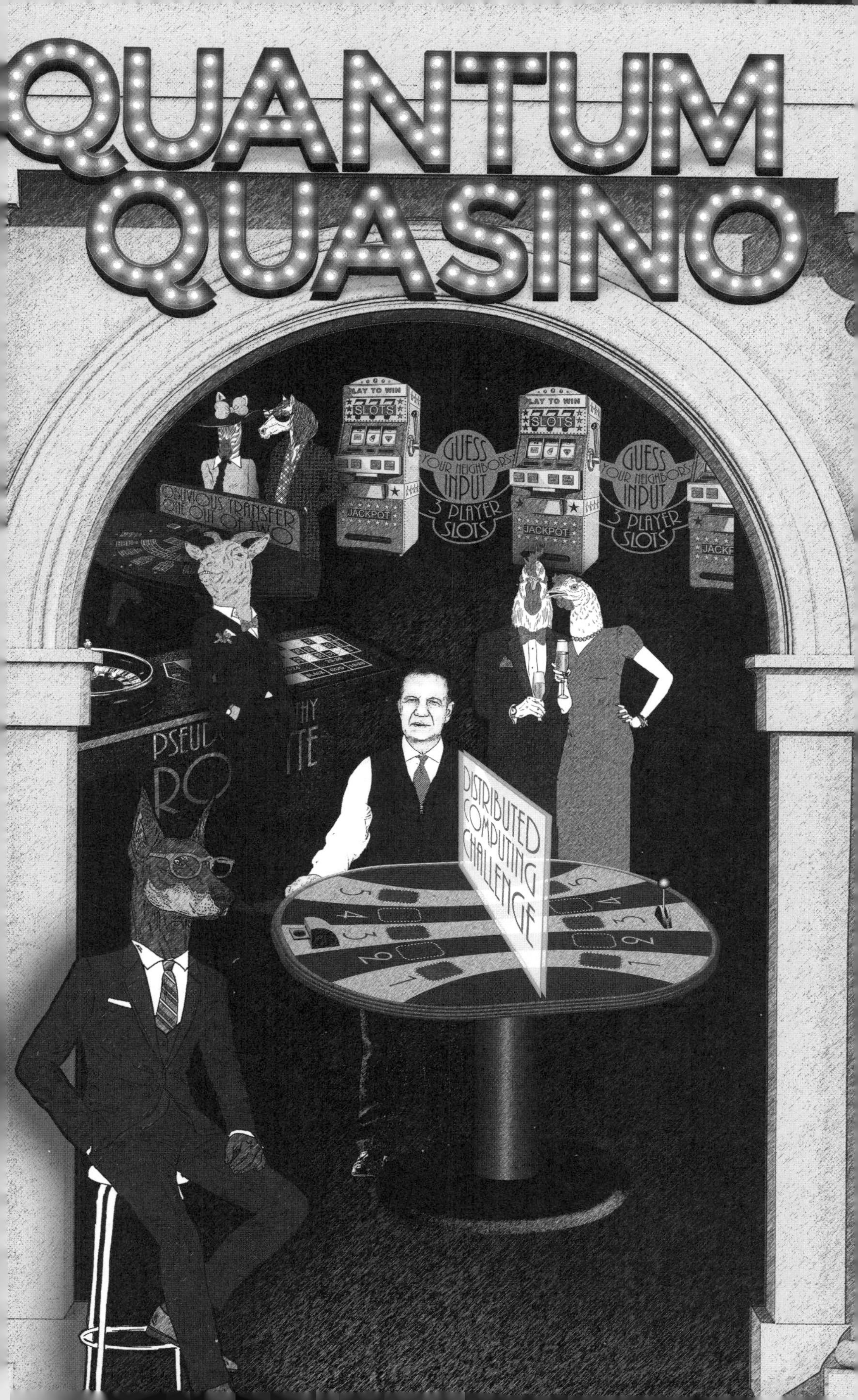
QUANTUM
QUASINO
PLAY TO WIN
SLOTS
JACKPOT
GUESS YOUR NEIGHBORS INPUT
3 PLAYER SLOTS
OBLIVIOUS TRANSFER ONE OUT OF TWO
DISTRIBUTED COMPUTING CHALLENGE

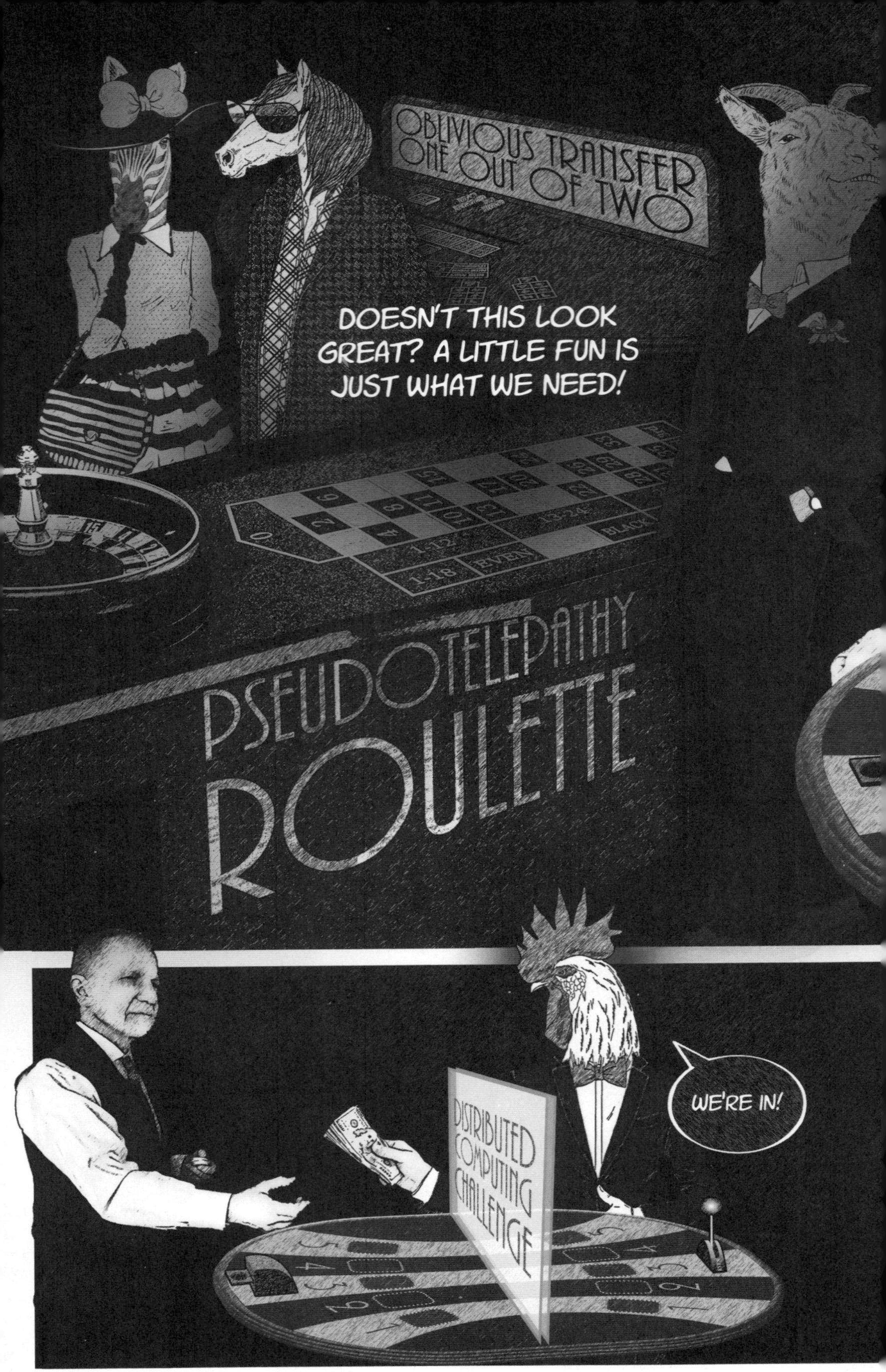
OBLIVIOUS TRANSFER ONE OUT OF TWO
DOESN'T THIS LOOK GREAT? A LITTLE FUN IS JUST WHAT WE NEED!
PSEUDOTELEPATHY ROULETTE
DISTRIBUTED COMPUTING CHALLENGE
WE'RE IN!

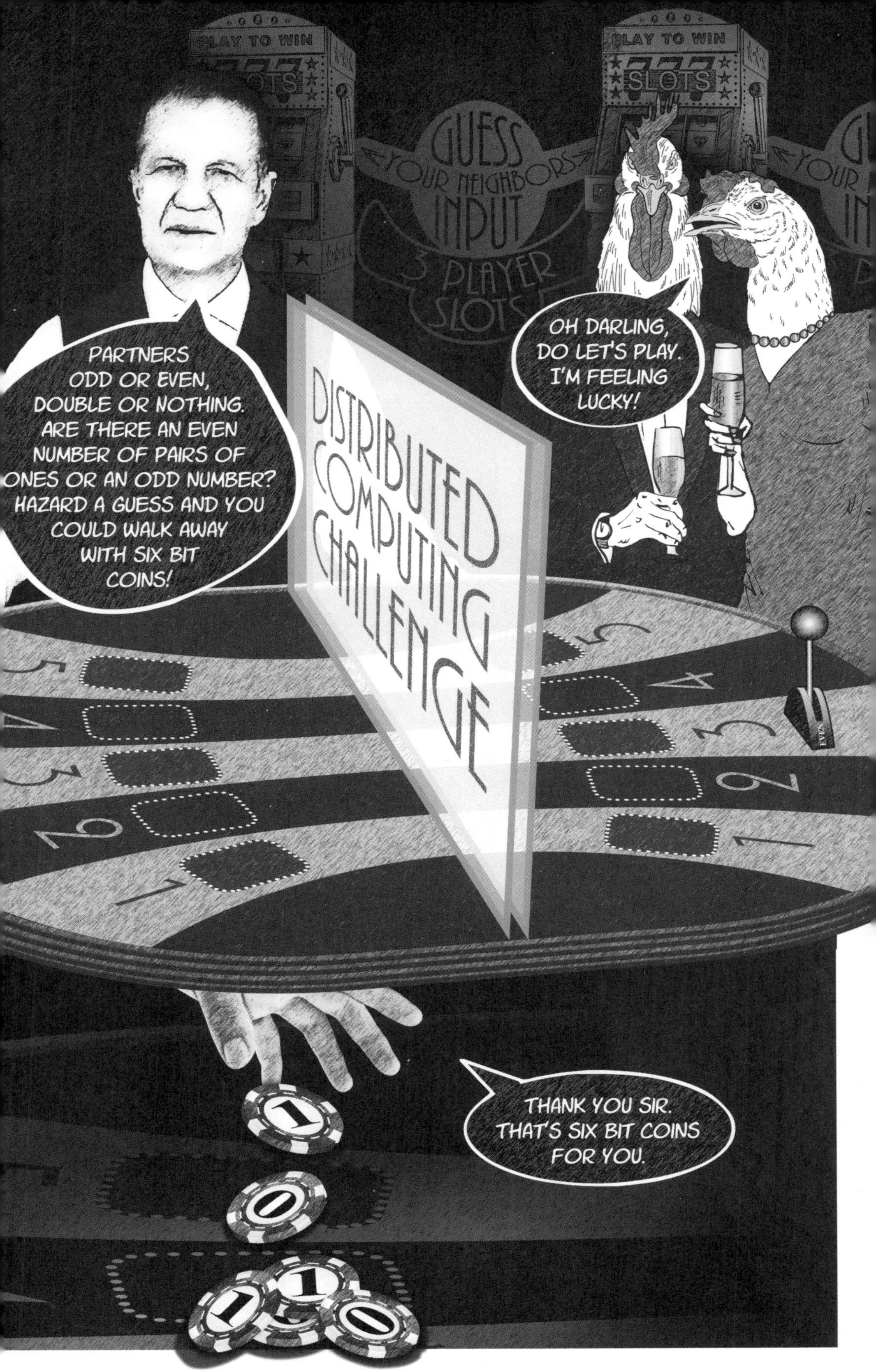
PLAY TO WIN
SLOTS
GUESS YOUR NEIGHBORS INPUT
3 PLAYER SLOTS
PARTNERS ODD OR EVEN, DOUBLE OR NOTHING. ARE THERE AN EVEN NUMBER OF PAIRS OF ONES OR AN ODD NUMBER? HAZARD A GUESS AND YOU COULD WALK AWAY WITH SIX BIT COINS!
DISTRIBUTED COMPUTING CHALLENGE
OH DARLING, DO LET'S PLAY. I'M FEELING LUCKY!
THANK YOU SIR. THAT'S SIX BIT COINS FOR YOU.

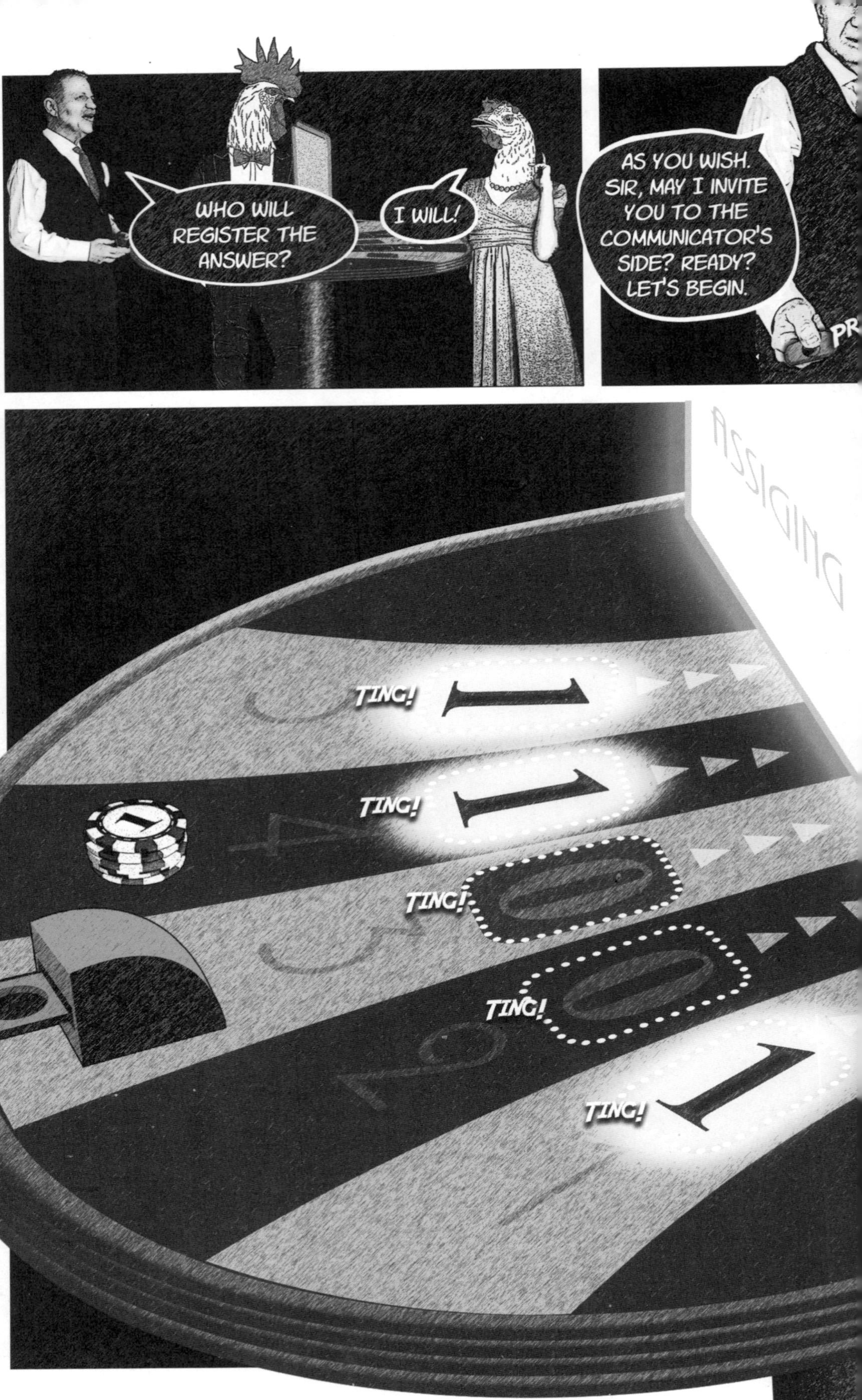
WHO WILL REGISTER THE ANSWER?
I WILL!
AS YOU WISH. SIR, MAY I INVITE YOU TO THE COMMUNICATOR'S SIDE? READY? LET'S BEGIN.
PR
ASSIGNING
TING!
TING!
TING!
TING!
TING!

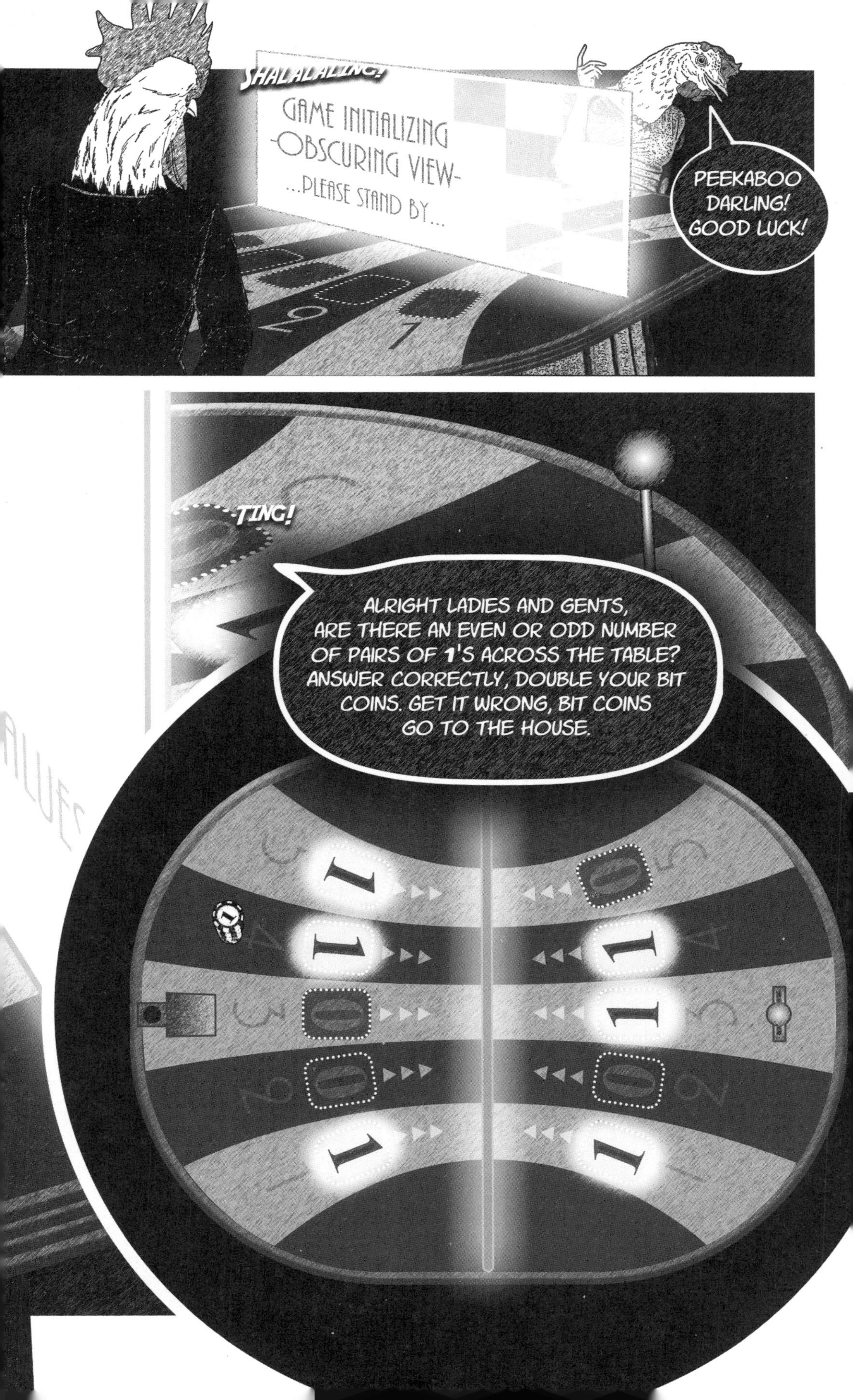
SHALALALING!
GAME INITIALIZING
-OBSCURING VIEW-
...PLEASE STAND BY...
PEEKABOO DARLING! GOOD LUCK!
TING!
ALRIGHT LADIES AND GENTS, ARE THERE AN EVEN OR ODD NUMBER OF PAIRS OF 1'S ACROSS THE TABLE? ANSWER CORRECTLY, DOUBLE YOUR BIT COINS. GET IT WRONG, BIT COINS GO TO THE HOUSE.

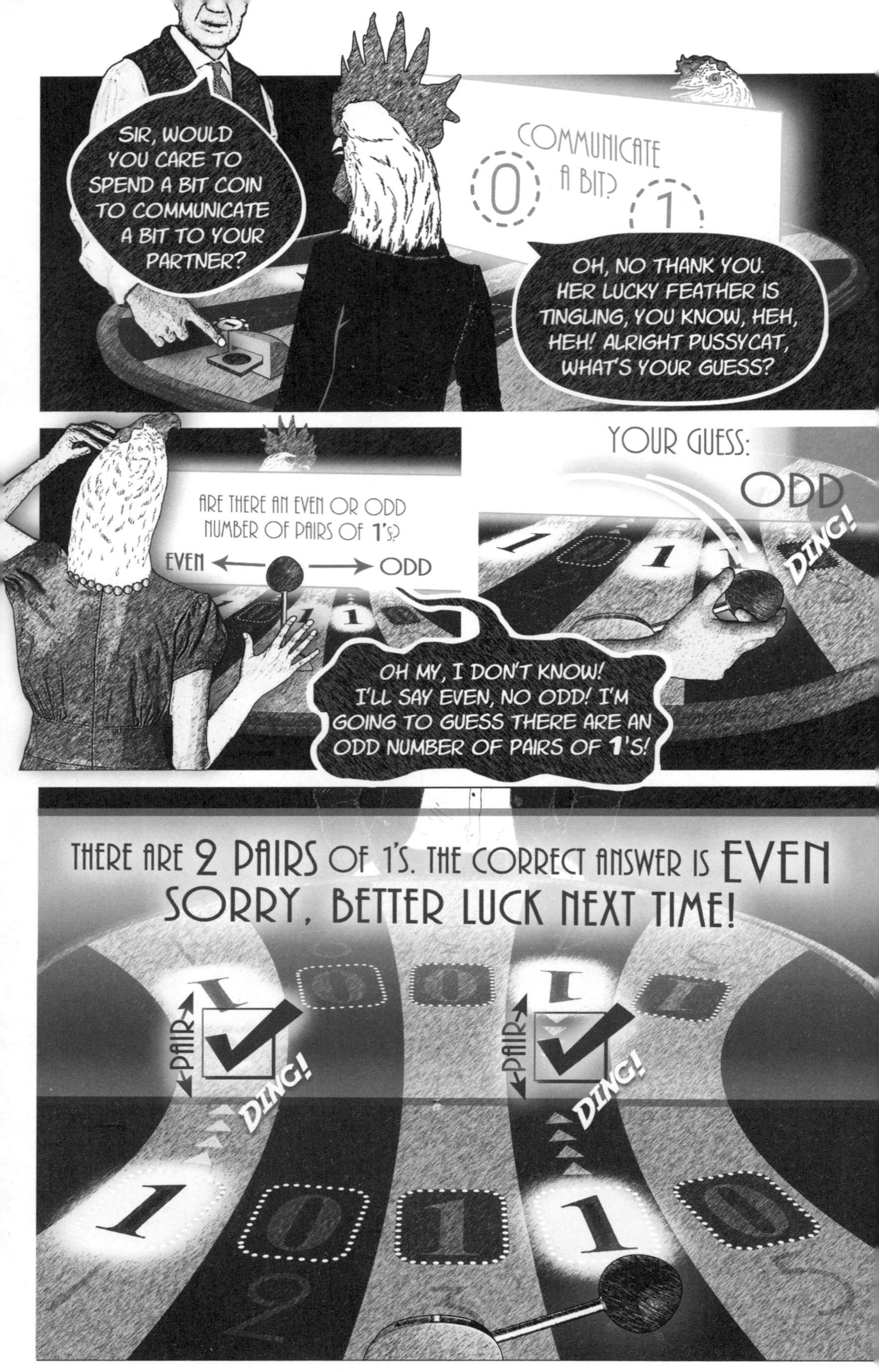
SIR, WOULD YOU CARE TO SPEND A BIT COIN TO COMMUNICATE A BIT TO YOUR PARTNER?
COMMUNICATE A BIT?
0
1
OH, NO THANK YOU. HER LUCKY FEATHER IS TINGLING, YOU KNOW, HEH, HEH! ALRIGHT PUSSYCAT, WHAT'S YOUR GUESS?
ARE THERE AN EVEN OR ODD NUMBER OF PAIRS OF 1'S?
EVEN
ODD
OH MY, I DON'T KNOW! I'LL SAY EVEN, NO ODD! I'M GOING TO GUESS THERE ARE AN ODD NUMBER OF PAIRS OF 1'S!
YOUR GUESS:
ODD
DING!
THERE ARE 2 PAIRS OF 1'S. THE CORRECT ANSWER IS EVEN
SORRY, BETTER LUCK NEXT TIME!
PAIR
DING!
PAIR
DING!
1
0
1
1
0

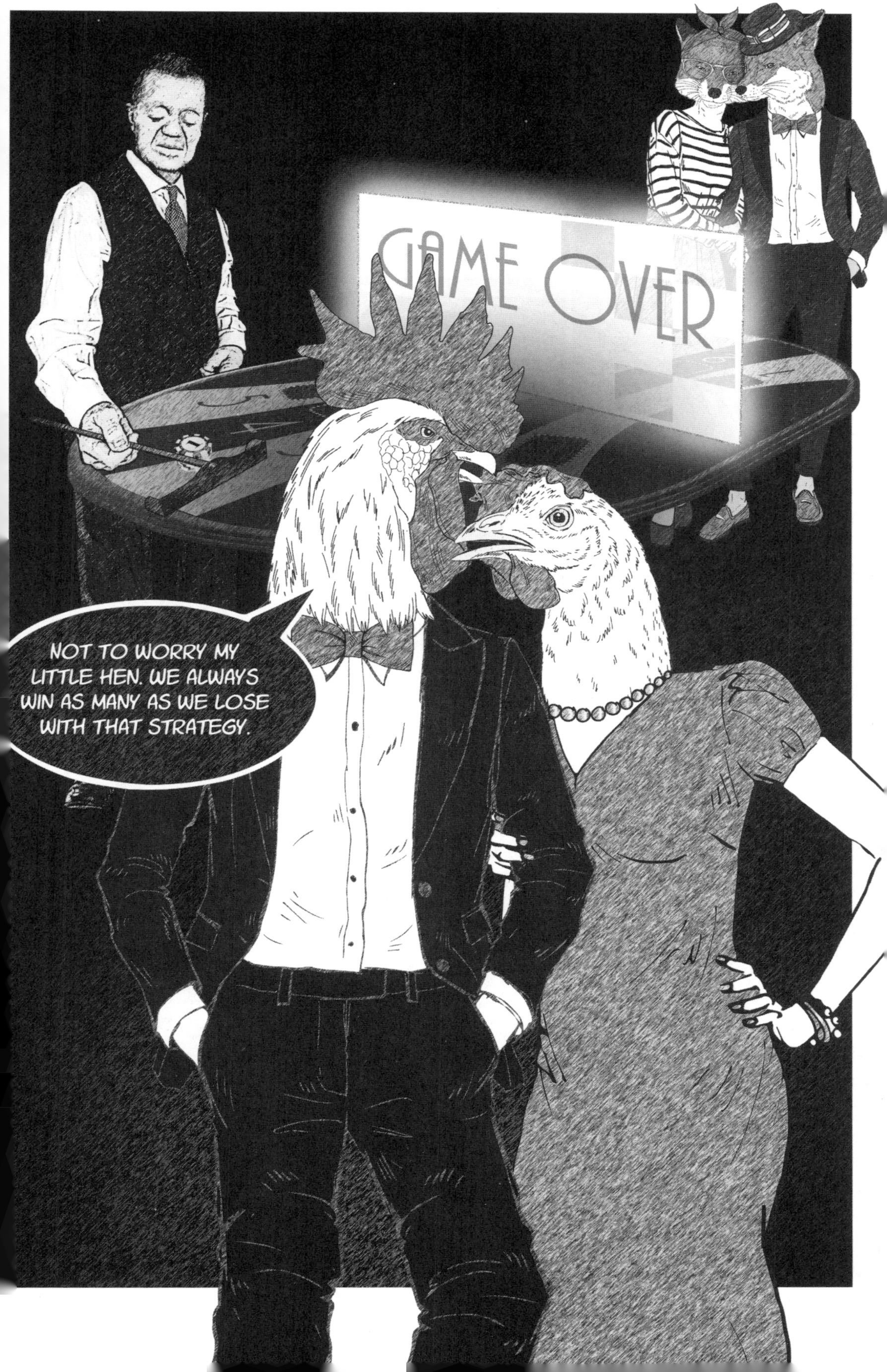
GAME OVER
NOT TO WORRY MY LITTLE HEN. WE ALWAYS WIN AS MANY AS WE LOSE WITH THAT STRATEGY.

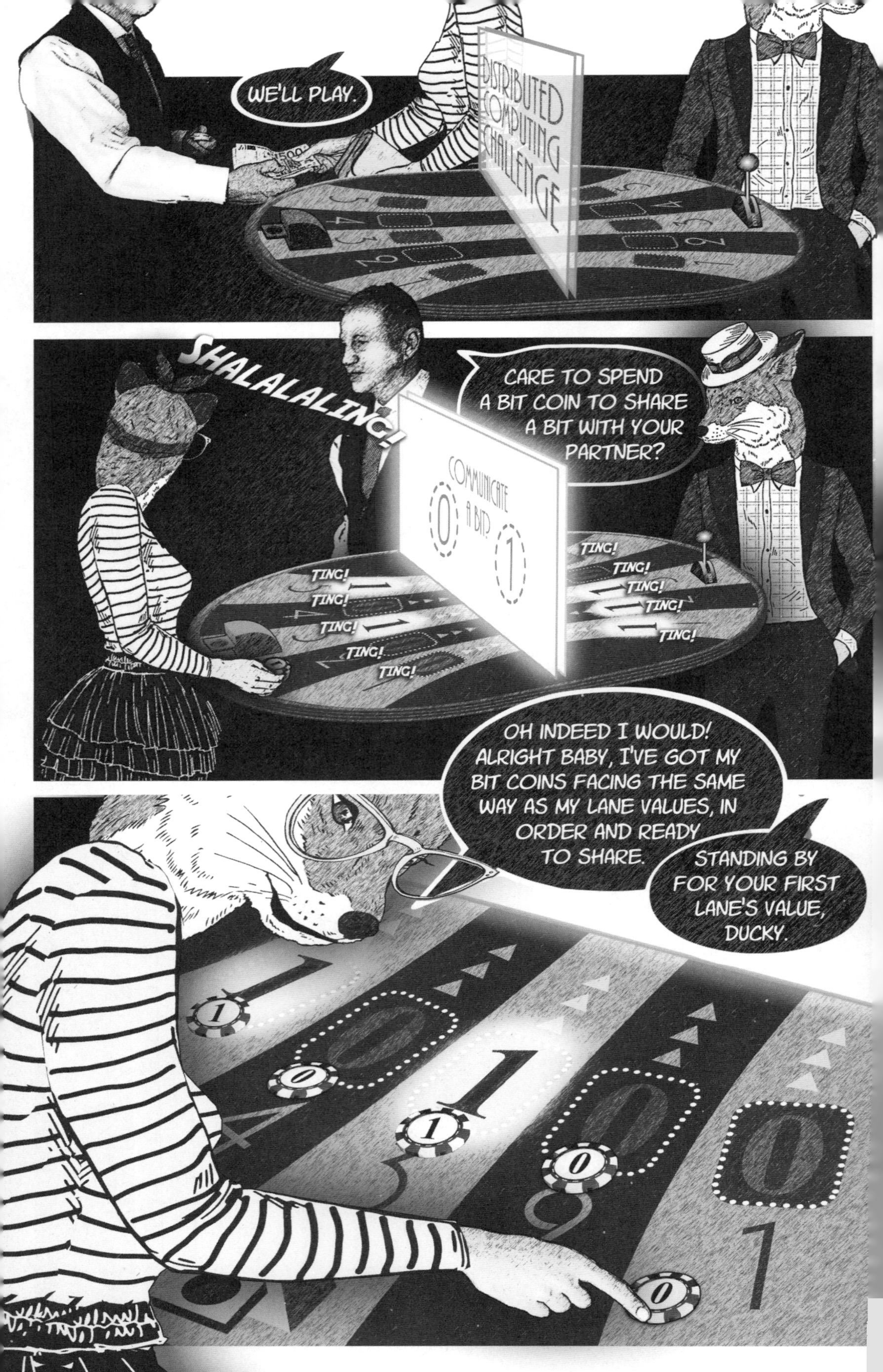
WE'LL PLAY.
DISTRIBUTED COMPUTING CHALLENGE
SHALALALING!
CARE TO SPEND A BIT COIN TO SHARE A BIT WITH YOUR PARTNER?
COMMUNICATE A BIT?
0
1
TING!
TING!
TING!
TING!
TING!
TING!
TING!
TING!
TING!
TING!
OH INDEED I WOULD! ALRIGHT BABY, I'VE GOT MY BIT COINS FACING THE SAME WAY AS MY LANE VALUES, IN ORDER AND READY TO SHARE.
STANDING BY FOR YOUR FIRST LANE'S VALUE, DUCKY.

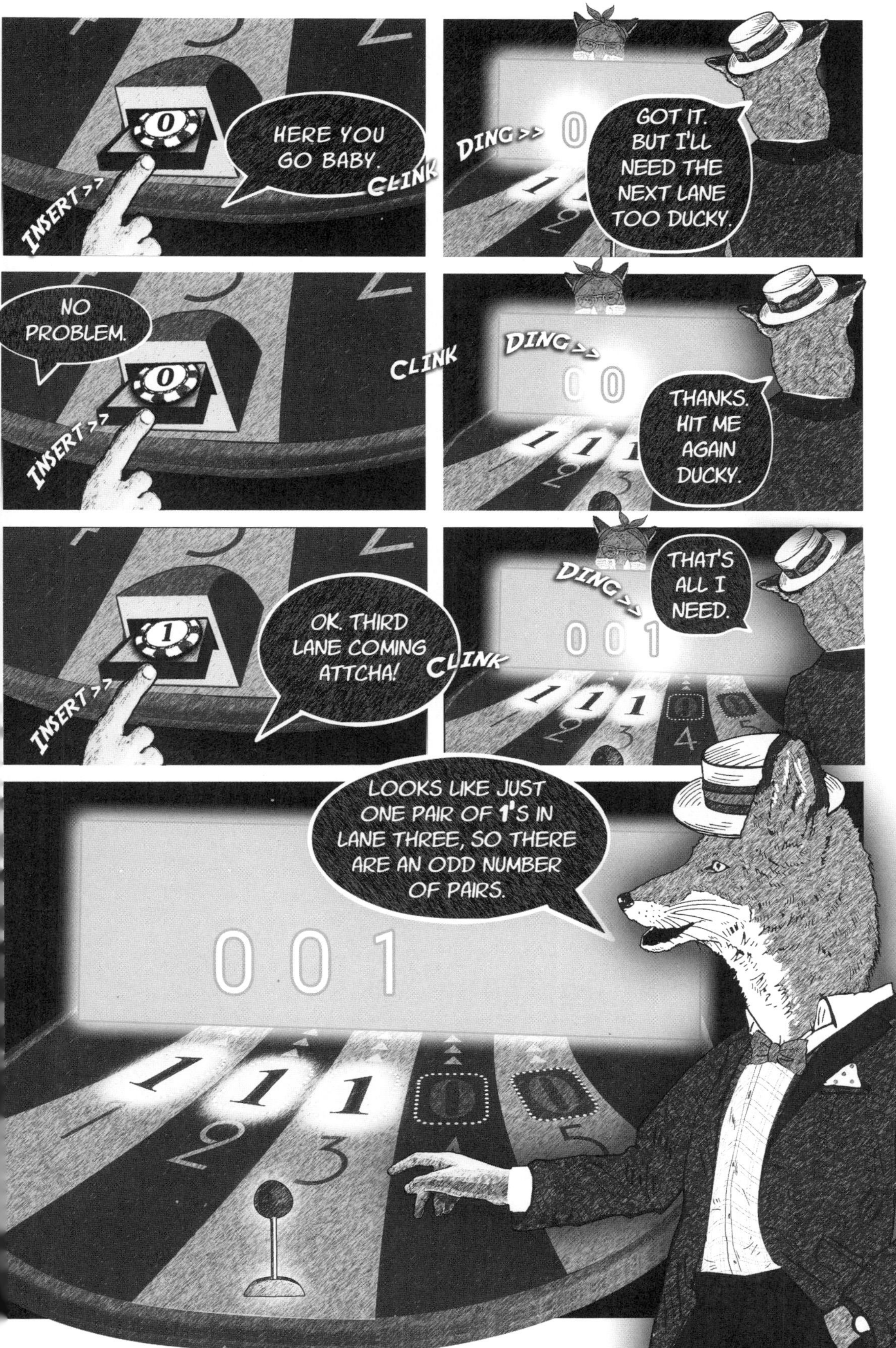
HERE YOU GO BABY.
INSERT>>
CLINK
DING>>
0
GOT IT. BUT I'LL NEED THE NEXT LANE TOO DUCKY.
NO PROBLEM.
INSERT>>
CLINK
DING>>
0 0
THANKS. HIT ME AGAIN DUCKY.
OK. THIRD LANE COMING ATTCHA!
INSERT>>
CLINK
DING>>
THAT'S ALL I NEED.
0 0 1
LOOKS LIKE JUST ONE PAIR OF 1'S IN LANE THREE, SO THERE ARE AN ODD NUMBER OF PAIRS.
0 0 1

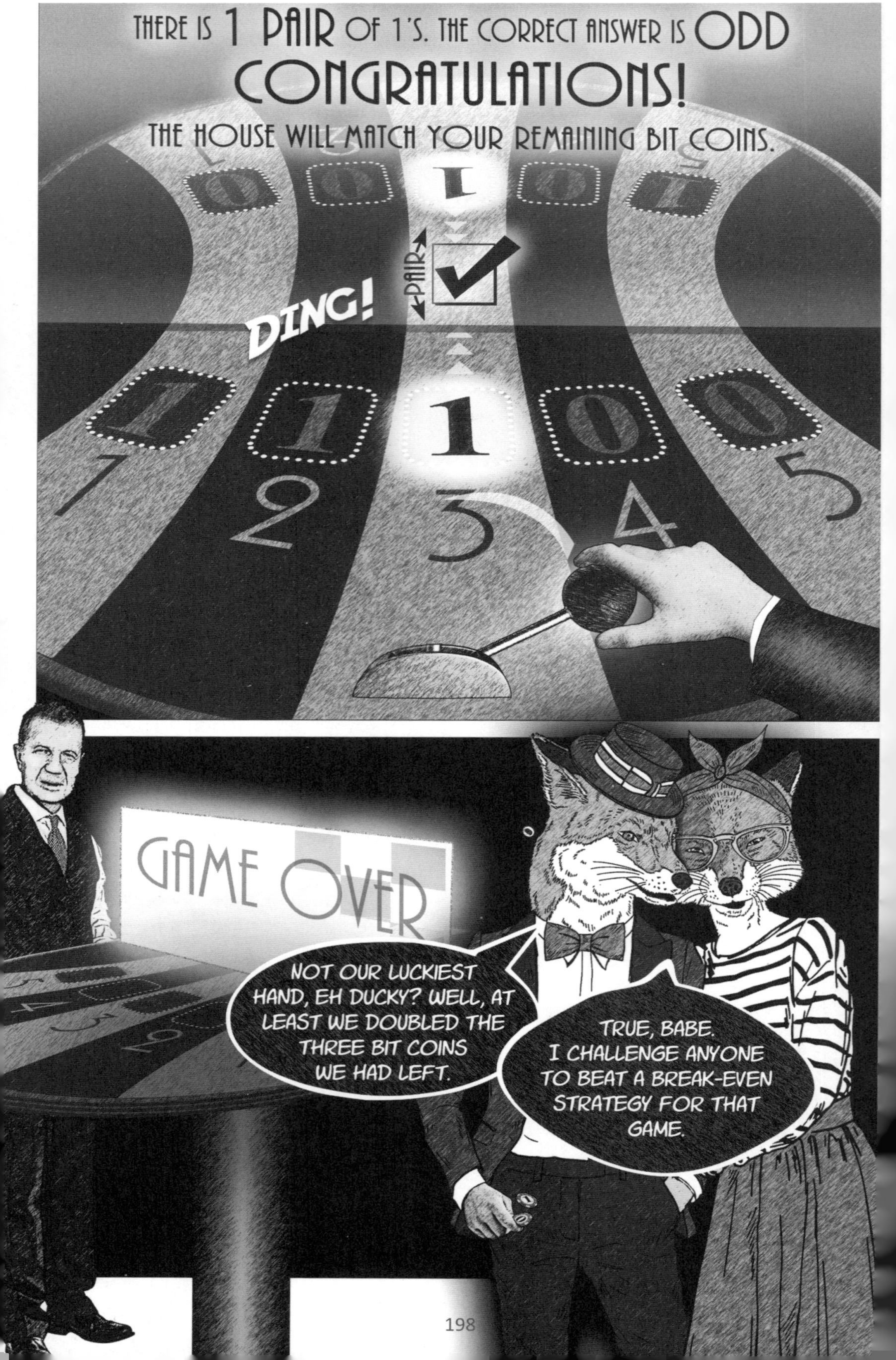
THERE IS 1 PAIR OF 1'S. THE CORRECT ANSWER IS ODD
CONGRATULATIONS!
THE HOUSE WILL MATCH YOUR REMAINING BIT COINS.
PAIR
DING!
GAME OVER
NOT OUR LUCKIEST HAND, EH DUCKY? WELL, AT LEAST WE DOUBLED THE THREE BIT COINS WE HAD LEFT.
TRUE, BABE. I CHALLENGE ANYONE TO BEAT A BREAK-EVEN STRATEGY FOR THAT GAME.

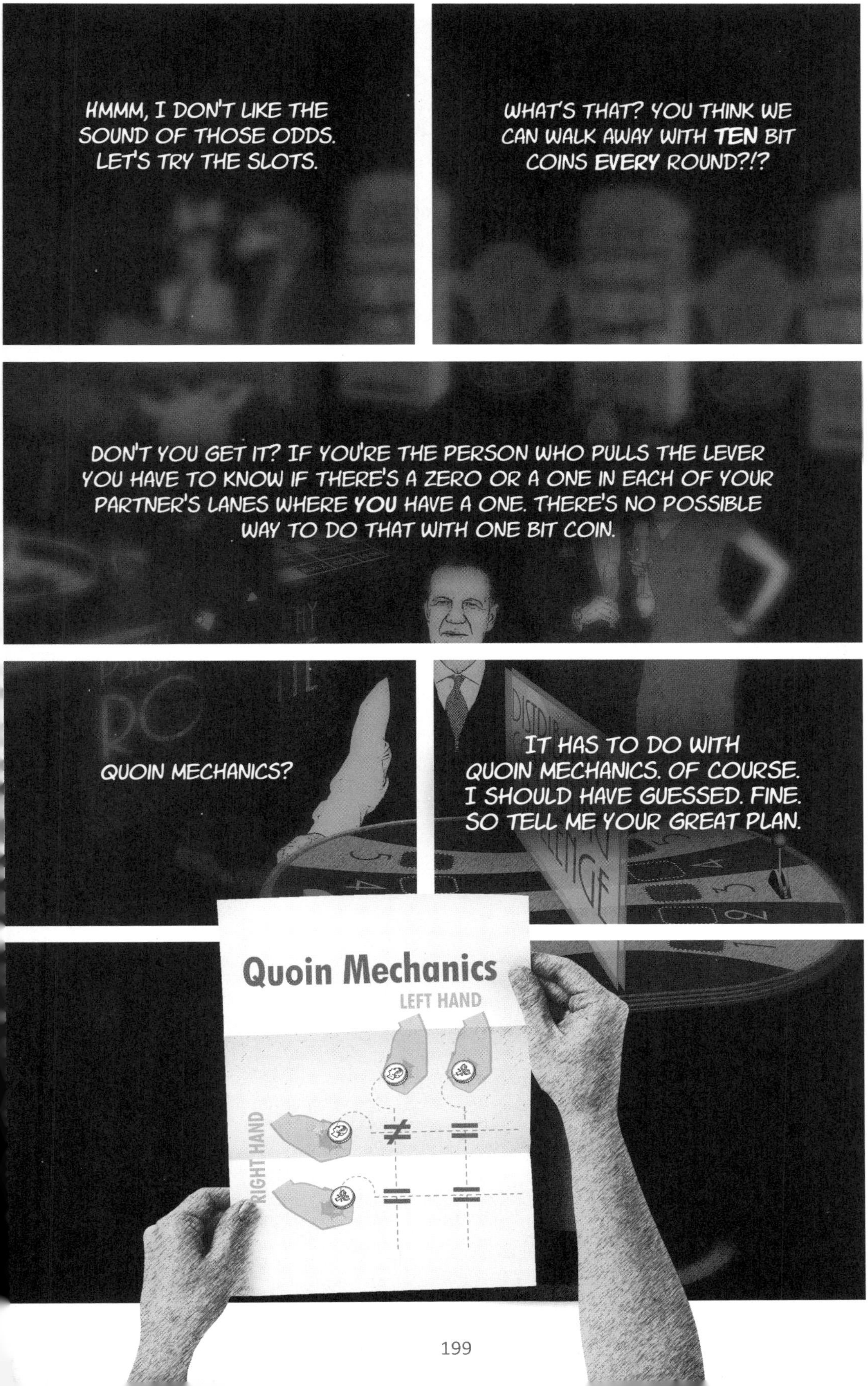
HMMM, I DON'T LIKE THE SOUND OF THOSE ODDS. LET'S TRY THE SLOTS.
WHAT'S THAT? YOU THINK WE CAN WALK AWAY WITH TEN BIT COINS EVERY ROUND?!?
DON'T YOU GET IT? IF YOU'RE THE PERSON WHO PULLS THE LEVER YOU HAVE TO KNOW IF THERE'S A ZERO OR A ONE IN EACH OF YOUR PARTNER'S LANES WHERE YOU HAVE A ONE. THERE'S NO POSSIBLE WAY TO DO THAT WITH ONE BIT COIN.
QUOIN MECHANICS?
IT HAS TO DO WITH QUOIN MECHANICS. OF COURSE. I SHOULD HAVE GUESSED. FINE. SO TELL ME YOUR GREAT PLAN.
Quoin Mechanics
LEFT HAND
RIGHT HAND
≠
=
=
=

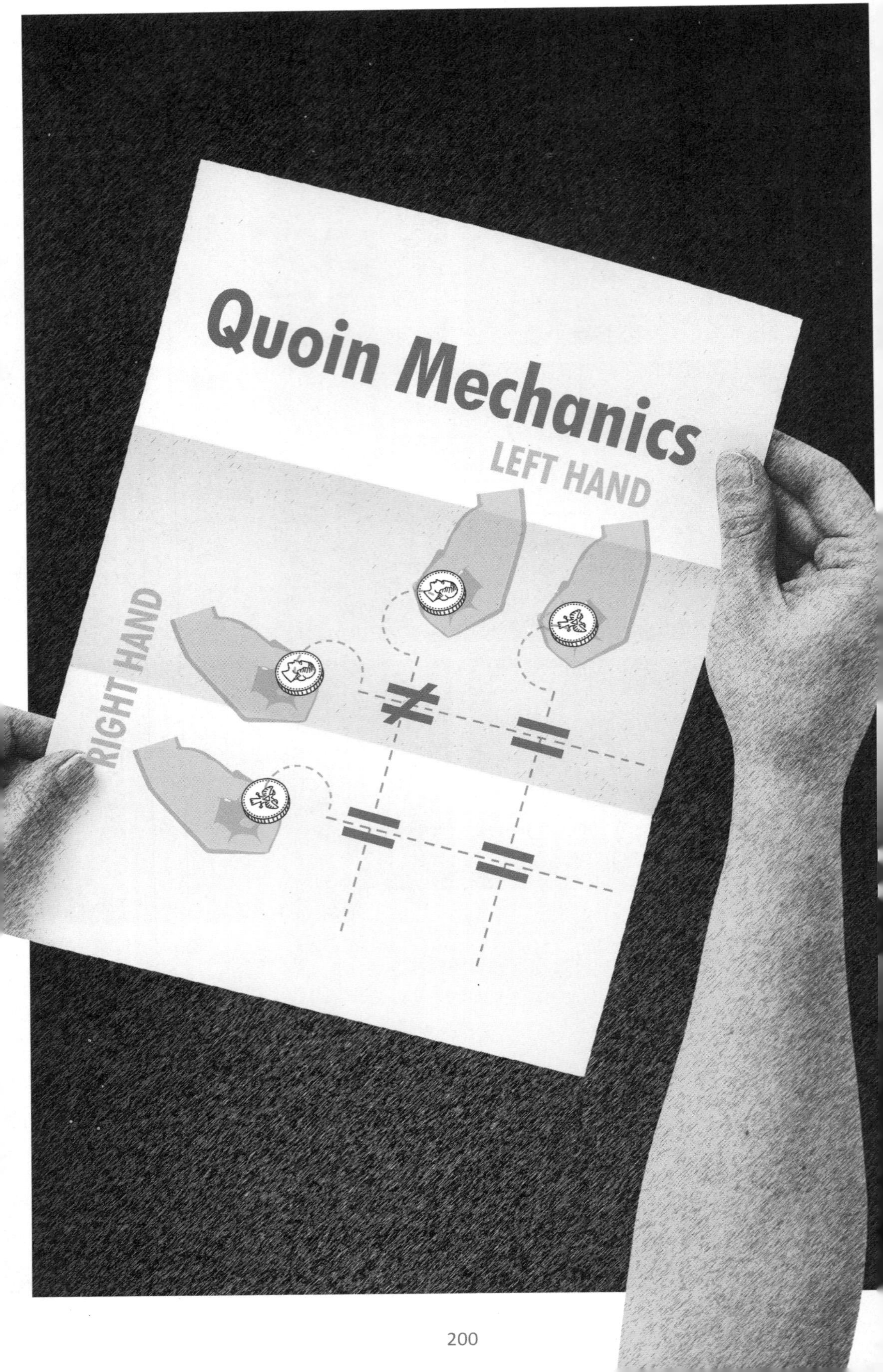
Quoin Mechanics
LEFT HAND
RIGHT HAND

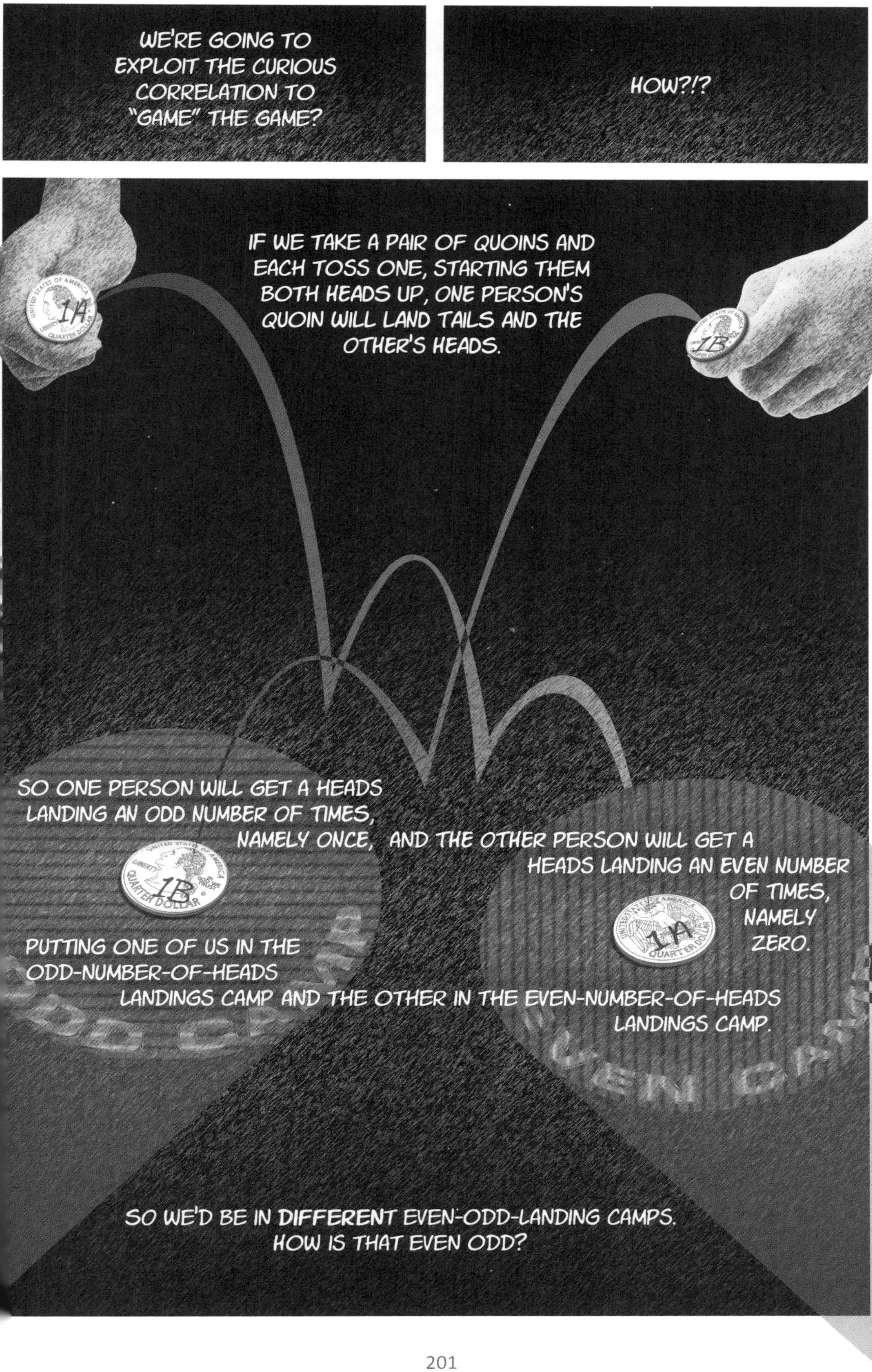

WE'RE GOING TO
EXPLOIT THE CURIOUS
CORRELATION TO
"GAME" THE GAME?

HOW?!?

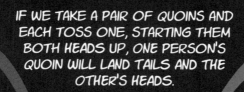

IF WE TAKE A PAIR OF QUOINS AND
EACH TOSS ONE, STARTING THEM
BOTH HEADS UP, ONE PERSON'S
QUOIN WILL LAND TAILS AND THE
OTHER'S HEADS.

SO ONE PERSON WILL GET A HEADS
LANDING AN ODD NUMBER OF TIMES,
NAMELY ONCE, AND THE OTHER PERSON WILL GET A
HEADS LANDING AN EVEN NUMBER
OF TIMES,
NAMELY
ZERO.

PUTTING ONE OF US IN THE
ODD-NUMBER-OF-HEADS
LANDINGS CAMP AND THE OTHER IN THE EVEN-NUMBER-OF-HEADS
LANDINGS CAMP.

SO WE'D BE IN **DIFFERENT** EVEN'-ODD-LANDING CAMPS.
HOW IS THAT EVEN ODD?

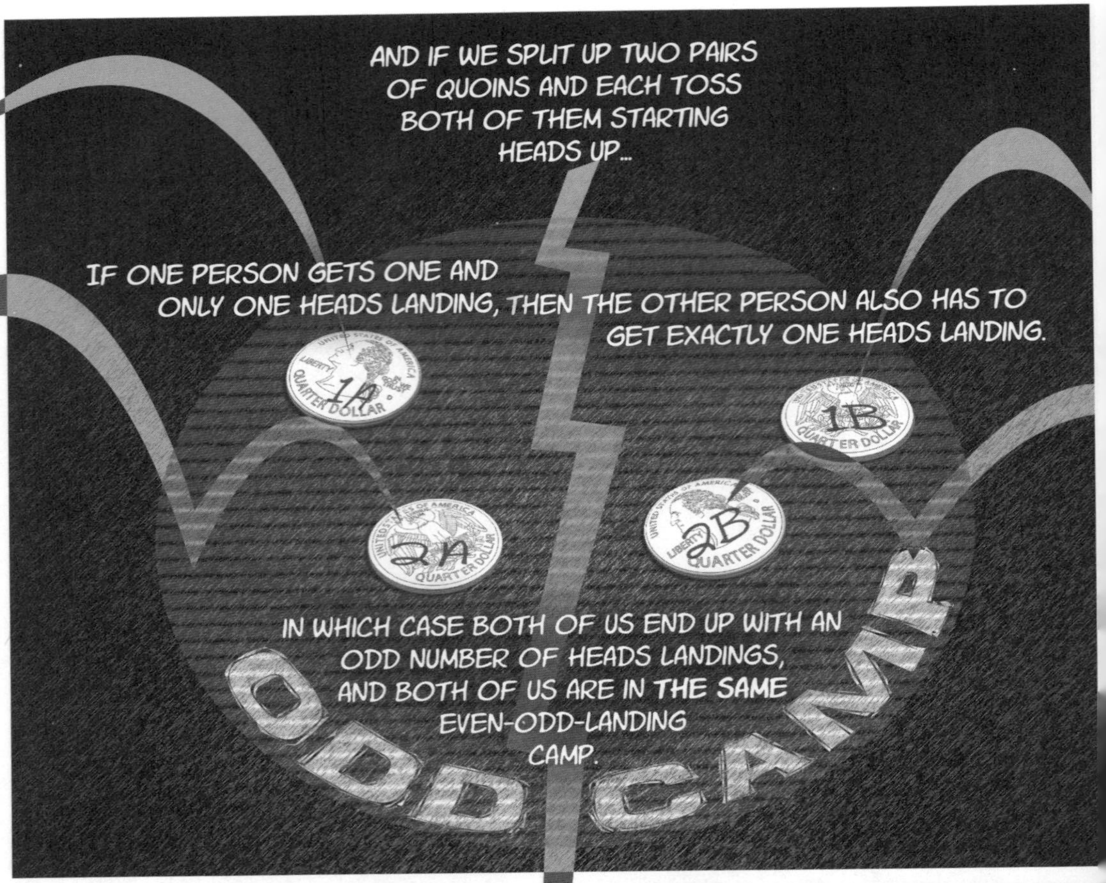
AND IF WE SPLIT UP TWO PAIRS OF QUOINS AND EACH TOSS BOTH OF THEM STARTING HEADS UP...
IF ONE PERSON GETS ONE AND ONLY ONE HEADS LANDING, THEN THE OTHER PERSON ALSO HAS TO GET EXACTLY ONE HEADS LANDING.
1A
1B
2A
2B
IN WHICH CASE BOTH OF US END UP WITH AN ODD NUMBER OF HEADS LANDINGS, AND BOTH OF US ARE IN THE SAME EVEN-ODD-LANDING CAMP.
ODD CAMP

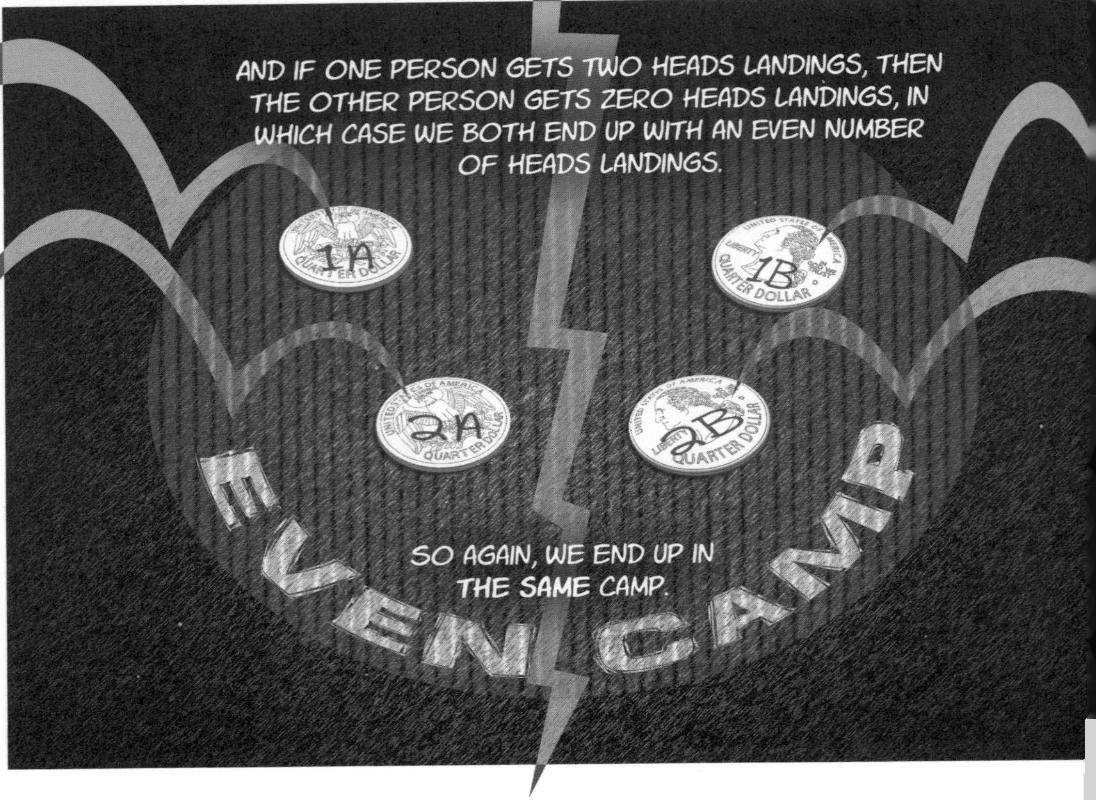
AND IF ONE PERSON GETS TWO HEADS LANDINGS, THEN THE OTHER PERSON GETS ZERO HEADS LANDINGS, IN WHICH CASE WE BOTH END UP WITH AN EVEN NUMBER OF HEADS LANDINGS.
1A
1B
2A
2B
SO AGAIN, WE END UP IN THE SAME CAMP.
EVEN CAMP

THIS WORKS NO MATTER HOW MANY TIMES
WE TOSS PAIRS OF QUOINS STARTING THEM
BOTH HEADS UP.

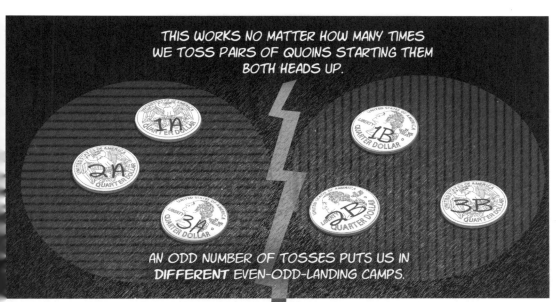

AN ODD NUMBER OF TOSSES PUTS US IN
DIFFERENT EVEN-ODD-LANDING CAMPS.

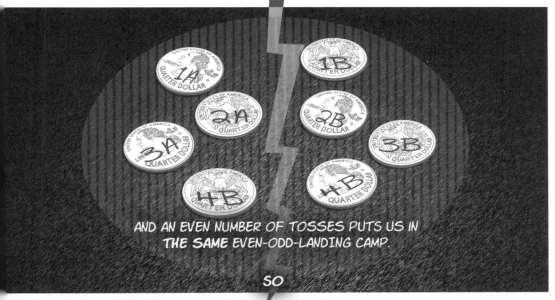

AND AN EVEN NUMBER OF TOSSES PUTS US IN
THE SAME EVEN-ODD-LANDING CAMP.

SO

IF WE FIND OURSELVES IN **THE
SAME** CAMP, THEN WE KNOW WE
MUST HAVE MADE AN **EVEN** NUMBER
OF HEADS/HEADS TOSSES...

AND IF WE ARE IN **DIFFERENT**
CAMPS, THEN THERE MUST HAVE
BEEN AN **ODD** NUMBER.

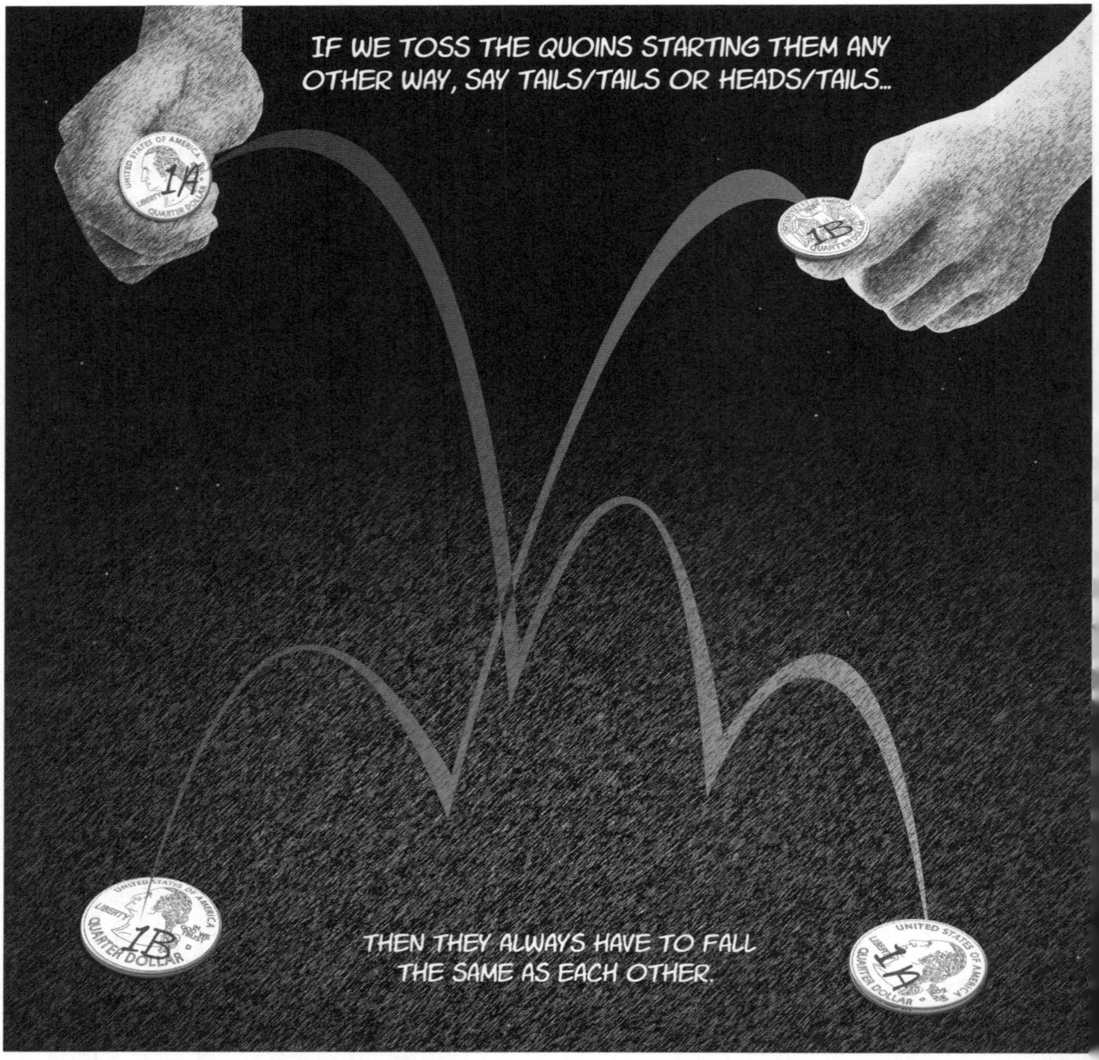
IF WE TOSS THE QUOINS STARTING THEM ANY OTHER WAY, SAY TAILS/TAILS OR HEADS/TAILS...
1A
1B
1B
THEN THEY ALWAYS HAVE TO FALL THE SAME AS EACH OTHER.
1A

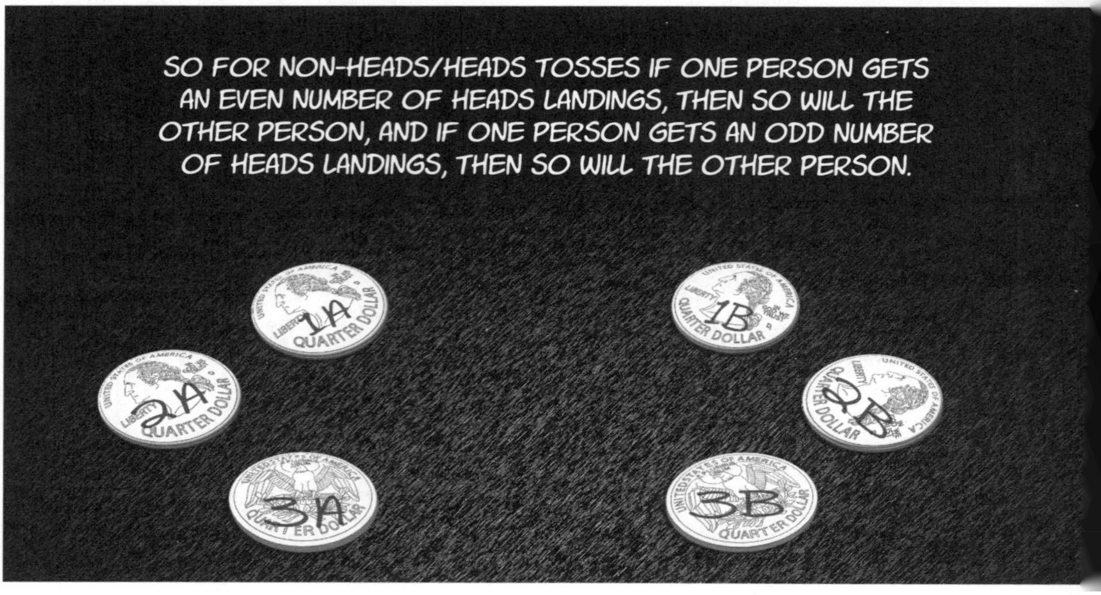
SO FOR NON-HEADS/HEADS TOSSES IF ONE PERSON GETS AN EVEN NUMBER OF HEADS LANDINGS, THEN SO WILL THE OTHER PERSON, AND IF ONE PERSON GETS AN ODD NUMBER OF HEADS LANDINGS, THEN SO WILL THE OTHER PERSON.
1A
1B
2A
2B
3A
3B

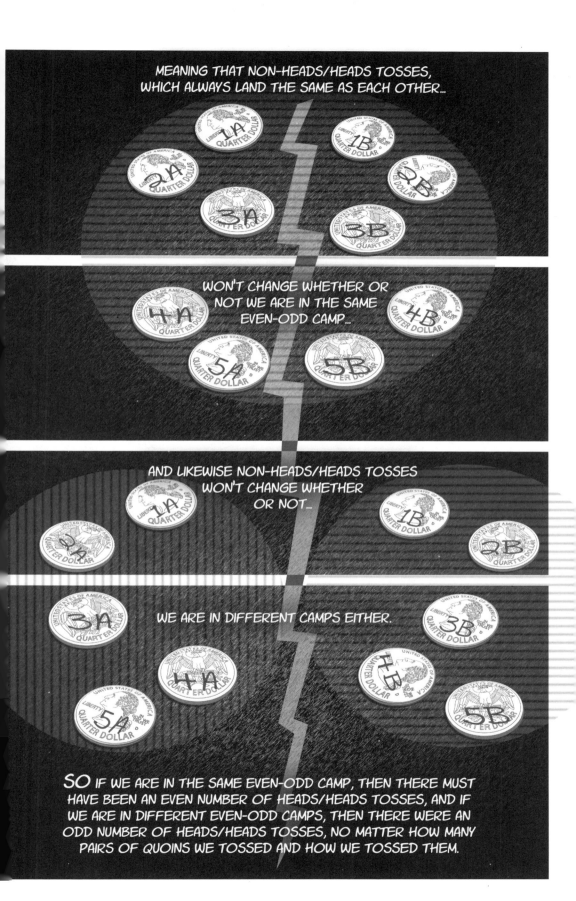

MEANING THAT NON-HEADS/HEADS TOSSES, WHICH ALWAYS LAND THE SAME AS EACH OTHER...

WON'T CHANGE WHETHER OR NOT WE ARE IN THE SAME EVEN-ODD CAMP...

AND LIKEWISE NON-HEADS/HEADS TOSSES WON'T CHANGE WHETHER OR NOT...

WE ARE IN DIFFERENT CAMPS EITHER.

SO IF WE ARE IN THE SAME EVEN-ODD CAMP, THEN THERE MUST HAVE BEEN AN EVEN NUMBER OF HEADS/HEADS TOSSES, AND IF WE ARE IN DIFFERENT EVEN-ODD CAMPS, THEN THERE WERE AN ODD NUMBER OF HEADS/HEADS TOSSES, NO MATTER HOW MANY PAIRS OF QUOINS WE TOSSED AND HOW WE TOSSED THEM.

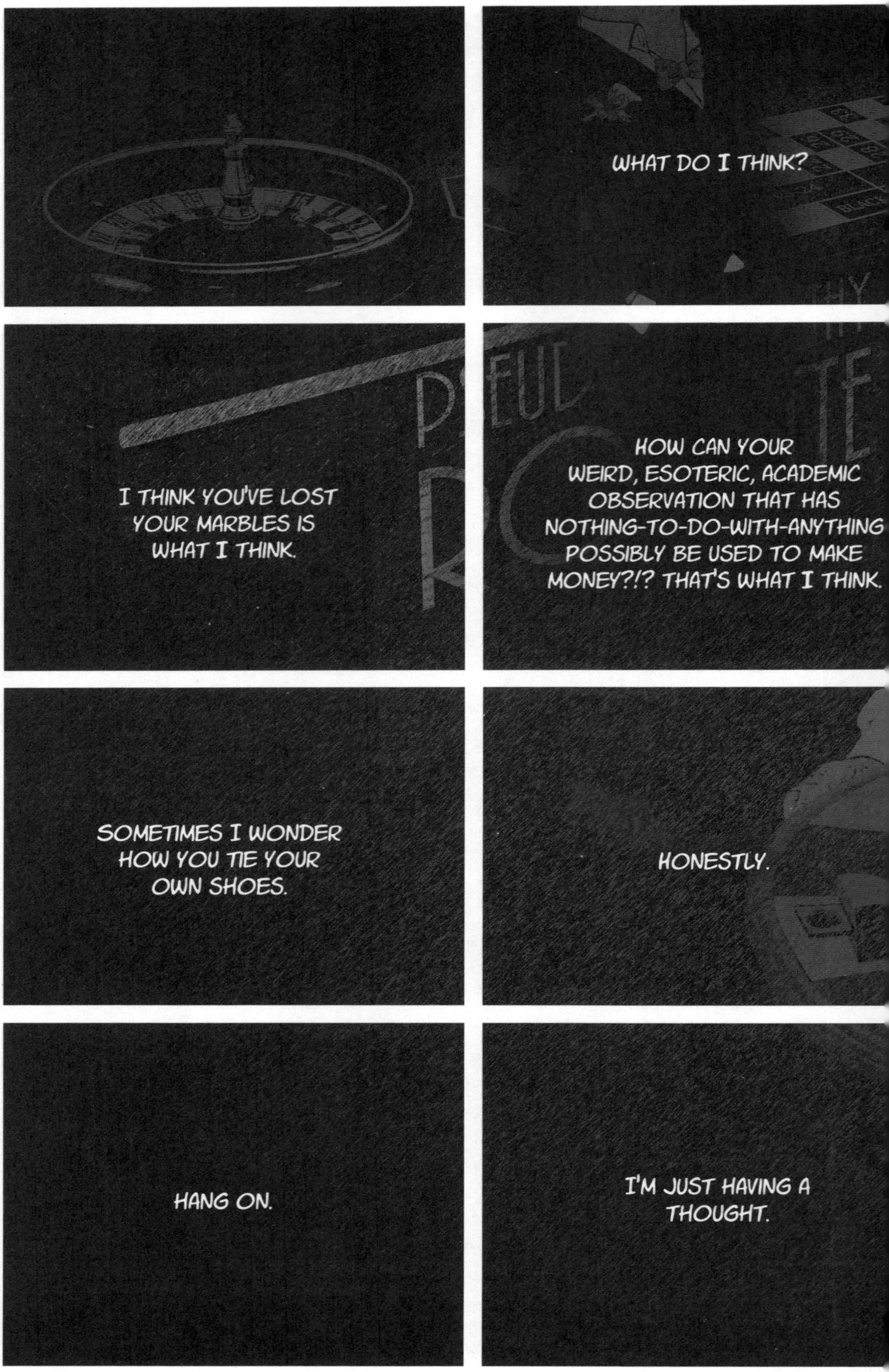
WHAT DO I THINK?
I THINK YOU'VE LOST YOUR MARBLES IS WHAT I THINK.
HOW CAN YOUR WEIRD, ESOTERIC, ACADEMIC OBSERVATION THAT HAS NOTHING-TO-DO-WITH-ANYTHING POSSIBLY BE USED TO MAKE MONEY?!? THAT'S WHAT I THINK.
SOMETIMES I WONDER HOW YOU TIE YOUR OWN SHOES.
HONESTLY.
HANG ON.
I'M JUST HAVING A THOUGHT.

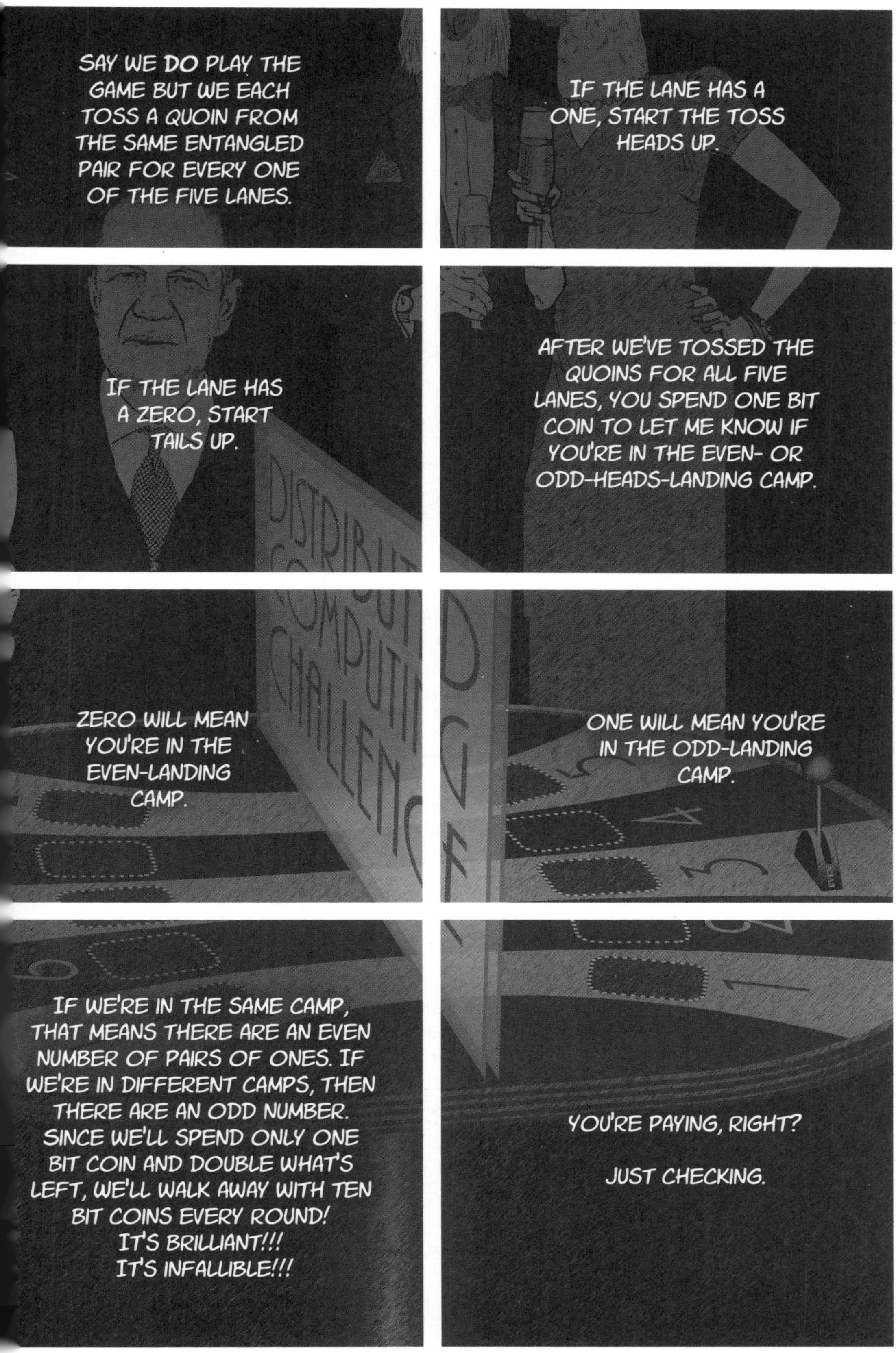
SAY WE DO PLAY THE GAME BUT WE EACH TOSS A QUOIN FROM THE SAME ENTANGLED PAIR FOR EVERY ONE OF THE FIVE LANES.
IF THE LANE HAS A ONE, START THE TOSS HEADS UP.
IF THE LANE HAS A ZERO, START TAILS UP.
AFTER WE'VE TOSSED THE QUOINS FOR ALL FIVE LANES, YOU SPEND ONE BIT COIN TO LET ME KNOW IF YOU'RE IN THE EVEN- OR ODD-HEADS-LANDING CAMP.
DISTRIBUTED COMPUTING CHALLENGE
ZERO WILL MEAN YOU'RE IN THE EVEN-LANDING CAMP.
ONE WILL MEAN YOU'RE IN THE ODD-LANDING CAMP.
IF WE'RE IN THE SAME CAMP, THAT MEANS THERE ARE AN EVEN NUMBER OF PAIRS OF ONES. IF WE'RE IN DIFFERENT CAMPS, THEN THERE ARE AN ODD NUMBER. SINCE WE'LL SPEND ONLY ONE BIT COIN AND DOUBLE WHAT'S LEFT, WE'LL WALK AWAY WITH TEN BIT COINS EVERY ROUND! IT'S BRILLIANT!!! IT'S INFALLIBLE!!!
YOU'RE PAYING, RIGHT?
JUST CHECKING.

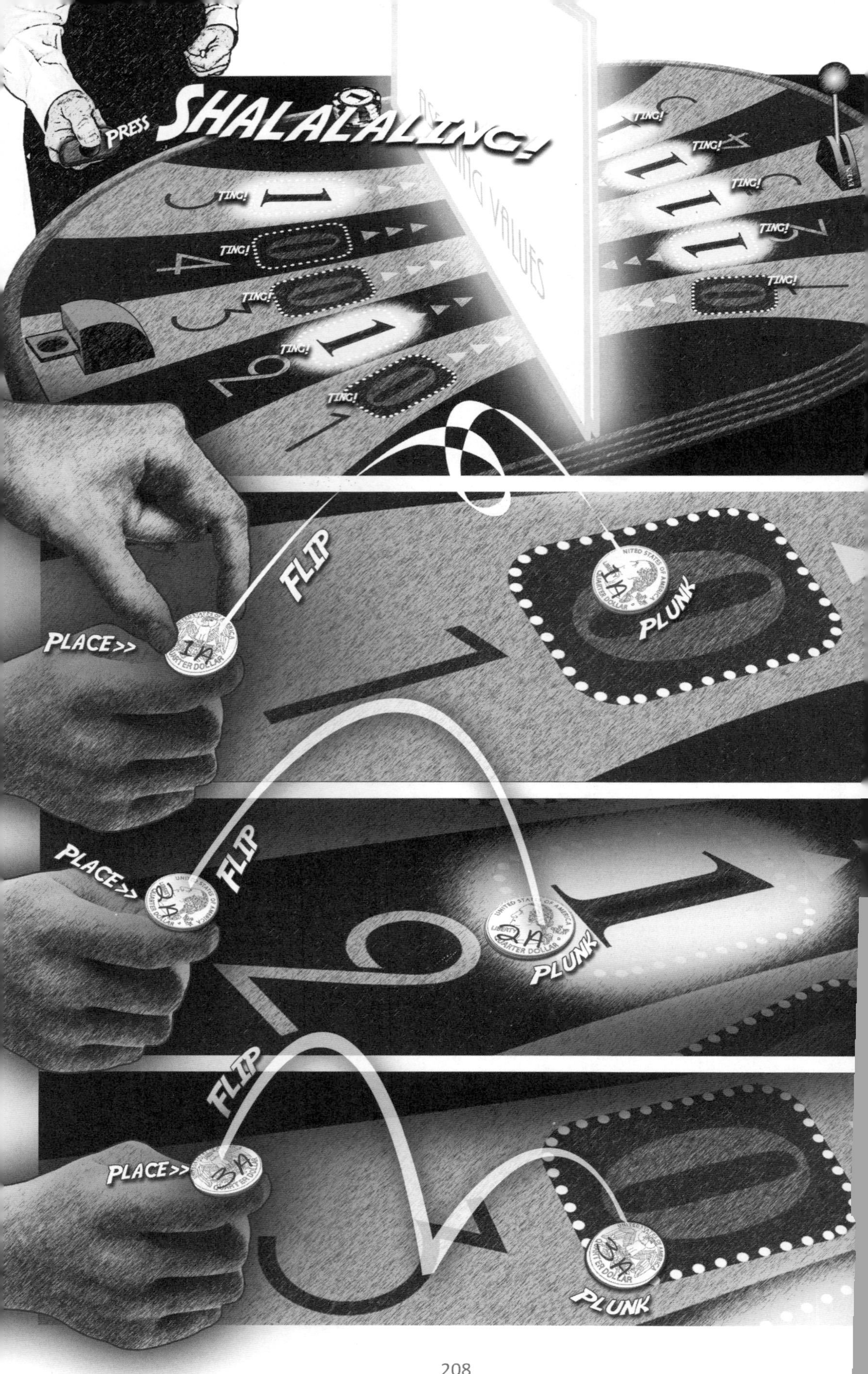
PRESS
SHALALALING!
TING!
TING!
TING!
TING!
TING!
TING!
TING!
TING!
TING!
TING!
PLACE>>
1A
FLIP
1A
PLUNK
PLACE>>
2A
FLIP
2A
PLUNK
FLIP
PLACE>>
3A
3A
PLUNK

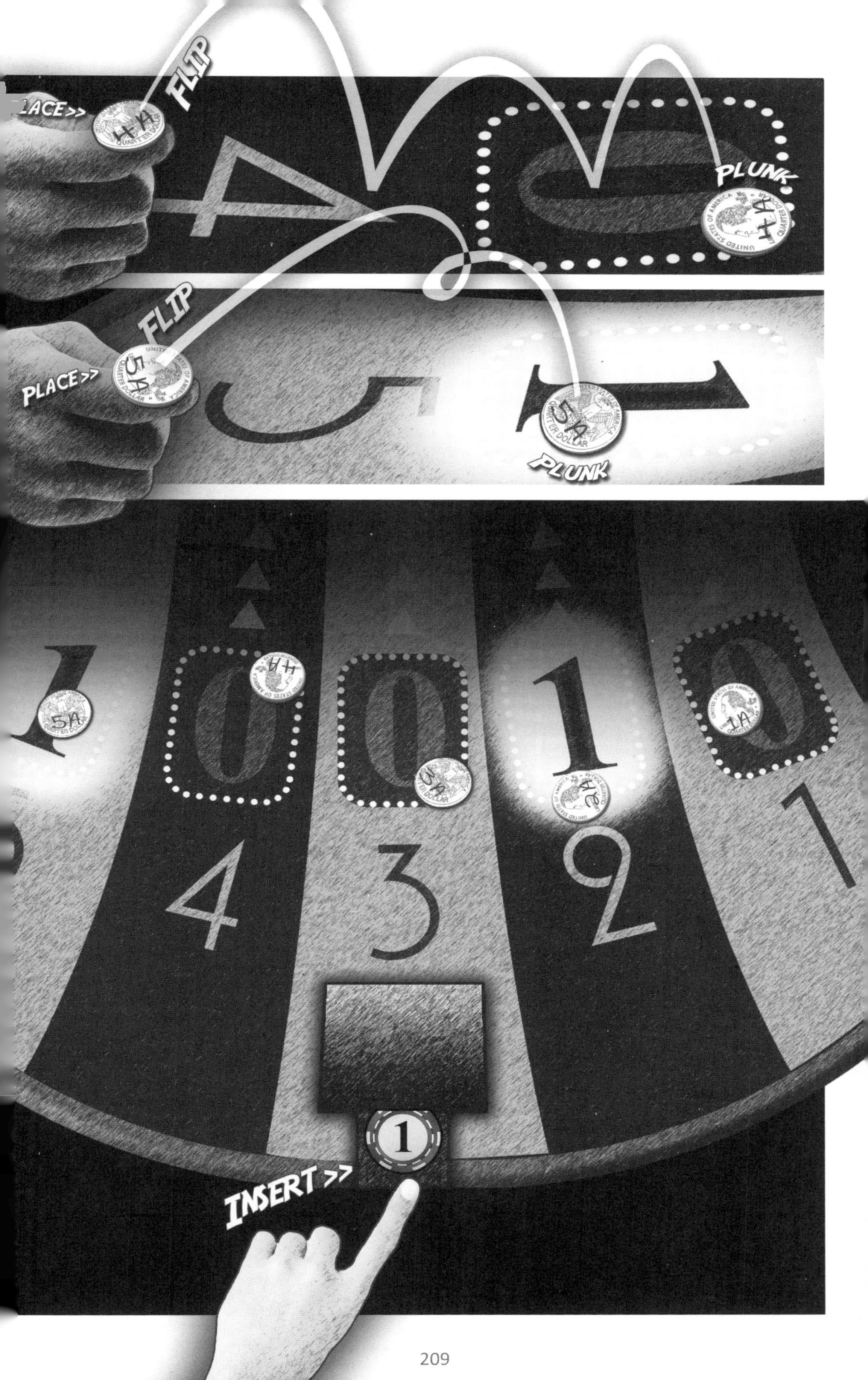

PLACE>>
FLIP
PLUNK
FLIP
PLACE>>
PLUNK
INSERT>>

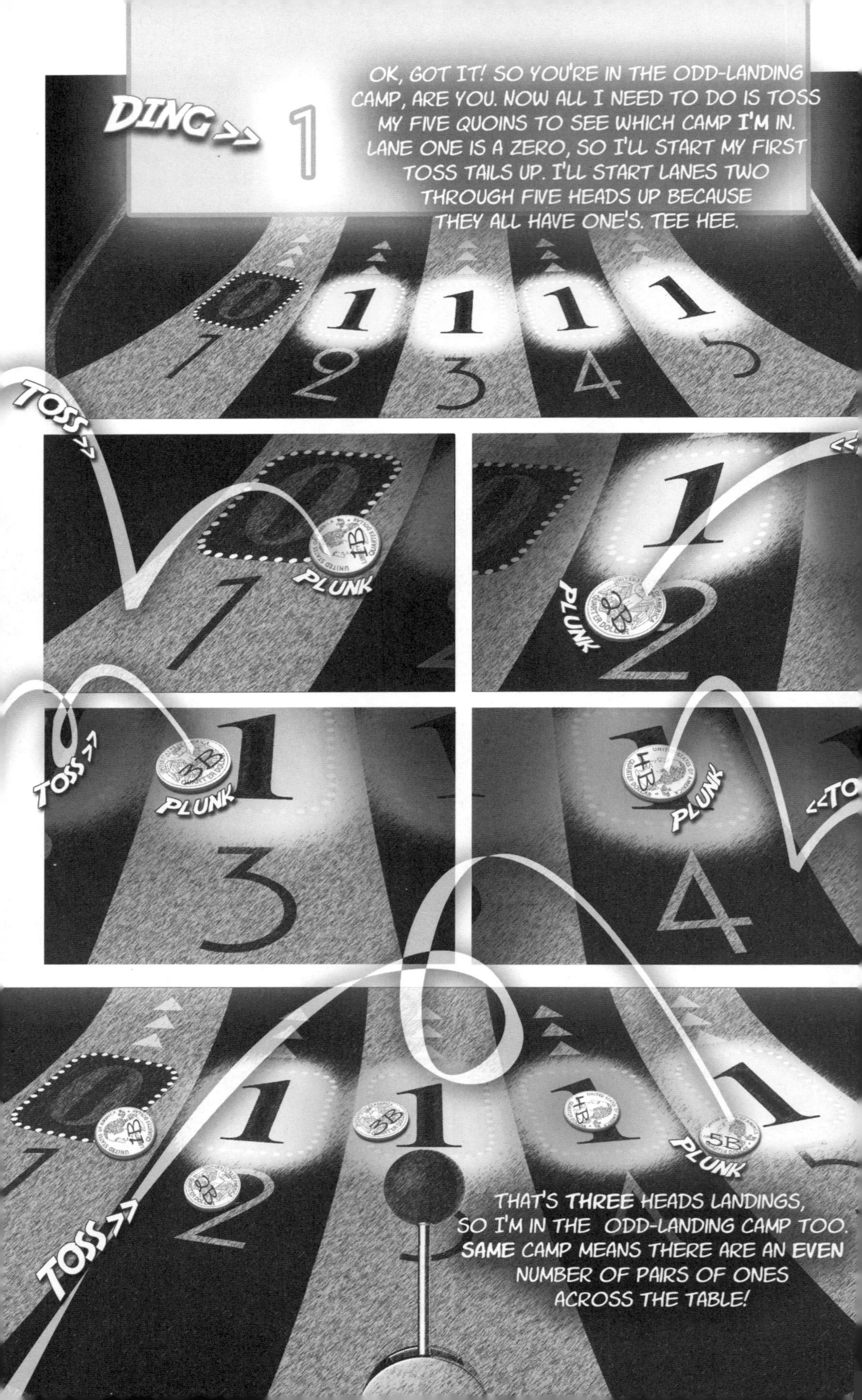
DING >> 1
OK, GOT IT! SO YOU'RE IN THE ODD-LANDING CAMP, ARE YOU. NOW ALL I NEED TO DO IS TOSS MY FIVE QUOINS TO SEE WHICH CAMP **I'M** IN. LANE ONE IS A ZERO, SO I'LL START MY FIRST TOSS TAILS UP. I'LL START LANES TWO THROUGH FIVE HEADS UP BECAUSE THEY ALL HAVE ONE'S. TEE HEE.
0 1 1 1 1
1 2 3 4 5
TOSS>>
PLUNK
PLUNK
TOSS >>
PLUNK
PLUNK
PLUNK
TOSS>>
THAT'S **THREE** HEADS LANDINGS, SO I'M IN THE ODD-LANDING CAMP TOO. **SAME** CAMP MEANS THERE ARE AN **EVEN** NUMBER OF PAIRS OF ONES ACROSS THE TABLE!

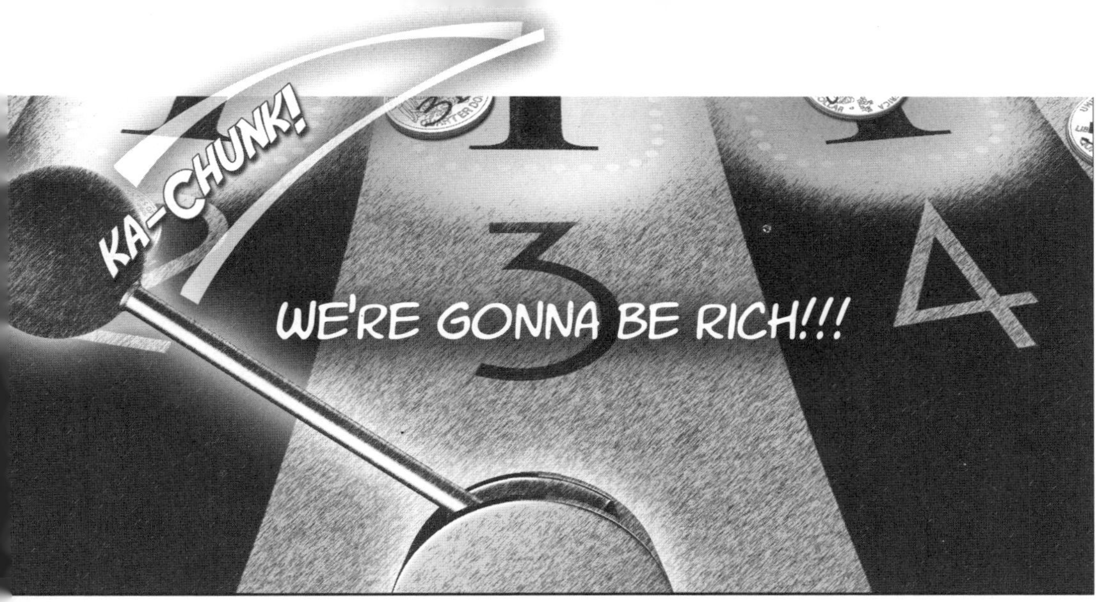
KA-CHUNK!
WE'RE GONNA BE RICH!!!

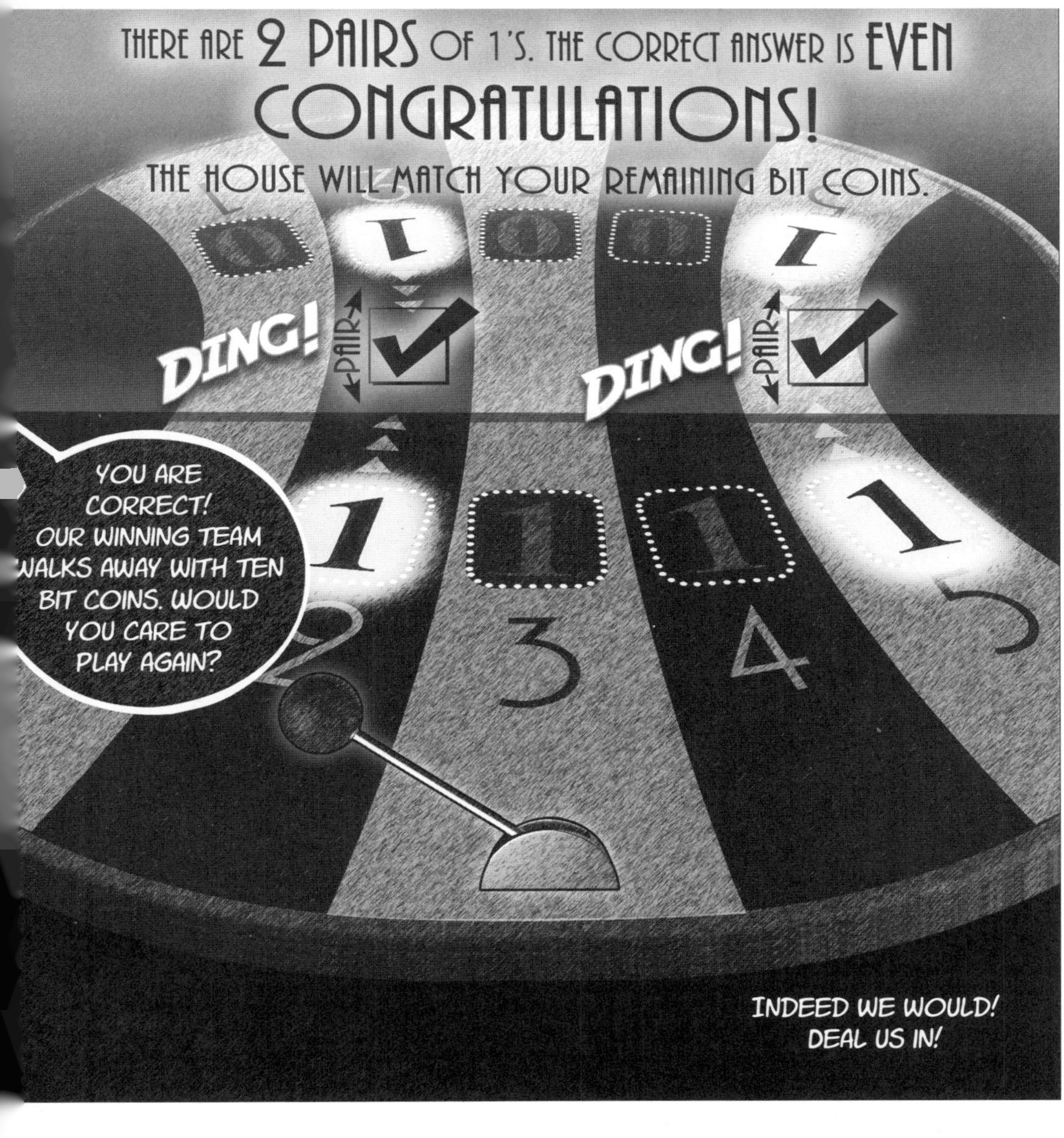
THERE ARE 2 PAIRS OF 1'S. THE CORRECT ANSWER IS EVEN
CONGRATULATIONS!
THE HOUSE WILL MATCH YOUR REMAINING BIT COINS.
DING!
PAIR
DING!
PAIR
YOU ARE CORRECT! OUR WINNING TEAM WALKS AWAY WITH TEN BIT COINS. WOULD YOU CARE TO PLAY AGAIN?
INDEED WE WOULD! DEAL US IN!

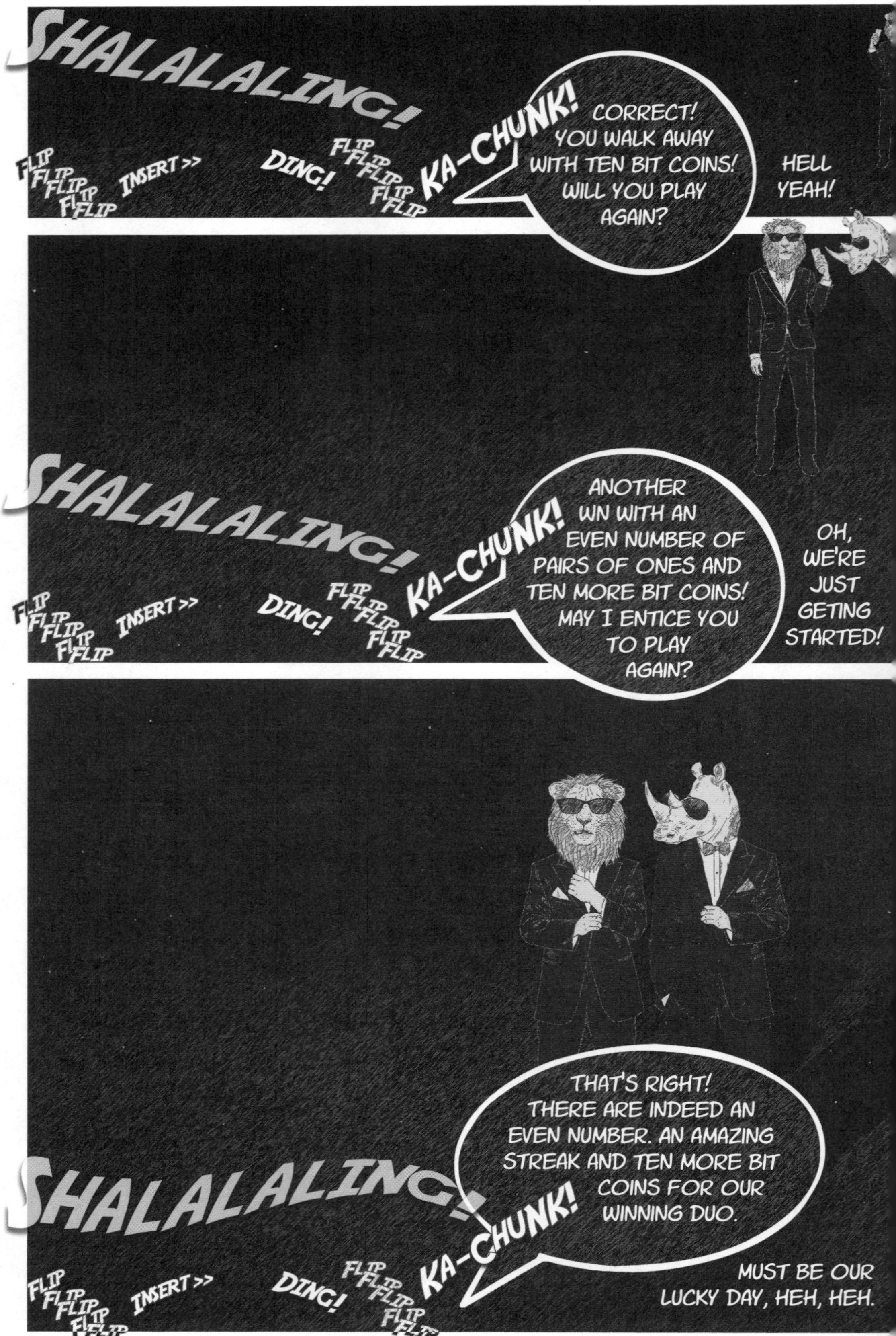
SHALALALING!
FLIP FLIP FLIP FLIP FLIP INSERT >> DING! FLIP FLIP FLIP FLIP FLIP
KA-CHUNK!
CORRECT! YOU WALK AWAY WITH TEN BIT COINS! WILL YOU PLAY AGAIN?
HELL YEAH!
SHALALALING!
FLIP FLIP FLIP FLIP FLIP INSERT >> DING! FLIP FLIP FLIP FLIP FLIP
KA-CHUNK!
ANOTHER WN WITH AN EVEN NUMBER OF PAIRS OF ONES AND TEN MORE BIT COINS! MAY I ENTICE YOU TO PLAY AGAIN?
OH, WE'RE JUST GETING STARTED!
SHALALALING!
FLIP FLIP FLIP FLIP FLIP INSERT >> DING! FLIP FLIP FLIP FLIP FLIP
KA-CHUNK!
THAT'S RIGHT! THERE ARE INDEED AN EVEN NUMBER. AN AMAZING STREAK AND TEN MORE BIT COINS FOR OUR WINNING DUO.
MUST BE OUR LUCKY DAY, HEH, HEH.

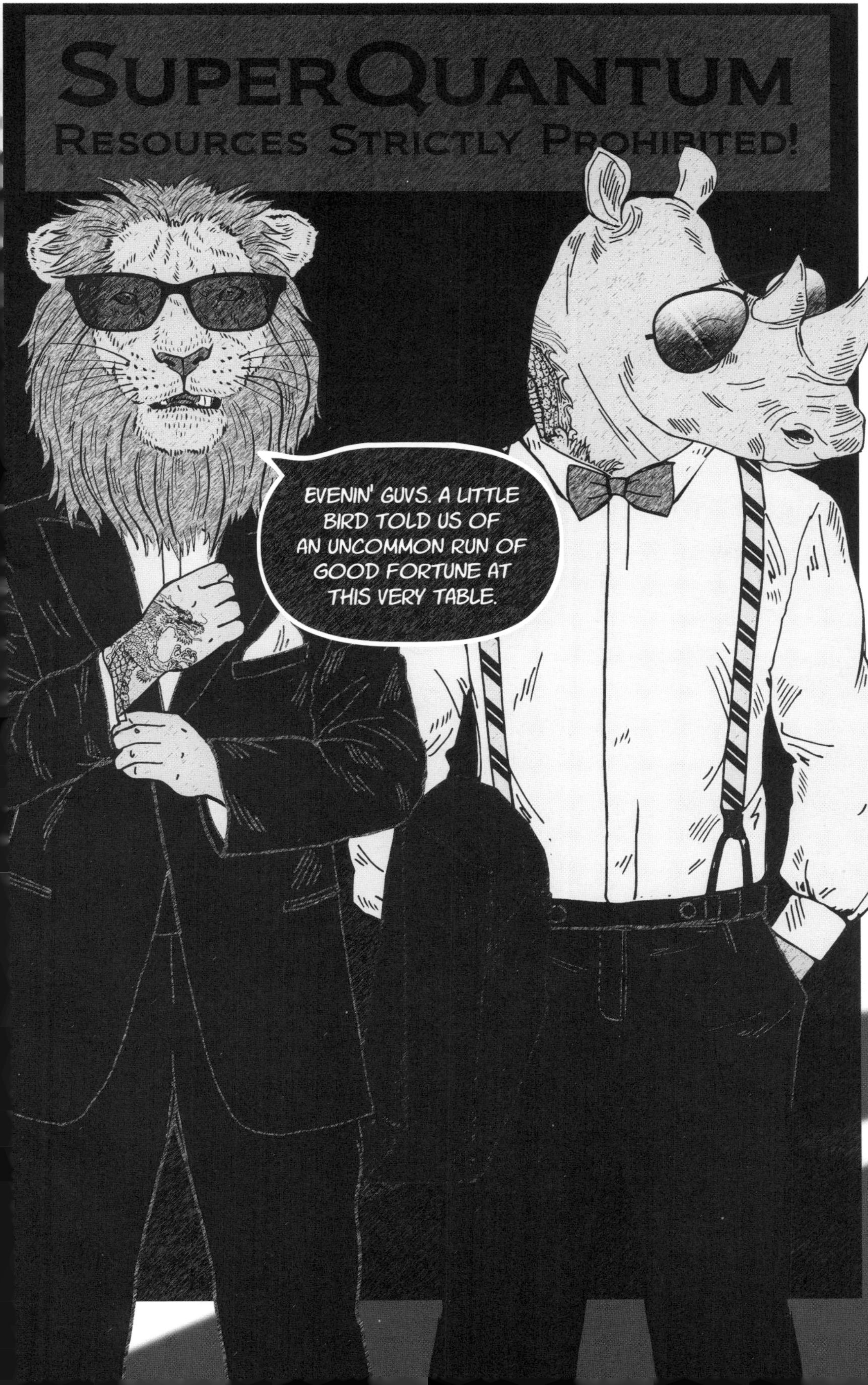
SuperQuantum
Resources Strictly Prohibited!
EVENIN' GUVS. A LITTLE BIRD TOLD US OF AN UNCOMMON RUN OF GOOD FORTUNE AT THIS VERY TABLE.

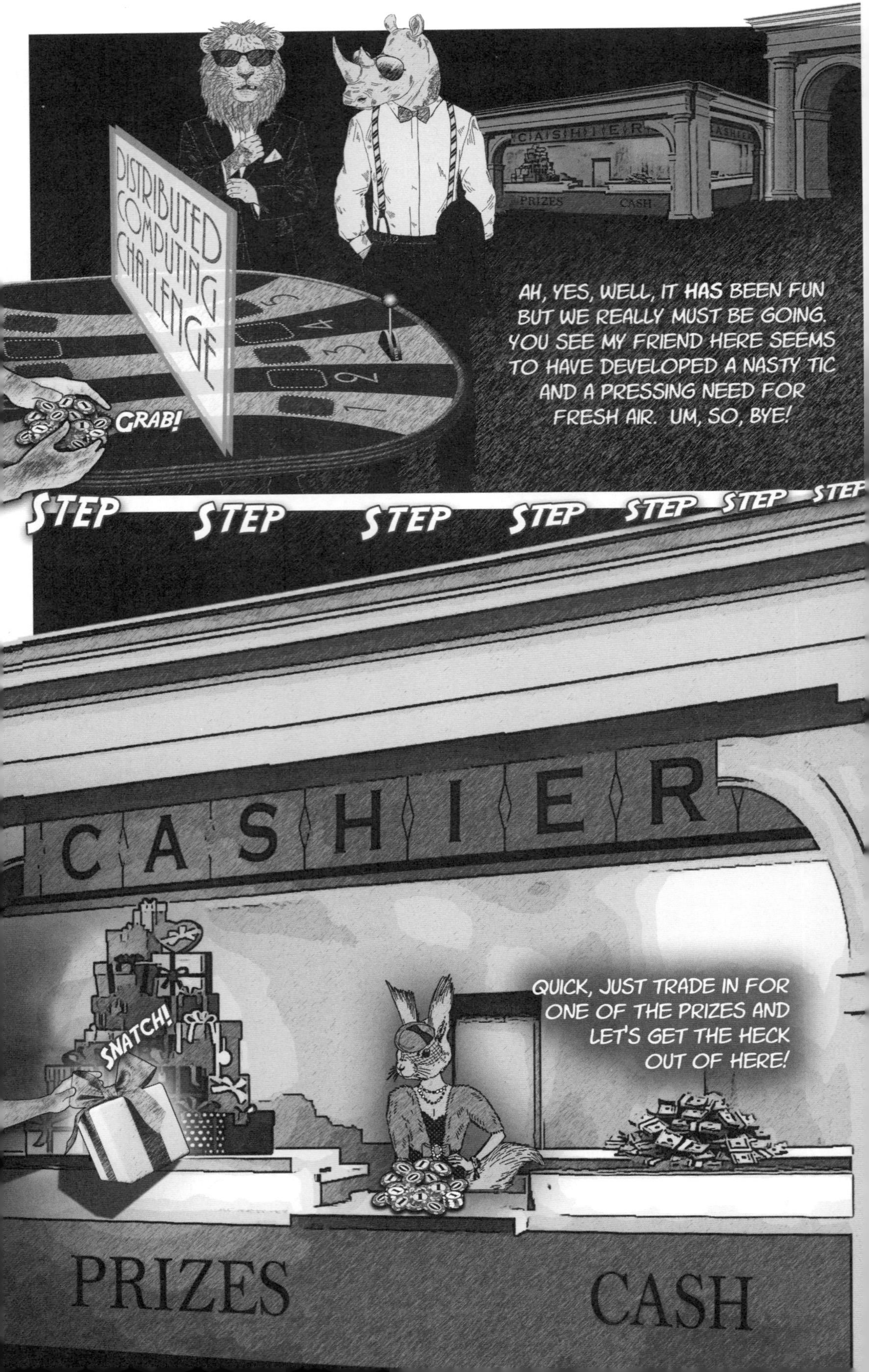
DISTRIBUTED COMPUTING CHALLENGE
CASHIER
PRIZES
CASH
GRAB!
AH, YES, WELL, IT **HAS** BEEN FUN BUT WE REALLY MUST BE GOING. YOU SEE MY FRIEND HERE SEEMS TO HAVE DEVELOPED A NASTY TIC AND A PRESSING NEED FOR FRESH AIR. UM, SO, BYE!
STEP STEP STEP STEP STEP STEP STEP
CASHIER
SNATCH!
QUICK, JUST TRADE IN FOR ONE OF THE PRIZES AND LET'S GET THE HECK OUT OF HERE!
PRIZES
CASH

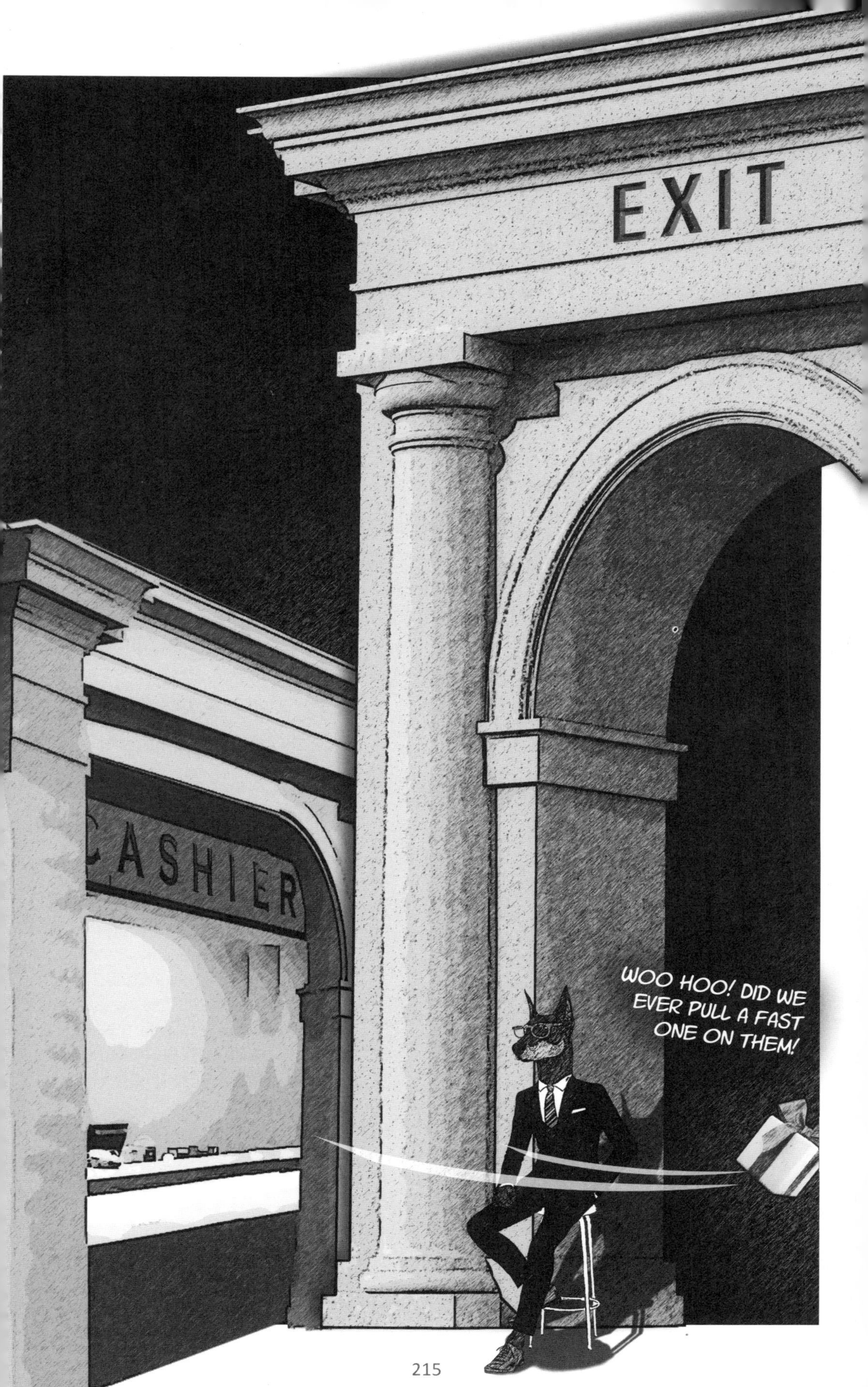
EXIT
CASHIER
WOO HOO! DID WE EVER PULL A FAST ONE ON THEM!

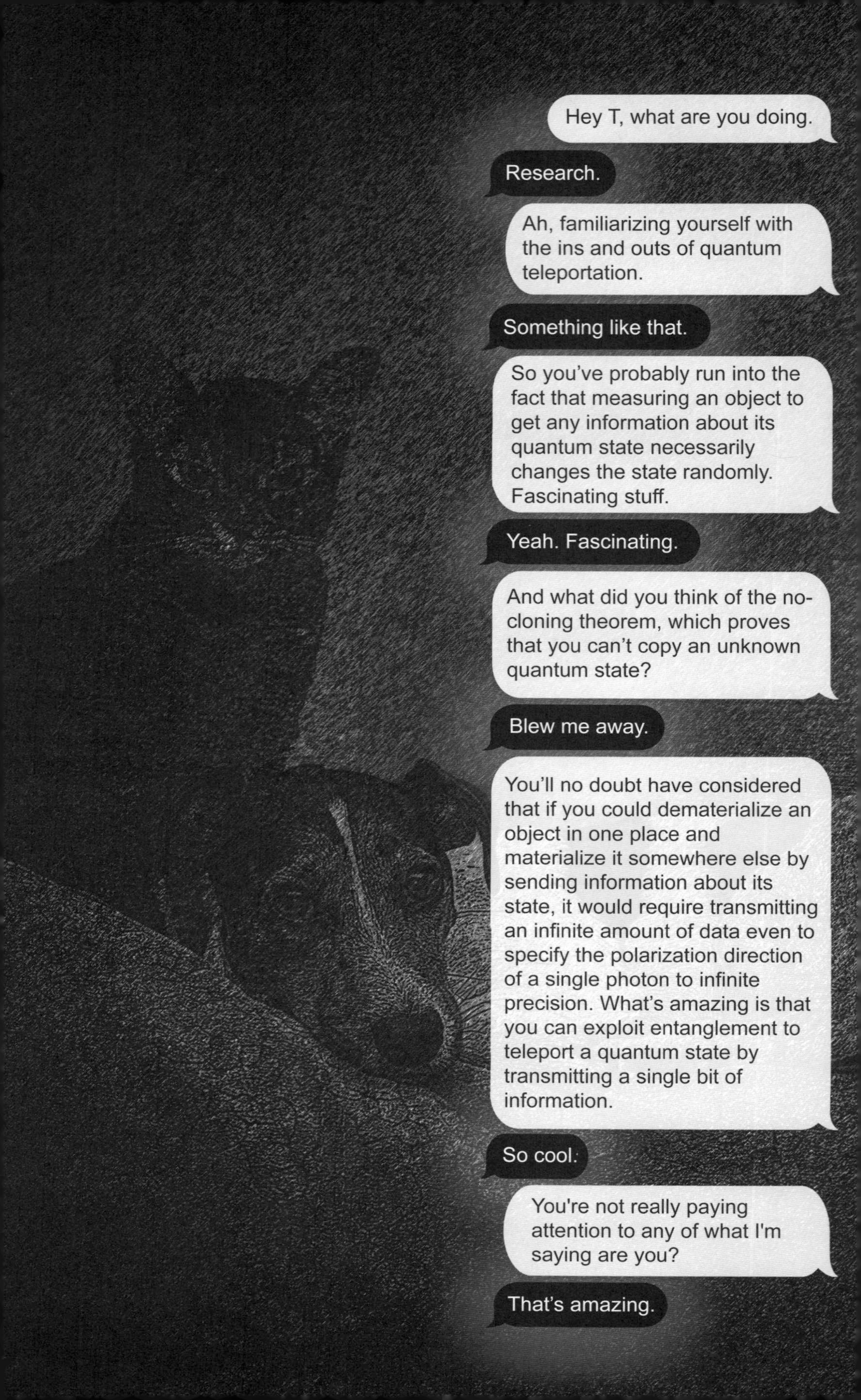
Hey T, what are you doing.
Research.
Ah, familiarizing yourself with the ins and outs of quantum teleportation.
Something like that.
So you’ve probably run into the fact that measuring an object to get any information about its quantum state necessarily changes the state randomly. Fascinating stuff.
Yeah. Fascinating.
And what did you think of the no-cloning theorem, which proves that you can’t copy an unknown quantum state?
Blew me away.
You’ll no doubt have considered that if you could dematerialize an object in one place and materialize it somewhere else by sending information about its state, it would require transmitting an infinite amount of data even to specify the polarization direction of a single photon to infinite precision. What’s amazing is that you can exploit entanglement to teleport a quantum state by transmitting a single bit of information.
So cool.
You're not really paying attention to any of what I'm saying are you?
That’s amazing.

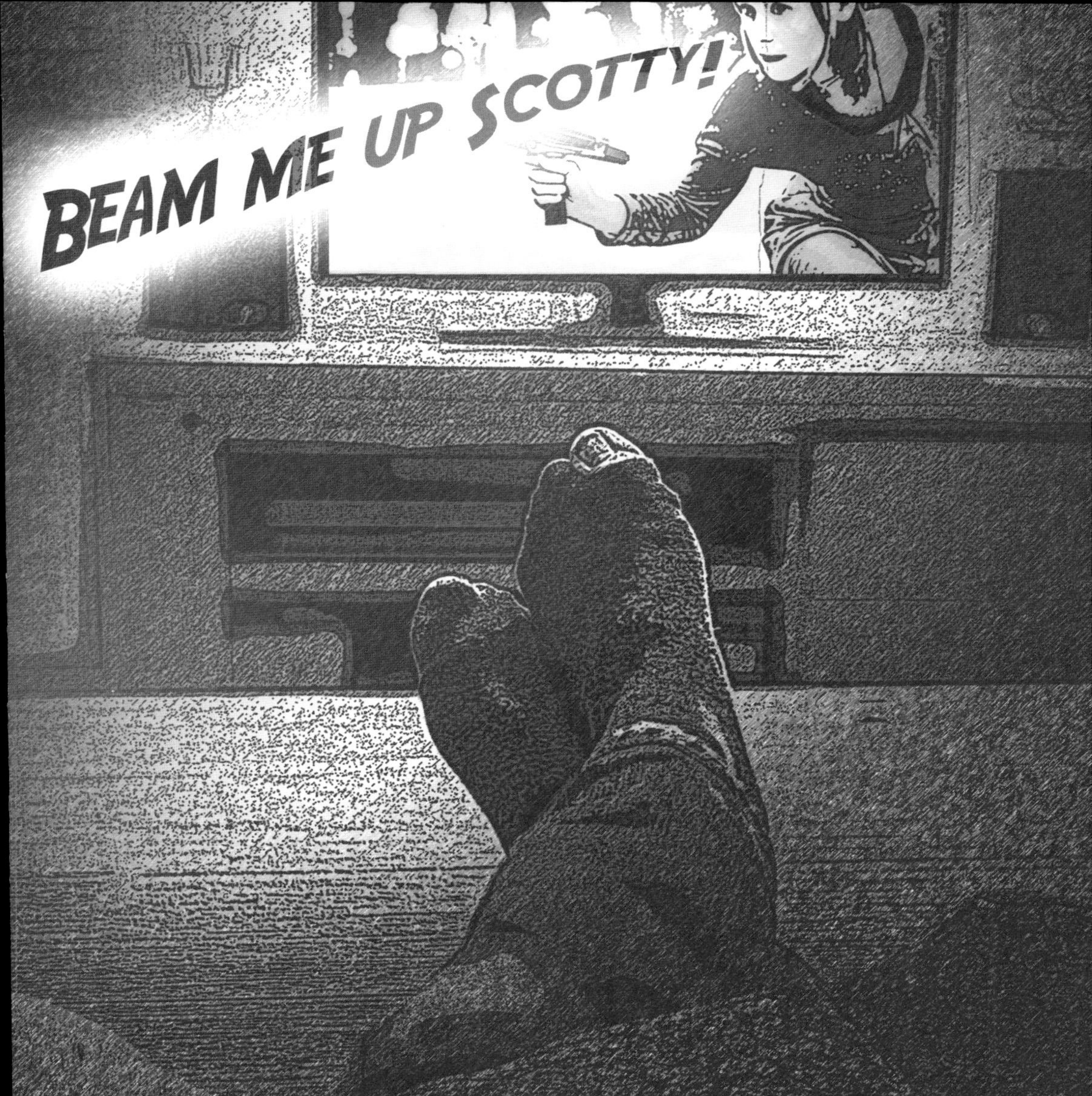
BEAM ME UP SCOTTY!

Soul (state) Teleportation

Later, when a newsman asked me whether it was possible to teleport not only the body but also the soul, I answered "only the soul." Even that is a gross oversimplification.

—Asher Peres

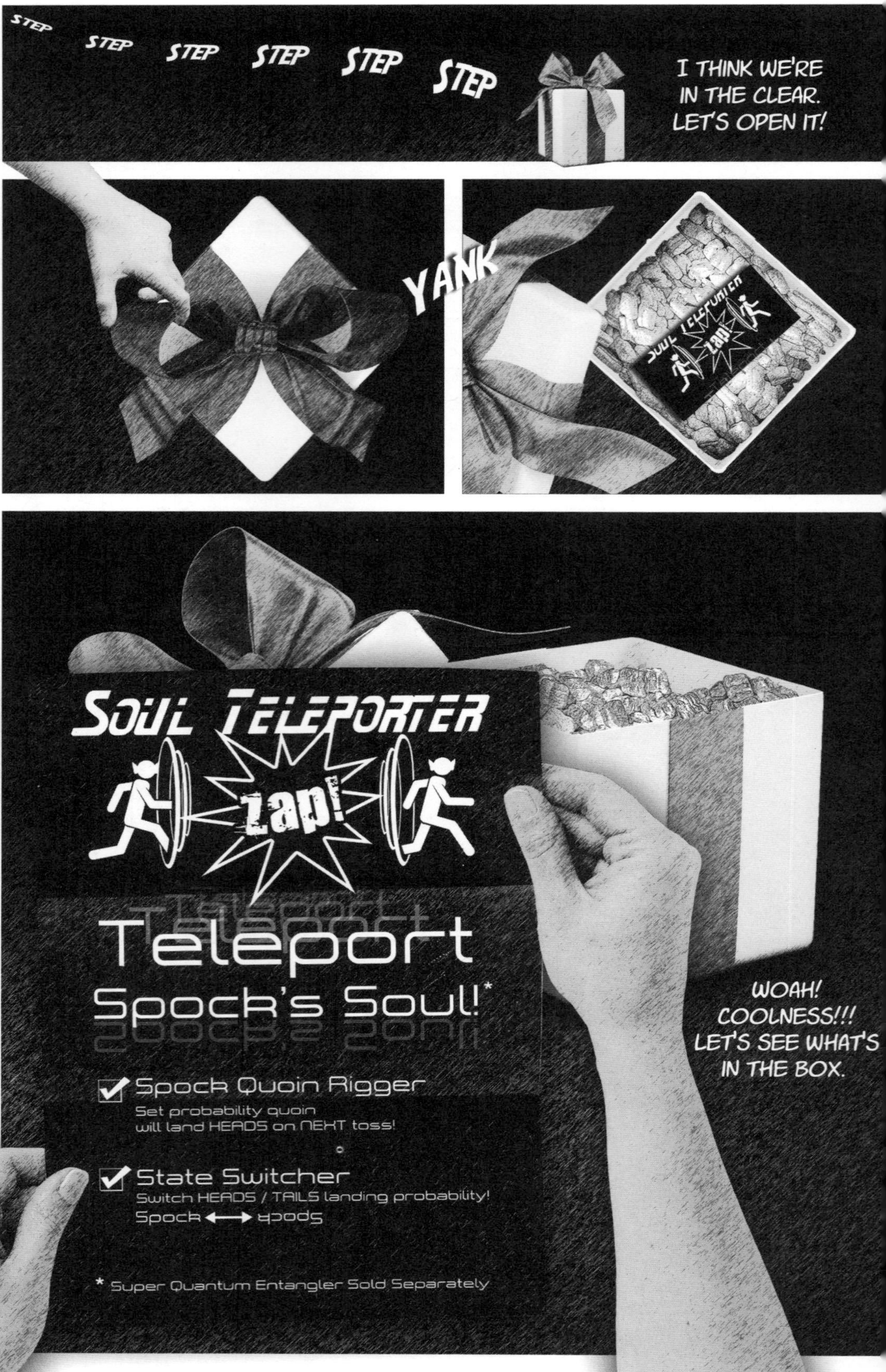
STEP
STEP
STEP
STEP
STEP
STEP
I THINK WE'RE IN THE CLEAR. LET'S OPEN IT!
YANK
SOUL TELEPORTER
zap!
Teleport
Spock's Soul!*
Spock Quoin Rigger
Set probability quoin will land HEADS on NEXT toss!
State Switcher
Switch HEADS / TAILS landing probability!
* Super Quantum Entangler Sold Separately
WOAH! COOLNESS!!! LET'S SEE WHAT'S IN THE BOX.

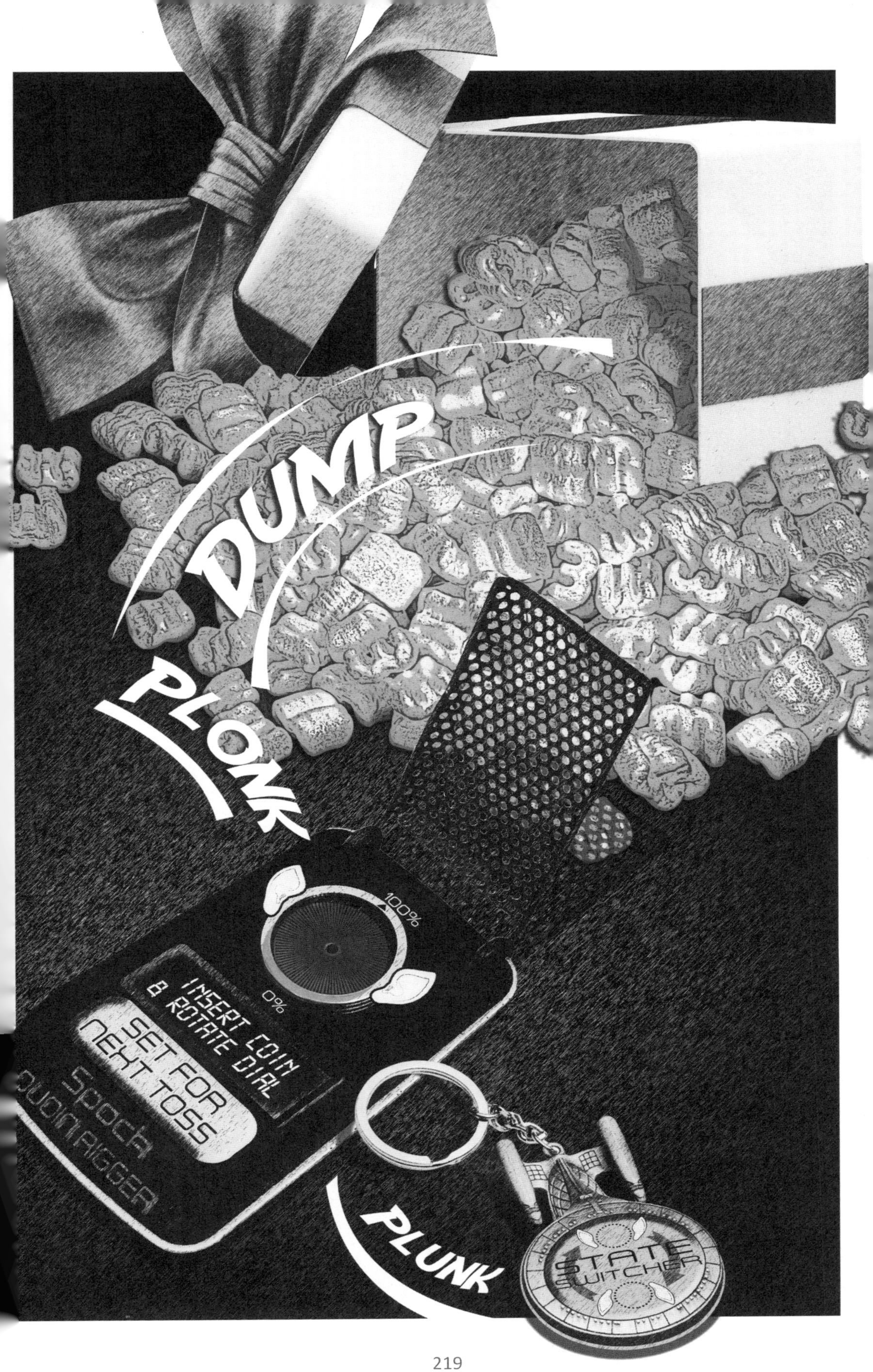
DUMP
PLONK
100%
0%
INSERT COIN
& ROTATE DIAL
SET FOR
NEXT TOSS
Spock
PLUNK
STATE
SWITCHER

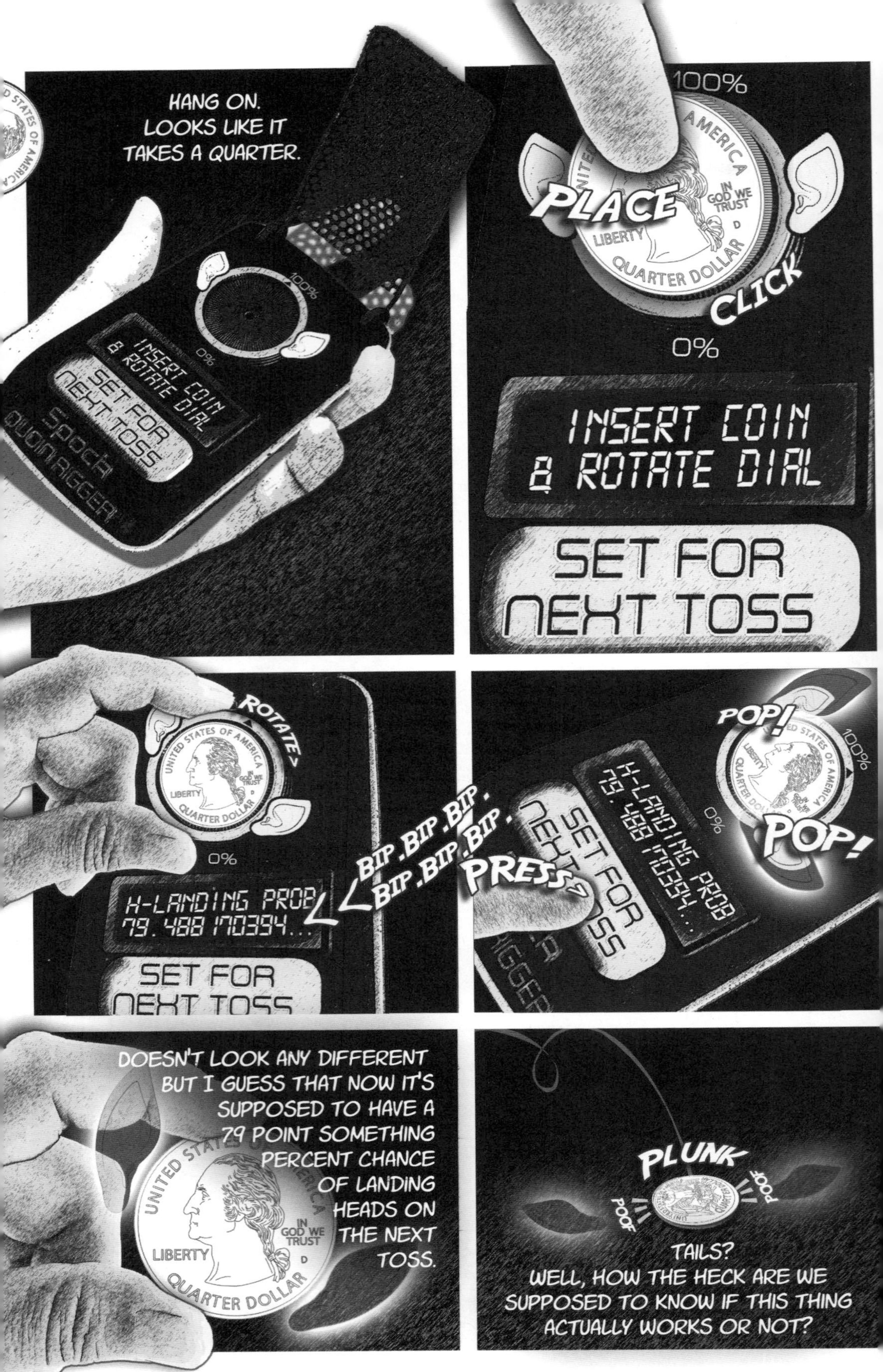
HANG ON. LOOKS LIKE IT TAKES A QUARTER.
100%
0%
INSERT COIN & ROTATE DIAL
SET FOR NEXT TOSS
100%
PLACE
UNITED STATES OF AMERICA
IN GOD WE TRUST
LIBERTY
QUARTER DOLLAR
CLICK
0%
INSERT COIN & ROTATE DIAL
SET FOR NEXT TOSS
ROTATE>
0%
H-LANDING PROB 79.488170394...
BIP.BIP.BIP.
BIP.BIP.BIP.
SET FOR NEXT TOSS
PRESS>
POP!
POP!
100%
0%
H-LANDING PROB 79.488170394...
SET FOR NEXT TOSS
DOESN'T LOOK ANY DIFFERENT BUT I GUESS THAT NOW IT'S SUPPOSED TO HAVE A 79 POINT SOMETHING PERCENT CHANCE OF LANDING HEADS ON THE NEXT TOSS.
PLUNK
POOF
POOF
TAILS?
WELL, HOW THE HECK ARE WE SUPPOSED TO KNOW IF THIS THING ACTUALLY WORKS OR NOT?

50 BILLION REPEATS LATER

YUP, IT WORKS.

SO I GUESS A **"SPOCK"** QUOIN IS JUST A COIN THAT'S RIGGED TO LAND HEADS WITH WHATEVER ODDS YOU SET ON THE DIAL.

WHAT'S THE STATE SWITCHER THING DO?

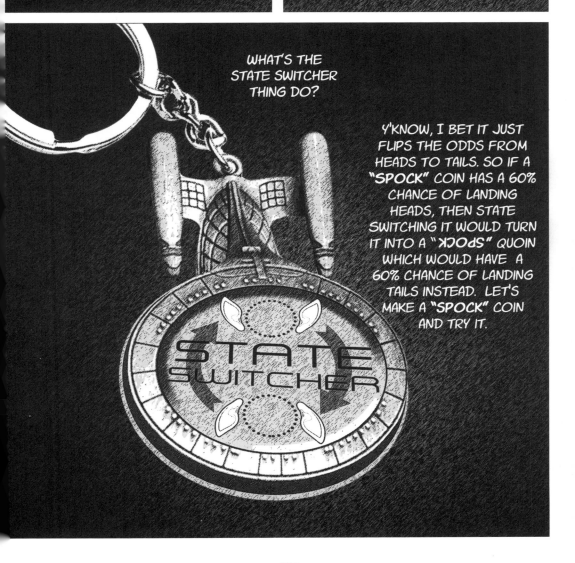

Y'KNOW, I BET IT JUST FLIPS THE ODDS FROM HEADS TO TAILS. SO IF A **"SPOCK"** COIN HAS A 60% CHANCE OF LANDING HEADS, THEN STATE SWITCHING IT WOULD TURN IT INTO A **"ꓘↃOꟼƧ"** QUOIN WHICH WOULD HAVE A 60% CHANCE OF LANDING TAILS INSTEAD. LET'S MAKE A **"SPOCK"** COIN AND TRY IT.

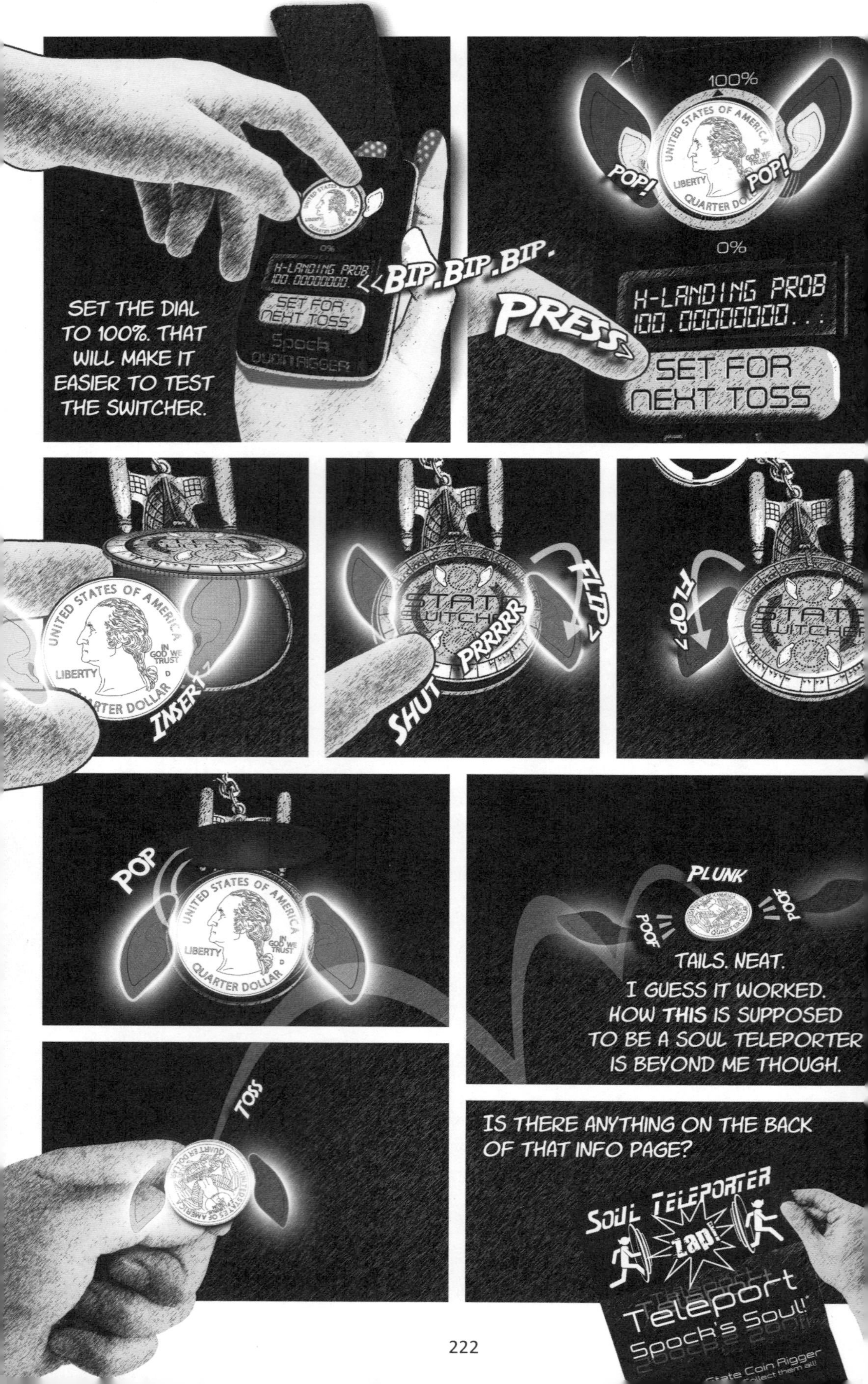

SET THE DIAL TO 100%. THAT WILL MAKE IT EASIER TO TEST THE SWITCHER.
0%
H-LANDING PROB 100.00000000.
BIP.BIP.BIP.
SET FOR NEXT TOSS
Spock
100%
POP!
POP!
0%
H-LANDING PROB 100.00000000...
PRESS
SET FOR NEXT TOSS
INSERT
SHUT
PRRRRR
FLIP
FLOP
STATE SWITCHER
POP
TOSS
PLUNK
POOF
POOF
TAILS. NEAT.
I GUESS IT WORKED.
HOW THIS IS SUPPOSED TO BE A SOUL TELEPORTER IS BEYOND ME THOUGH.
IS THERE ANYTHING ON THE BACK OF THAT INFO PAGE?
SOUL TELEPORTER
zap!
Teleport
Spock's Soul!
State Coin Rigger

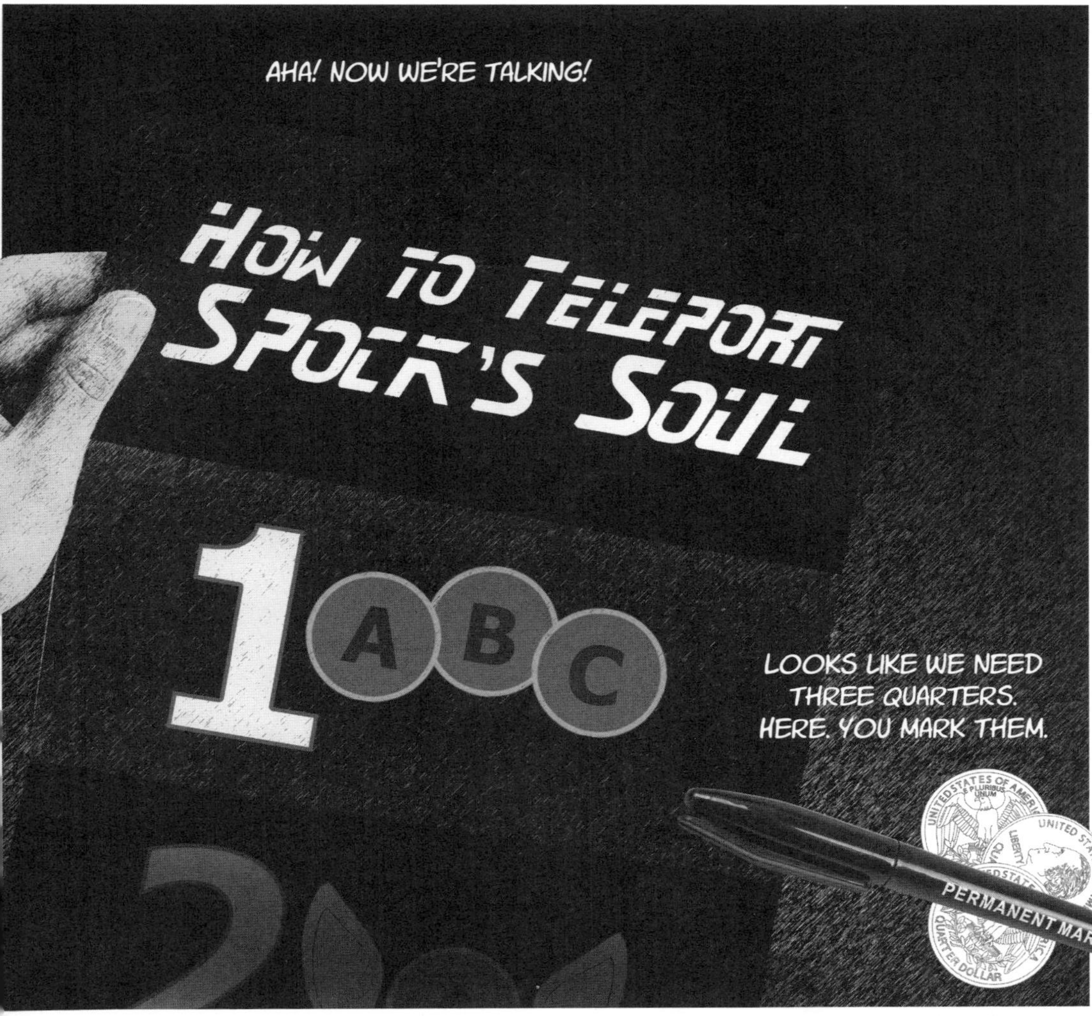
AHA! NOW WE'RE TALKING!
HOW TO TELEPORT SPOCK'S SOUL
1
A B C
LOOKS LIKE WE NEED THREE QUARTERS. HERE. YOU MARK THEM.
PERMANENT MARKER
2

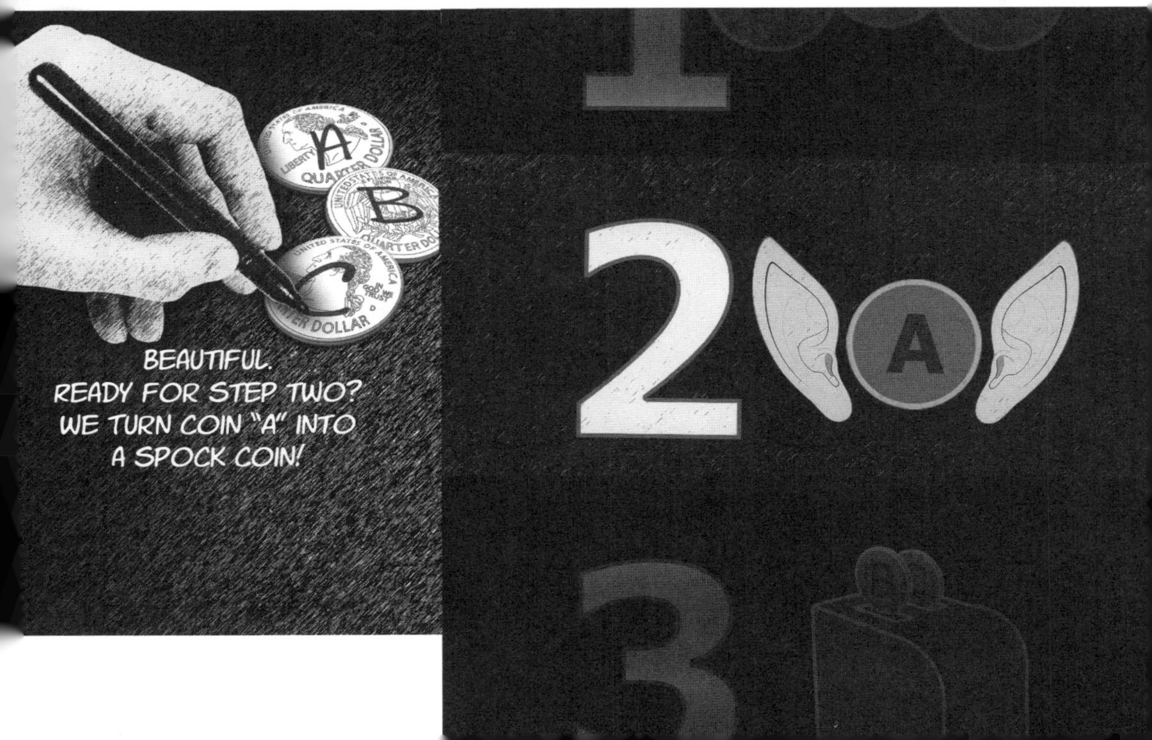
A
B
C
BEAUTIFUL. READY FOR STEP TWO? WE TURN COIN "A" INTO A SPOCK COIN!
1
2
A
3

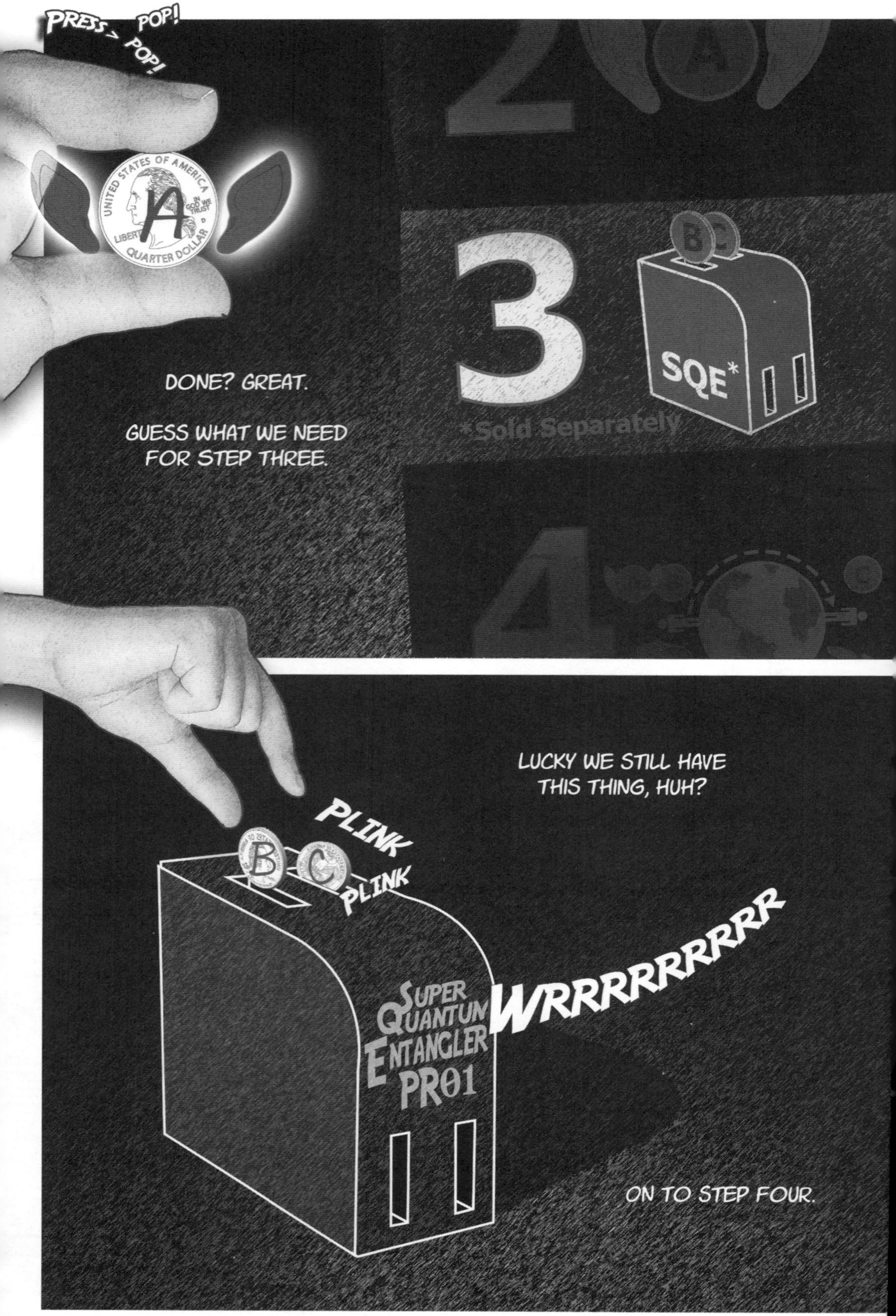
PRESS
POP!
POP!
UNITED STATES OF AMERICA
IN GOD WE TRUST
LIBERTY
QUARTER DOLLAR
A
2
A
3
B C
SQE*
*Sold Separately
4
DONE? GREAT.
GUESS WHAT WE NEED FOR STEP THREE.
LUCKY WE STILL HAVE THIS THING, HUH?
PLINK
PLINK
B C
SUPER QUANTUM ENTANGLER PRO1
WRRRRRRRRR
ON TO STEP FOUR.

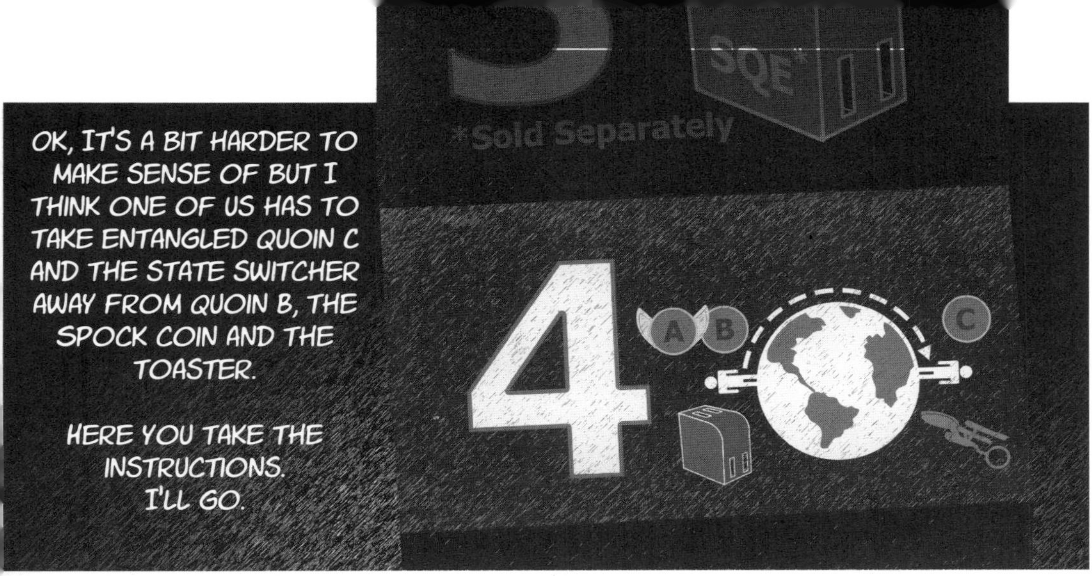

OK, IT'S A BIT HARDER TO MAKE SENSE OF BUT I THINK ONE OF US HAS TO TAKE ENTANGLED QUOIN C AND THE STATE SWITCHER AWAY FROM QUOIN B, THE SPOCK COIN AND THE TOASTER.
HERE YOU TAKE THE INSTRUCTIONS. I'LL GO.
SQE*
*Sold Separately
4
A
B
C

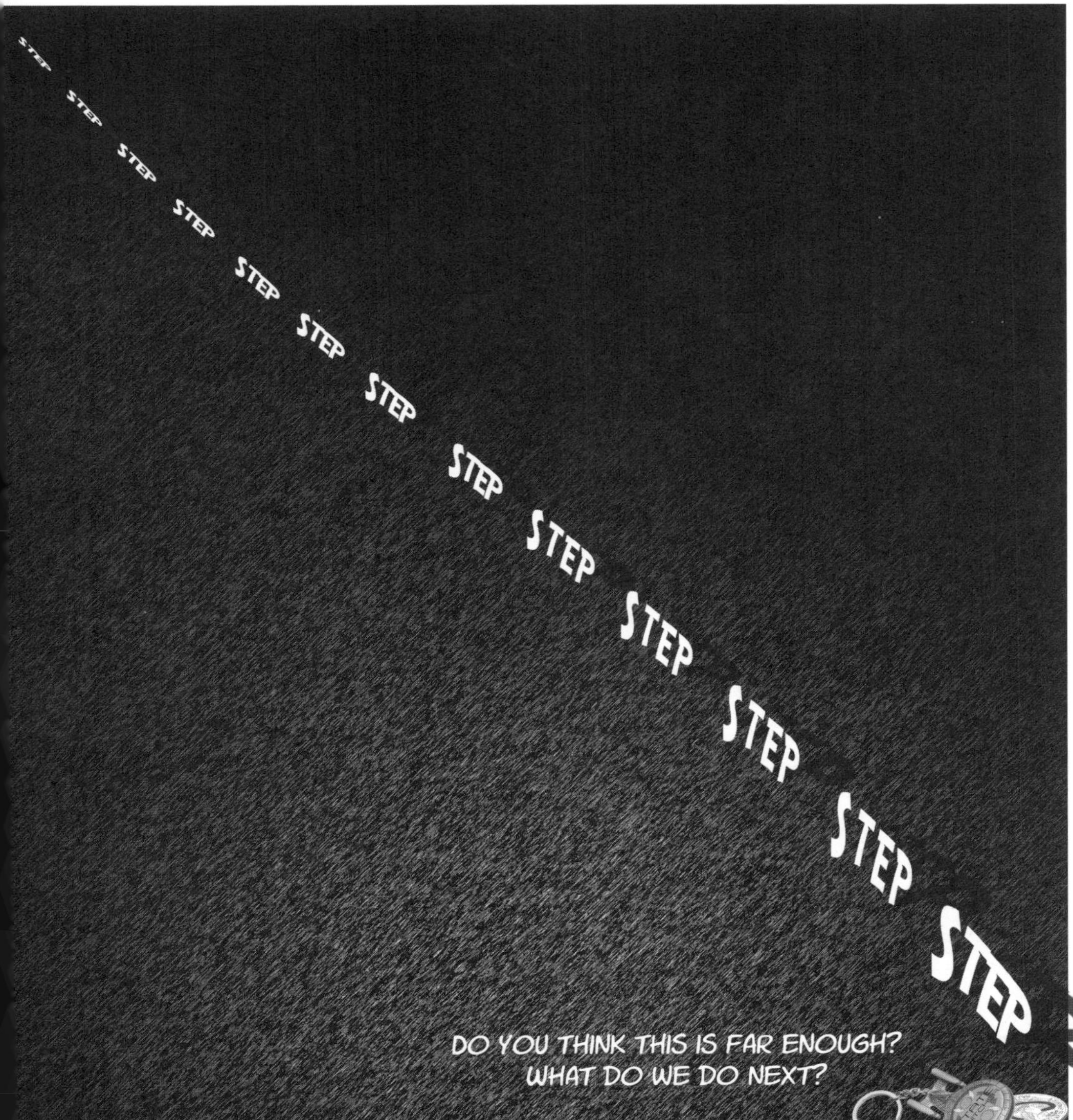

STEP STEP STEP STEP STEP STEP STEP STEP STEP STEP STEP STEP STEP
DO YOU THINK THIS IS FAR ENOUGH? WHAT DO WE DO NEXT?

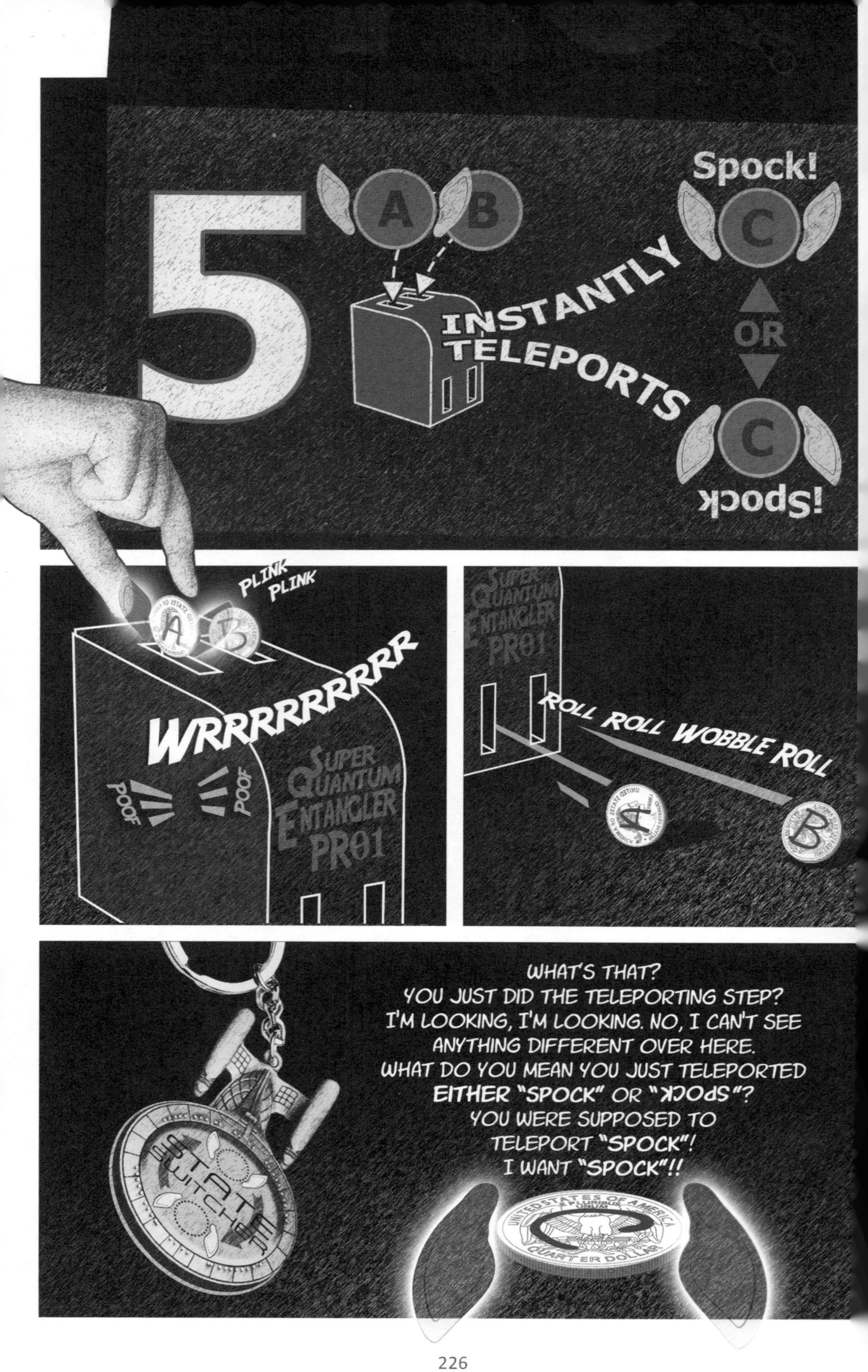
5
Spock!
A
B
C
INSTANTLY
TELEPORTS
OR
C
¡Spock
PLINK
PLINK
WRRRRRRRRR
POOF
POOF
SUPER QUANTUM ENTANGLER PRθ1
SUPER QUANTUM ENTANGLER PRθ1
ROLL ROLL WOBBLE ROLL
WHAT'S THAT?
YOU JUST DID THE TELEPORTING STEP?
I'M LOOKING, I'M LOOKING. NO, I CAN'T SEE
ANYTHING DIFFERENT OVER HERE.
WHAT DO YOU MEAN YOU JUST TELEPORTED
EITHER "SPOCK" OR "ꓘƆOԀS"?
YOU WERE SUPPOSED TO
TELEPORT "SPOCK"!
I WANT "SPOCK"!!
STATE SWITCHER
UNITED STATES OF AMERICA
QUARTER DOLLAR

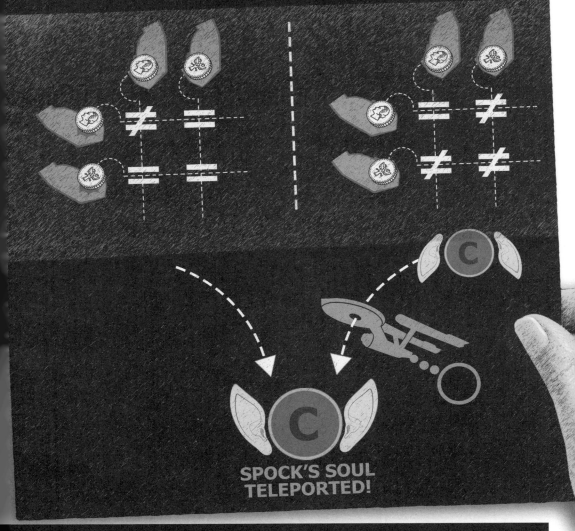

6 Ⓐ Ⓑ Are Now

Entangled OR Entangled

SPOCK'S SOUL TELEPORTED!

YOU'LL TELL ME IN A SECOND IF QUOINS "A" AND "B"
ARE "ENTANGLED" OR "ENTANGLED"?
WHAT DOES THAT HAVE TO DO WITH IT?
DO I WANT "SPOCK" OR NOT?
YOU KNOW I DO!
OK, OK, GO AHEAD.
I'LL WAIT.

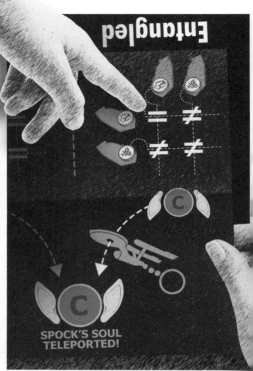

FLIP

PLONK

PLINK

Entangled

C

C

SPOCK'S SOUL
TELEPORTED!

YOU SAY "A" AND "B" WERE "ᗡƎ⅃ϼNA⊥NƎ"? SO? I HAVE TO PUT "C" THROUGH
THE STATE SWITCHER TO FINISH THE TELEPORTATION? OK.

INSERT >

SHUT

PRRRRRR

FLOP ?

STATE SWITCHER

FLIP ?

POP

ALRIGHT, I DID IT. SO SPOCK'S SOUL
SHOULD BE TELEPORTED INTO MY
COIN NOW, RIGHT? I'M GOING TO
TOSS IT TO SEE.

FLIP

PLUNK

POOF   POOF

HEADS.

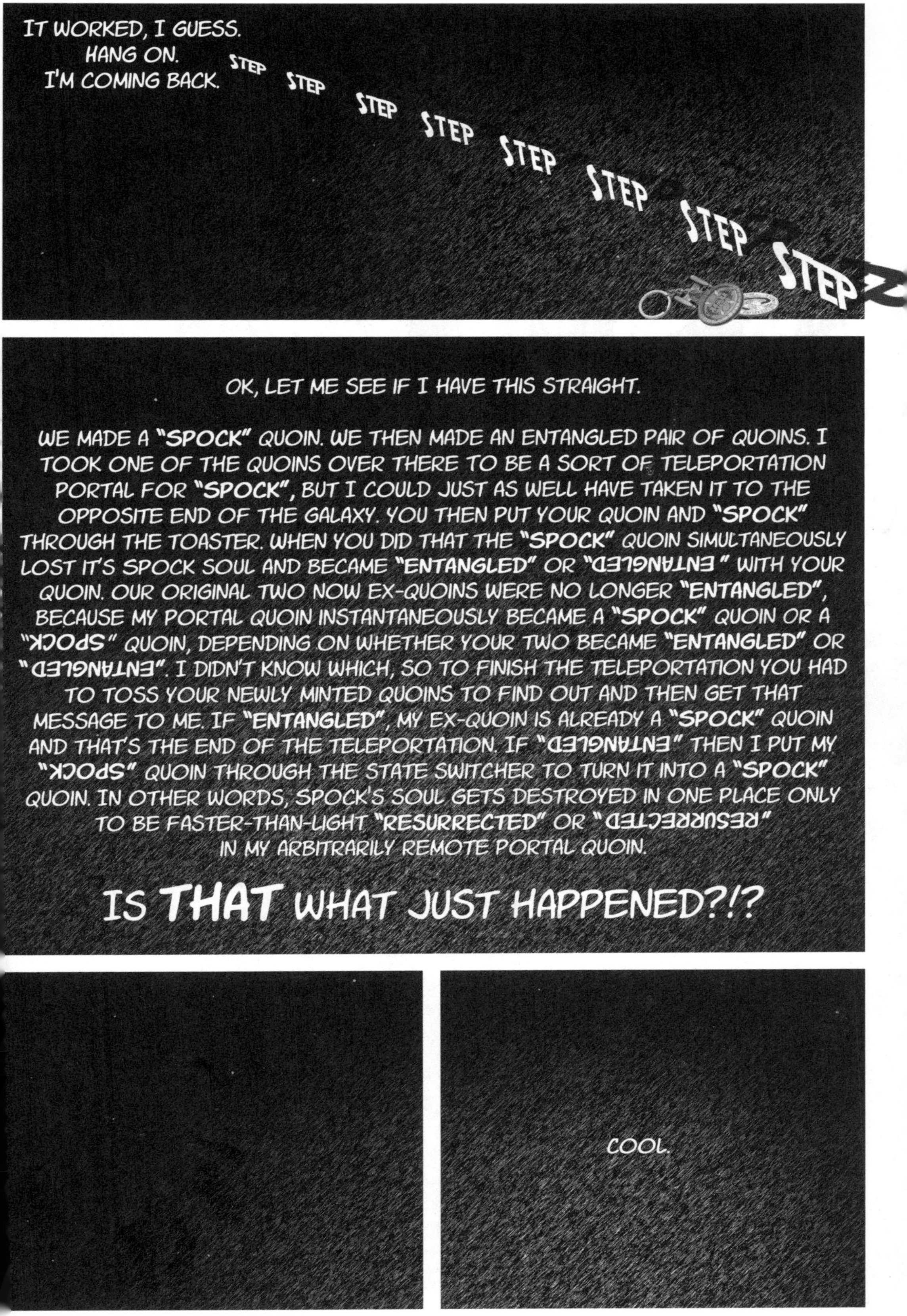

IT WORKED, I GUESS.
HANG ON.
I'M COMING BACK.
STEP STEP STEP STEP STEP STEP STEP STEP
OK, LET ME SEE IF I HAVE THIS STRAIGHT.
WE MADE A "SPOCK" QUOIN. WE THEN MADE AN ENTANGLED PAIR OF QUOINS. I TOOK ONE OF THE QUOINS OVER THERE TO BE A SORT OF TELEPORTATION PORTAL FOR "SPOCK", BUT I COULD JUST AS WELL HAVE TAKEN IT TO THE OPPOSITE END OF THE GALAXY. YOU THEN PUT YOUR QUOIN AND "SPOCK" THROUGH THE TOASTER. WHEN YOU DID THAT THE "SPOCK" QUOIN SIMULTANEOUSLY LOST IT'S SPOCK SOUL AND BECAME "ENTANGLED" OR "ENTANGLED" WITH YOUR QUOIN. OUR ORIGINAL TWO NOW EX-QUOINS WERE NO LONGER "ENTANGLED", BECAUSE MY PORTAL QUOIN INSTANTANEOUSLY BECAME A "SPOCK" QUOIN OR A "SPOCK" QUOIN, DEPENDING ON WHETHER YOUR TWO BECAME "ENTANGLED" OR "ENTANGLED". I DIDN'T KNOW WHICH, SO TO FINISH THE TELEPORTATION YOU HAD TO TOSS YOUR NEWLY MINTED QUOINS TO FIND OUT AND THEN GET THAT MESSAGE TO ME. IF "ENTANGLED", MY EX-QUOIN IS ALREADY A "SPOCK" QUOIN AND THAT'S THE END OF THE TELEPORTATION. IF "ENTANGLED" THEN I PUT MY "SPOCK" QUOIN THROUGH THE STATE SWITCHER TO TURN IT INTO A "SPOCK" QUOIN. IN OTHER WORDS, SPOCK'S SOUL GETS DESTROYED IN ONE PLACE ONLY TO BE FASTER-THAN-LIGHT "RESURRECTED" OR "RESURRECTED" IN MY ARBITRARILY REMOTE PORTAL QUOIN.
IS THAT WHAT JUST HAPPENED?!?
COOL.

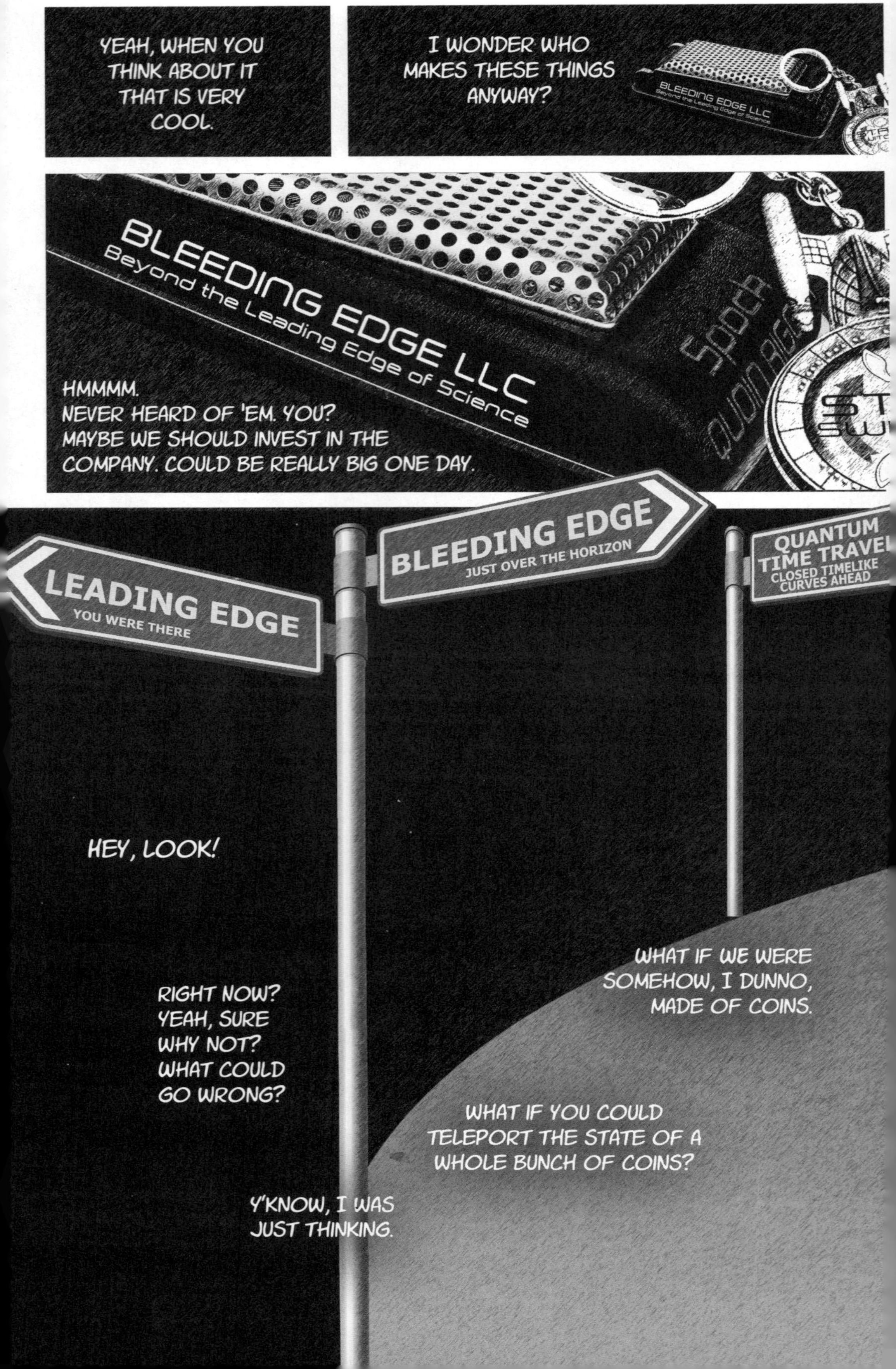
YEAH, WHEN YOU THINK ABOUT IT THAT IS VERY COOL.
I WONDER WHO MAKES THESE THINGS ANYWAY?
BLEEDING EDGE LLC
Beyond the Leading Edge of Science
Spock
HMMMM.
NEVER HEARD OF 'EM. YOU?
MAYBE WE SHOULD INVEST IN THE COMPANY. COULD BE REALLY BIG ONE DAY.
LEADING EDGE
YOU WERE THERE
BLEEDING EDGE
JUST OVER THE HORIZON
QUANTUM TIME TRAVEL
CLOSED TIMELIKE CURVES AHEAD
HEY, LOOK!
WHAT IF WE WERE SOMEHOW, I DUNNO, MADE OF COINS.
RIGHT NOW? YEAH, SURE WHY NOT? WHAT COULD GO WRONG?
WHAT IF YOU COULD TELEPORT THE STATE OF A WHOLE BUNCH OF COINS?
Y'KNOW, I WAS JUST THINKING.

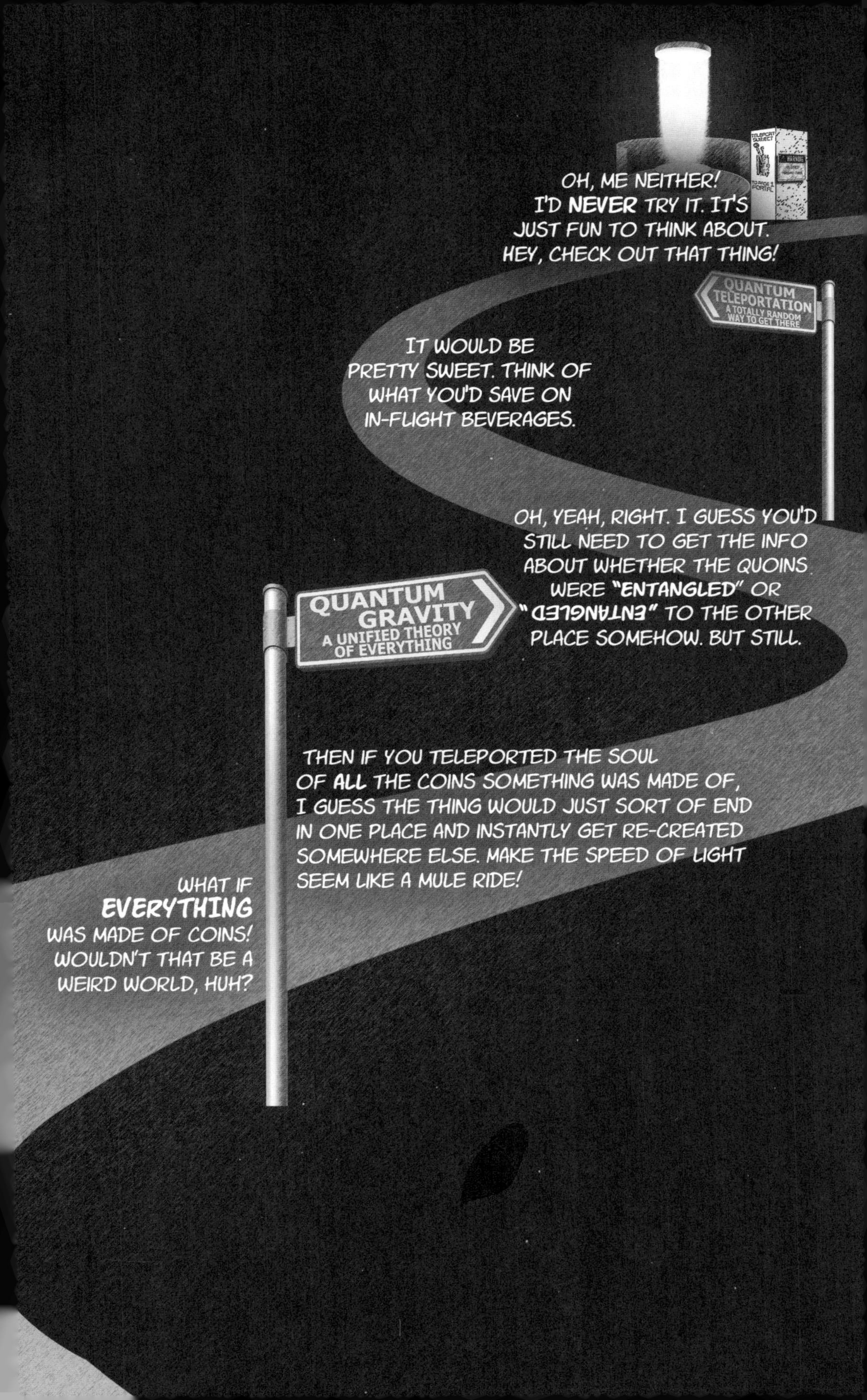
OH, ME NEITHER! I'D NEVER TRY IT. IT'S JUST FUN TO THINK ABOUT. HEY, CHECK OUT THAT THING!
QUANTUM TELEPORTATION A TOTALLY RANDOM WAY TO GET THERE
IT WOULD BE PRETTY SWEET. THINK OF WHAT YOU'D SAVE ON IN-FLIGHT BEVERAGES.
OH, YEAH, RIGHT. I GUESS YOU'D STILL NEED TO GET THE INFO ABOUT WHETHER THE QUOINS WERE "ENTANGLED" OR "ENTANGLED" TO THE OTHER PLACE SOMEHOW. BUT STILL.
QUANTUM GRAVITY A UNIFIED THEORY OF EVERYTHING
THEN IF YOU TELEPORTED THE SOUL OF ALL THE COINS SOMETHING WAS MADE OF, I GUESS THE THING WOULD JUST SORT OF END IN ONE PLACE AND INSTANTLY GET RE-CREATED SOMEWHERE ELSE. MAKE THE SPEED OF LIGHT SEEM LIKE A MULE RIDE!
WHAT IF EVERYTHING WAS MADE OF COINS! WOULDN'T THAT BE A WEIRD WORLD, HUH?

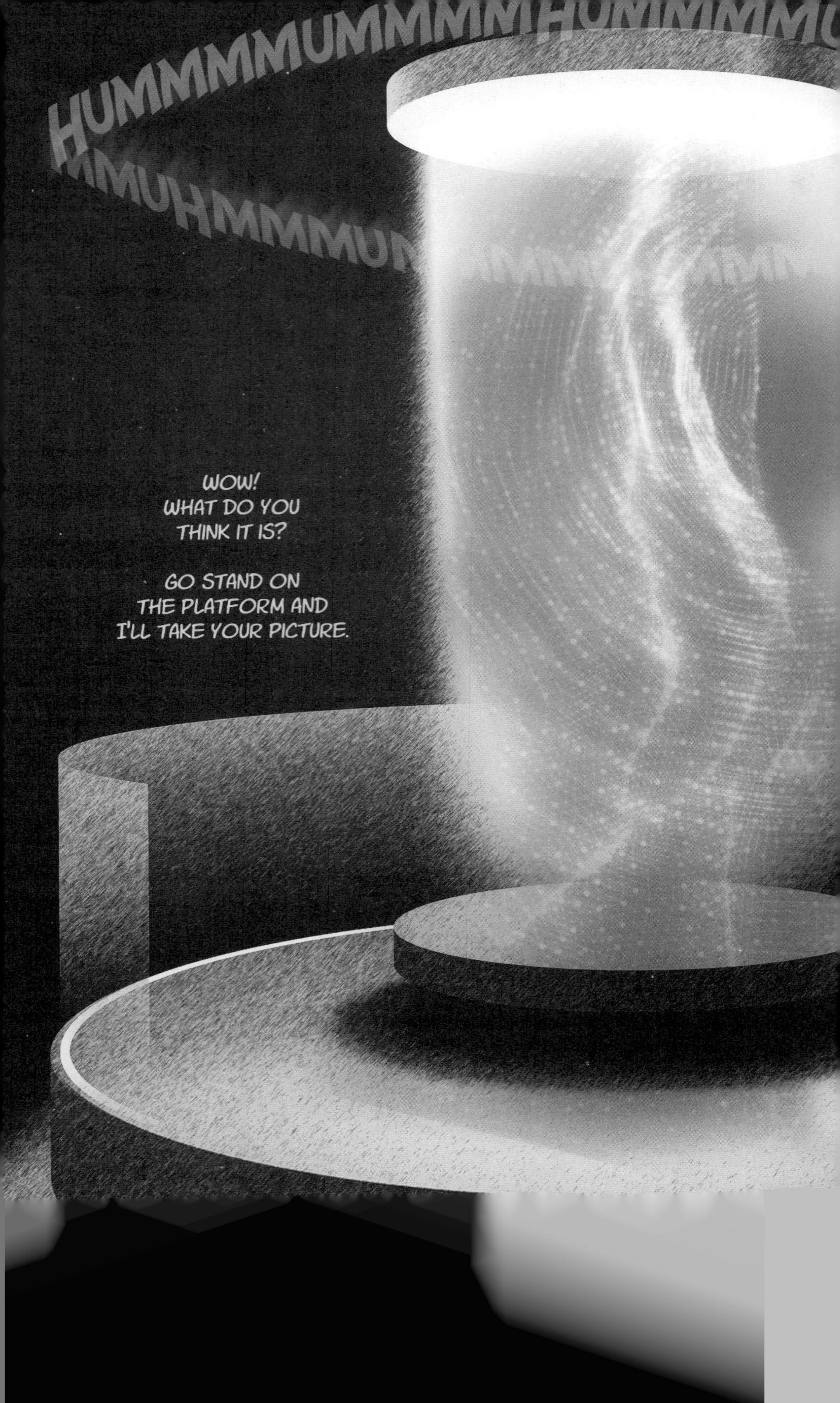
WOW!
WHAT DO YOU
THINK IT IS?
GO STAND ON
THE PLATFORM AND
I'LL TAKE YOUR PICTURE.

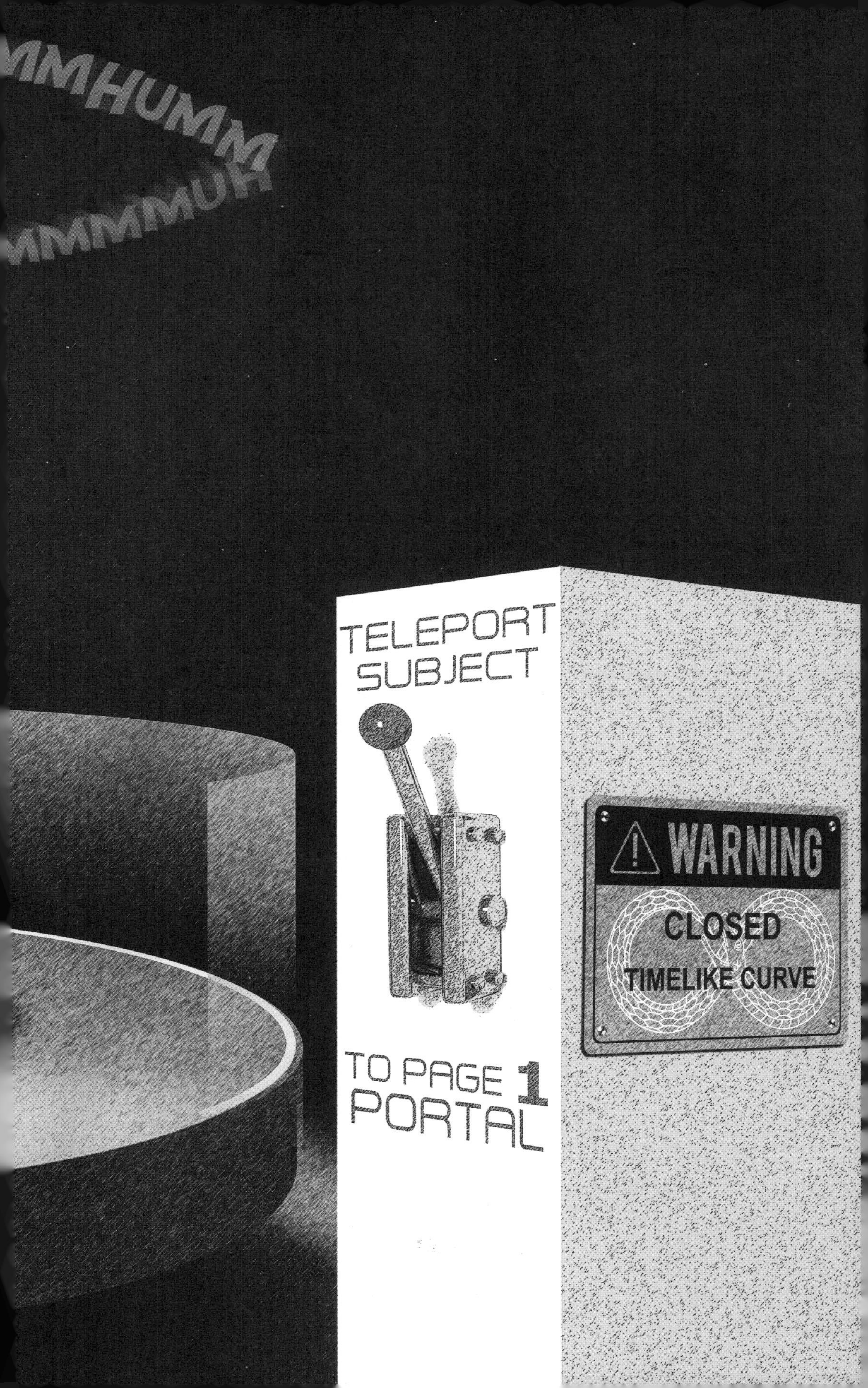
MMHUMM
MMMMUH
TELEPORT
SUBJECT
TO PAGE 1
PORTAL
WARNING
CLOSED
TIMELIKE CURVE

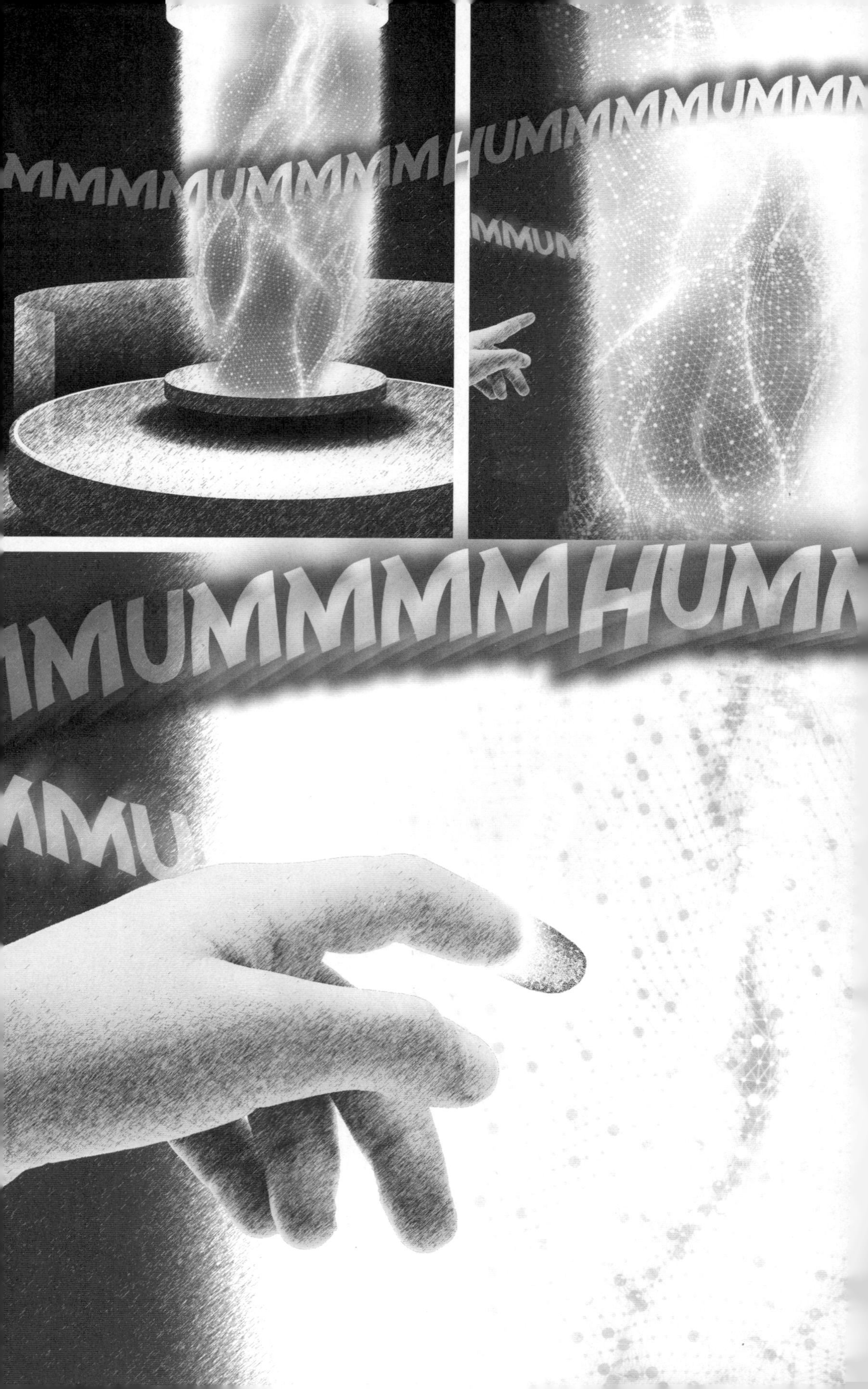

MMMMUMMMMMHUMMMMMUMMMM
MMMUM
MMUMMMMMHUMM
MMU

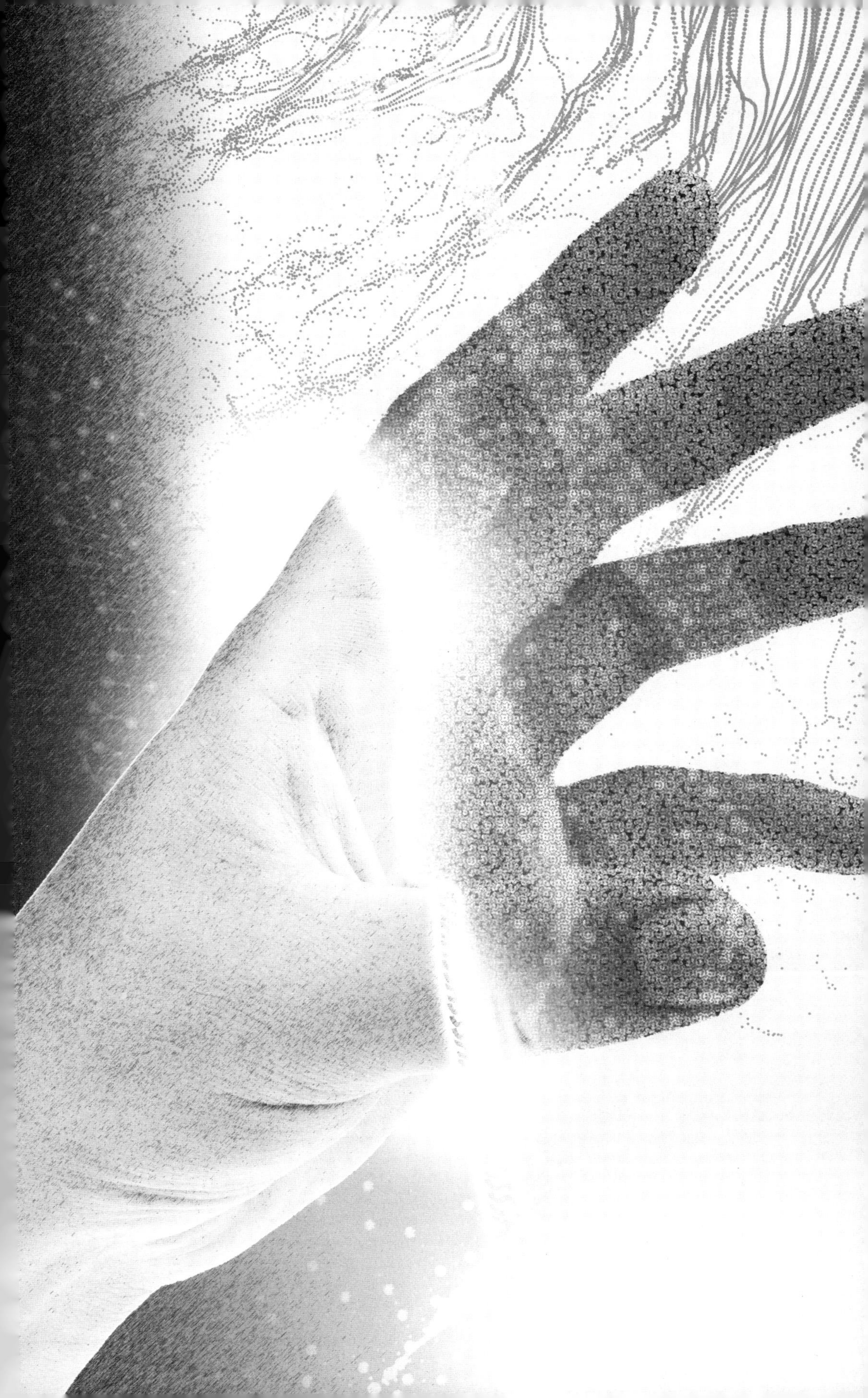

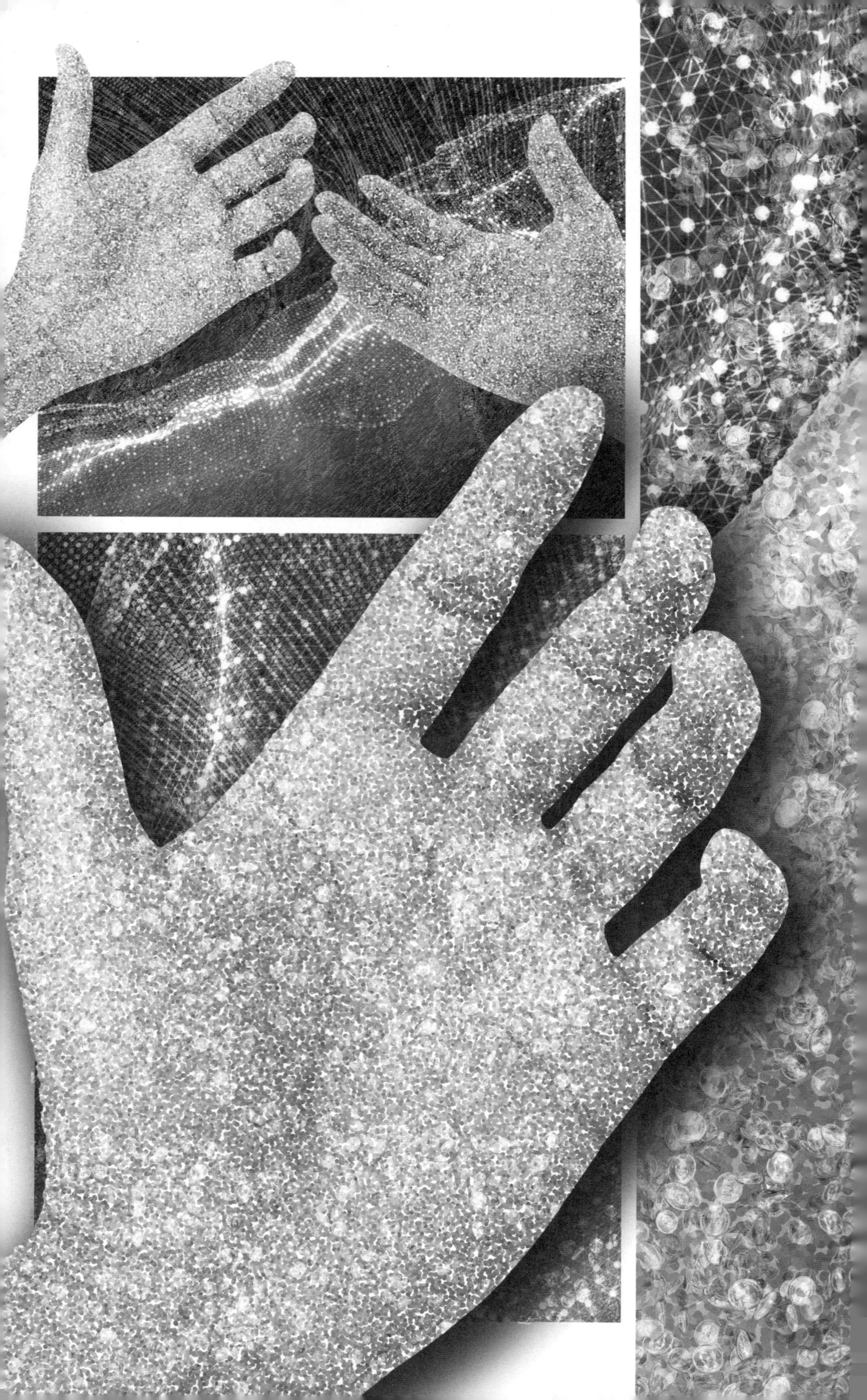

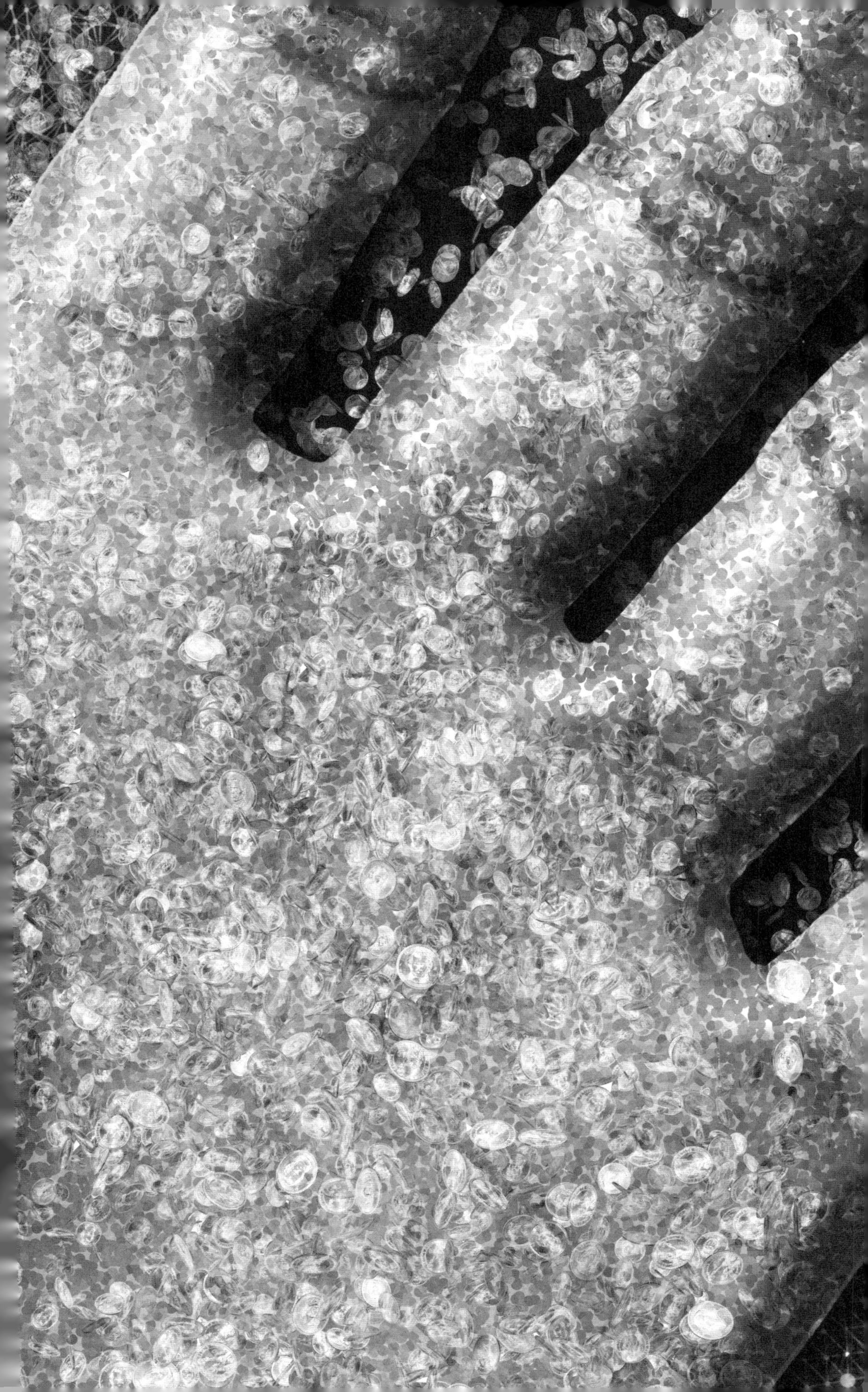

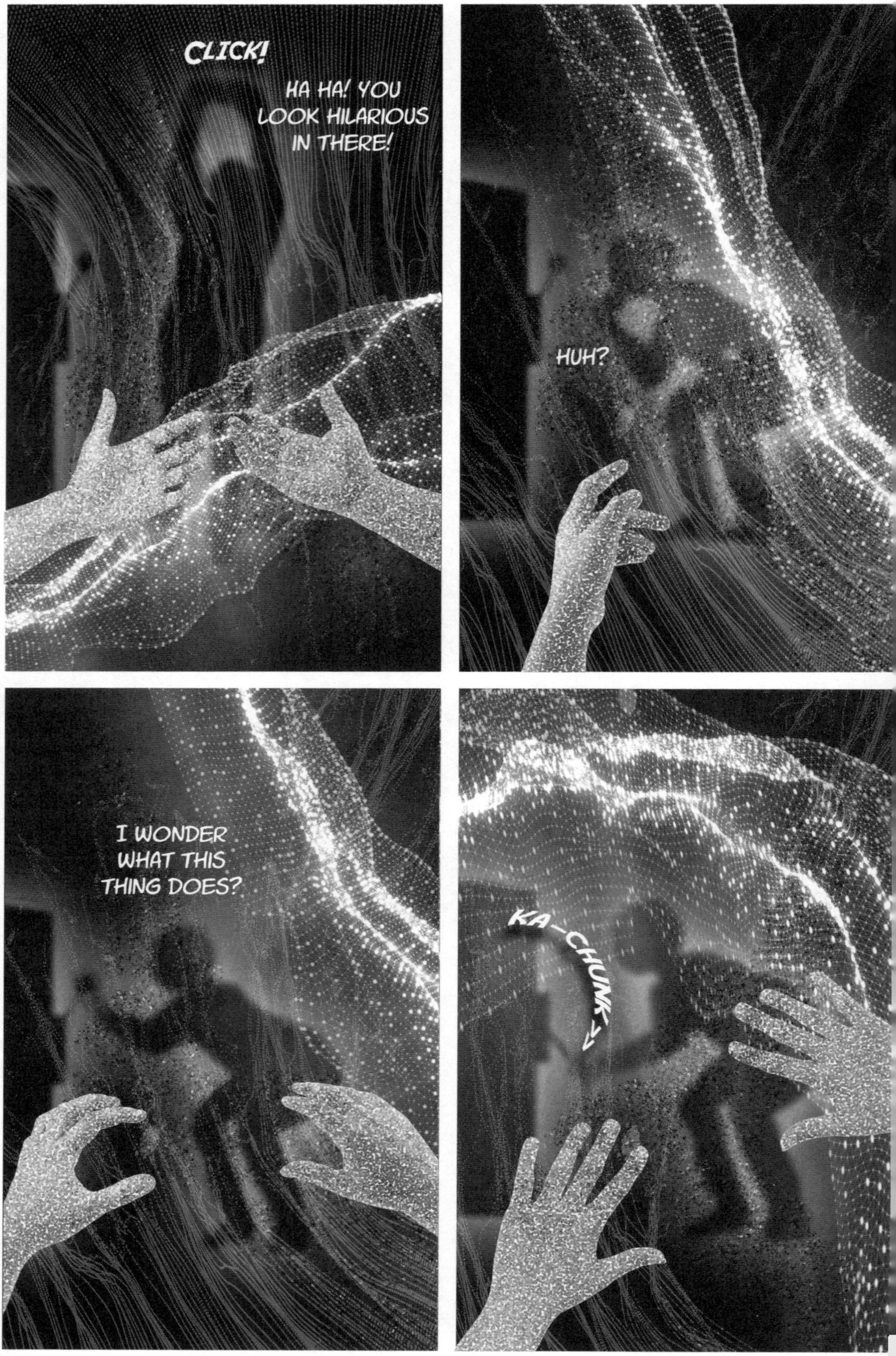
CLICK!
HA HA! YOU LOOK HILARIOUS IN THERE!
HUH?
I WONDER WHAT THIS THING DOES?
KA-CHUNK

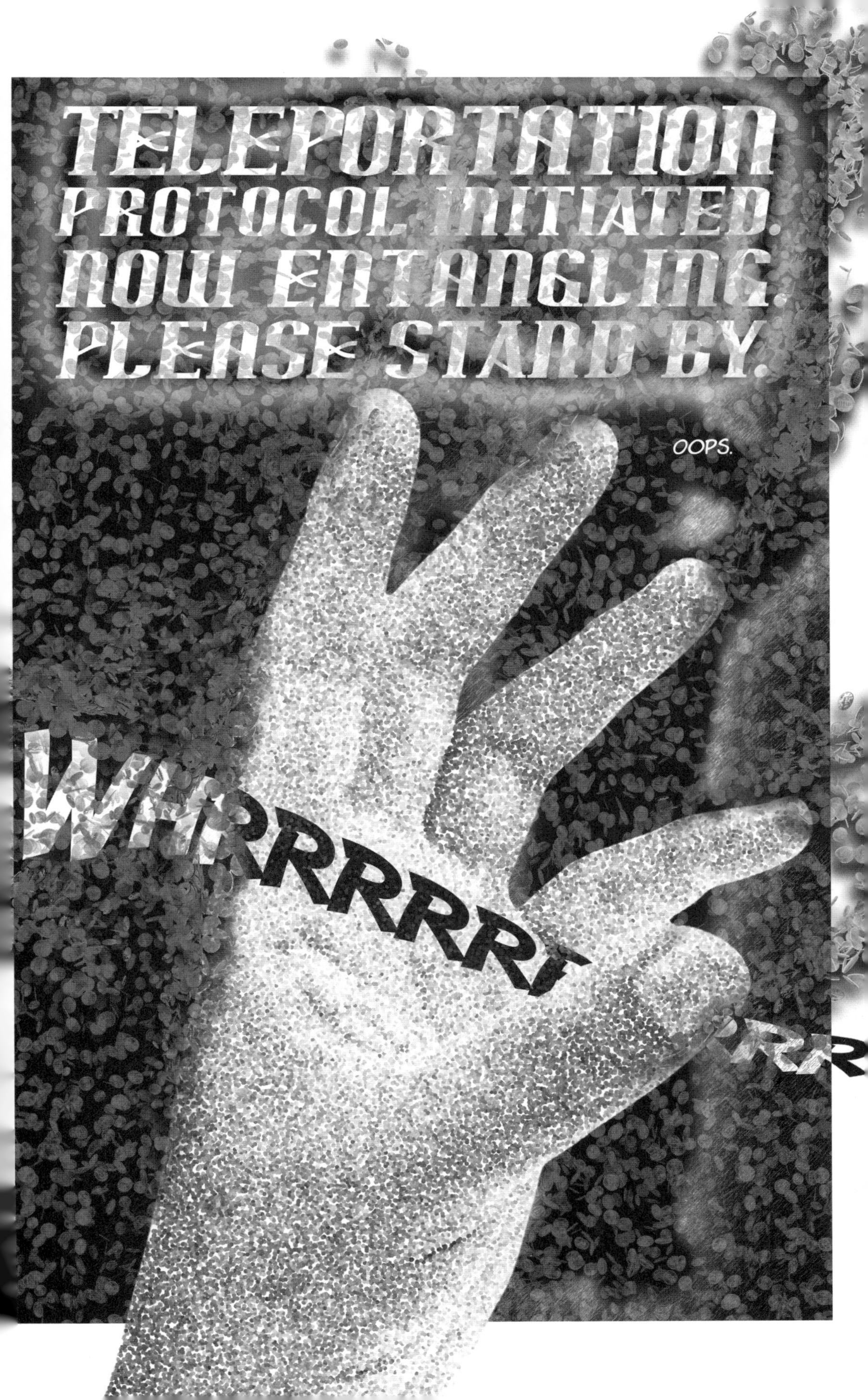
TELEPORTATION PROTOCOL INITIATED. NOW ENTANGLING. PLEASE STAND BY.
OOPS.
WHRRRRRRRRRR

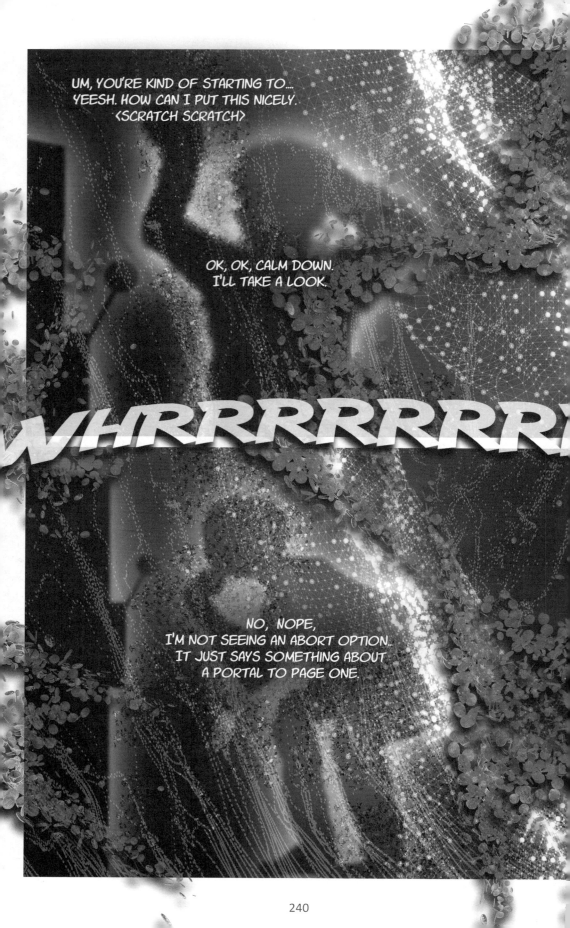

UM, YOU'RE KIND OF STARTING TO...
YEESH. HOW CAN I PUT THIS NICELY.
<SCRATCH SCRATCH>

OK, OK, CALM DOWN.
I'LL TAKE A LOOK.

WHRRRRRRRRRR

NO, NOPE,
I'M NOT SEEING AN ABORT OPTION..
IT JUST SAYS SOMETHING ABOUT
A PORTAL TO PAGE ONE.

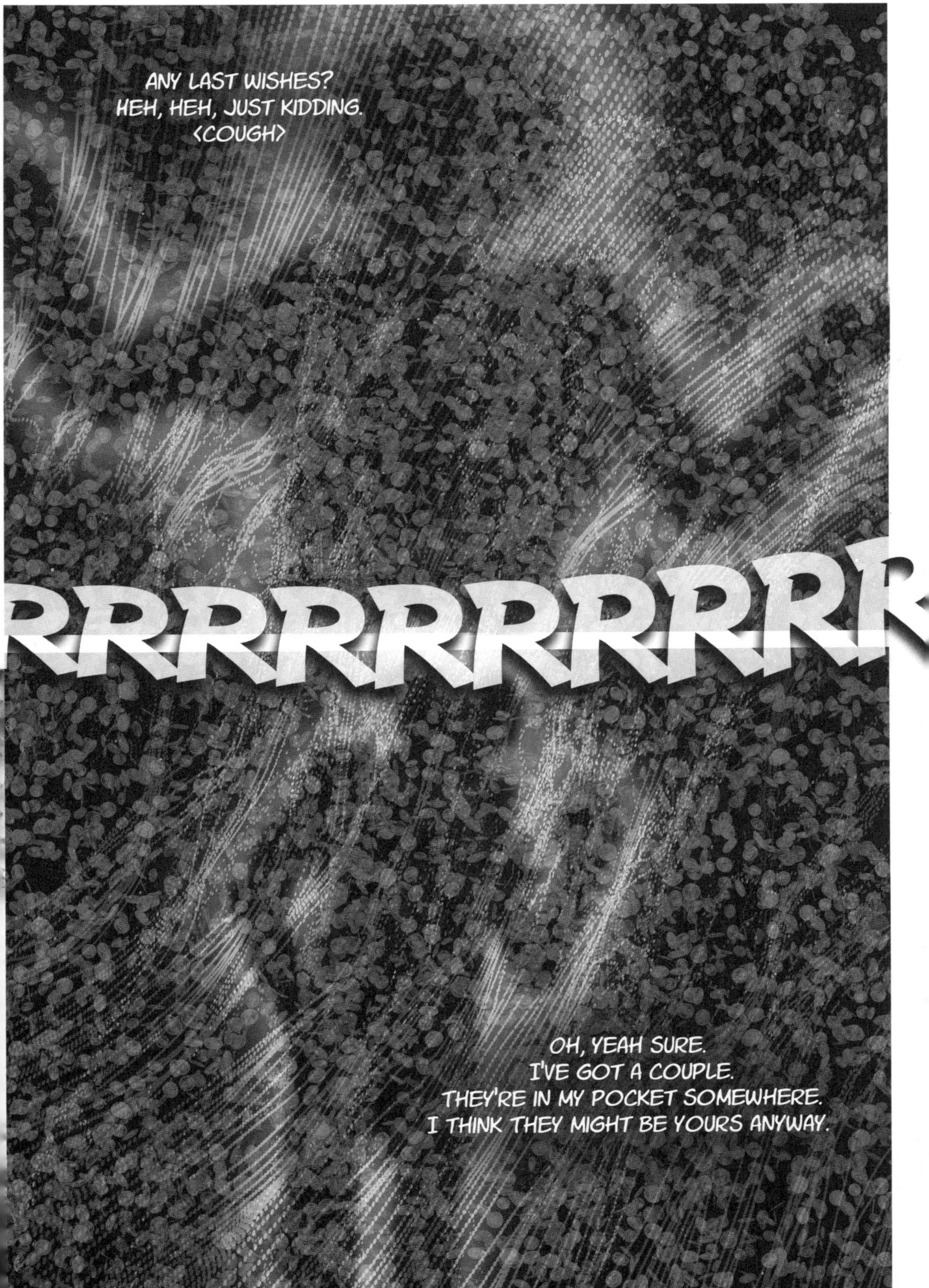
ANY LAST WISHES?
HEH, HEH, JUST KIDDING.
<COUGH>
RRRRRRRRRRR
OH, YEAH SURE.
I'VE GOT A COUPLE.
THEY'RE IN MY POCKET SOMEWHERE.
I THINK THEY MIGHT BE YOURS ANYWAY.

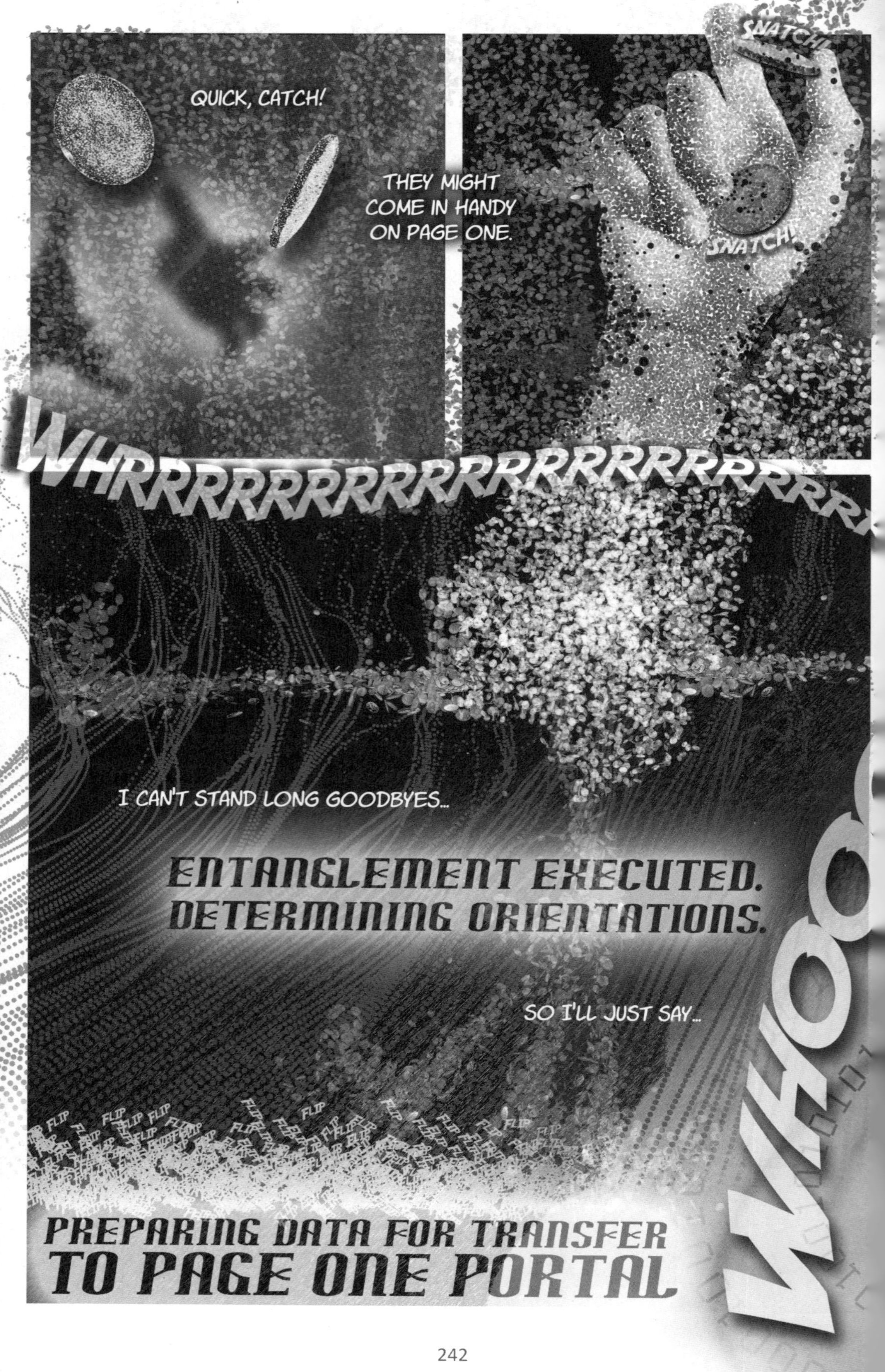
QUICK, CATCH!
THEY MIGHT COME IN HANDY ON PAGE ONE.
SNATCH!
SNATCH!
WHRRRRRRRRRRRRRRRRRRR
I CAN'T STAND LONG GOODBYES...
ENTANGLEMENT EXECUTED. DETERMINING ORIENTATIONS.
SO I'LL JUST SAY...
FLIP FLIP FLIP FLIP FLIP FLIP
PREPARING DATA FOR TRANSFER TO PAGE ONE PORTAL
WHOO

oooooooOOSH!

THIS HAS REALLY BEEN...

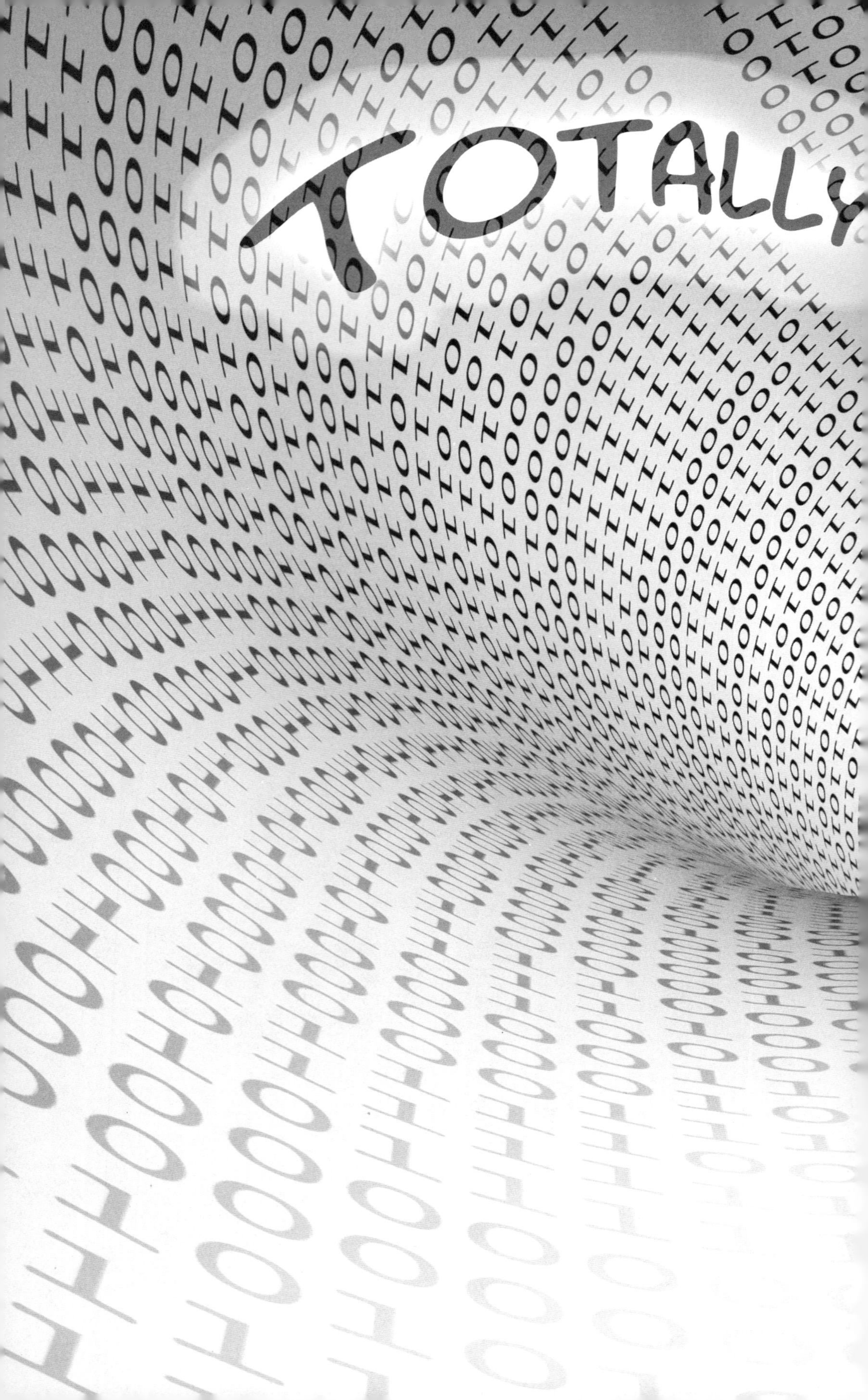
TOTALLY

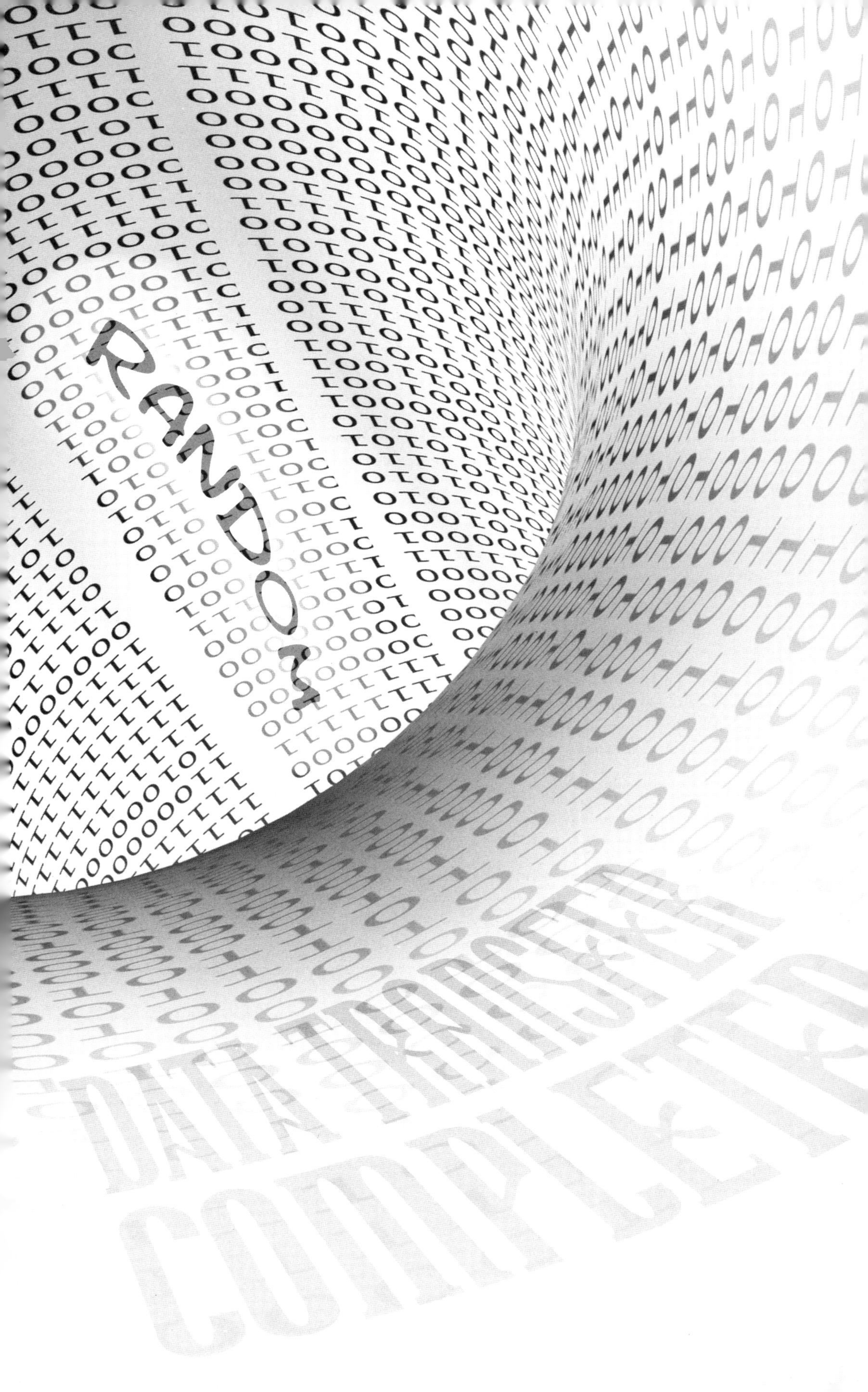
RANDOM

NOTES

For a link to an expanded version of these notes, see http://totallyrandom.info, where we also have lots of good stuff from the uncut version and more on how quoin entanglement relates to real-world quantum entanglement.

Most of the dialogue in the sequences with Einstein and Schrödinger is from the following sources, slightly tweaked to fit the narrative:

[1] Albert Einstein, Boris Podolsky, and Nathan Rosen, "Can quantum-mechanical description of physical reality be considered complete?" *Physical Review* 47, 777 (1935).

[2] Max Born (ed.), *The Born-Einstein Letters* (Walker and Co., New York, 1971).

[3] John D. Trimmer, "The present situation in quantum mechanics: A translation of Schrödinger's 'cat paradox' paper," *Proceedings of the American Philosophical Society* 124, 323–338 (1980). From Schrödinger's three-part paper "Die gegenwärtige Situation in der Quantenmechanik," *Die Naturwissenschaften* 48, 807–812; 49, 823–828; 50, 844–849 (1935).

Quotations from John Bell are mostly from his collection of papers:

[4] J. S. Bell, *Speakable and Unspeakable in Quantum Mechanics* (Cambridge University Press, Cambridge, 1987).

Page 7 The Superquantum Entangler PR01 takes two ordinary coins and turns them into "quoins" that produce the curious correlation when tossed. The correlation is the same as the correlation between inputs and outputs of a "nonlocal box" (generally referred to as a PR-box) introduced by Sandu Popescu and Daniel Rohrlich in "Quantum nonlocality as an axiom," *Foundations of Physics* 24, 370–385 (1994). Quantum systems like entangled photon pairs can produce something close to this correlation, but not quite.

Page 9 "I would not call . . . lines of thought." Erwin Schrödinger, "Discussion of probability relations between separated systems," *Mathematical Proceedings of the Cambridge Philosophical Society* 31, 555–563 (1935), p. 555.

Pages 16–17 Papers strewn on desk:

- Zeeya Merali, "Toughest test yet for quantum 'spookiness,'" *Nature* 525, 14–15 (September 2015).

- B. Hensen et al., "Loophole-free Bell inequality violation using electron spins separated by

246

1.3 kilometres," *Nature* 526, 682–686 (October 2015).

- Gregor Wehs et al., "Violations of Bell's inequality under strict Einstein locality conditions," *Physical Review Letters* 81, 5039–5043 (1998).

- K. Shalm et al., "Strong loophole-free test of local realism," *Physical Review Letters* 115,
250402–250412 (2015).

Page 22 "The scientific attitude . . . explanation." John Bell in [4], p. 152.

Page 37 John Bell derived the inequality that goes by his name in "On the Einstein-Podolsky-Rosen paradox," *Physics* 1, 195–200 (1964), reprinted in [4], pp. 14–21. Since Bell's original paper, there have been many versions of his theorem, with several related inequalities. One of the most useful inequalities was derived by J. Clauser, M. Horne, A. Shimony, and R. Holt in "Proposed experiment to test local hidden variable theories," *Physical Review Letters* 23, 880–884 (1969). The inequality on "The Punchline" page is a version of the Clauser-Horne-Simony-Holt inequality.

Page 38 This version of Bell's theorem is closely related to (and was stimulated by) the proof in Nicolas Gisin's book *Quantum Chance: Nonlocality, Teleportation and Other Quantum Marvels* (Copernicus, Göttingen, 2014).

Pages 42–43 "Conceivably, . . . velocity of light." John Bell in [4], p. 20; A. Aspect, J. Dalibard, and G. Roger, "Experimental test of Bell's inequality using time-varying analyzers," *Physical Review Letters A* 260, 323–327 (1999).

Pages 56–57 "Do not keep saying . . . like that." Richard Feynman, *The Character of Physical Law* (MIT Press, Cambridge, MA, 1967), p. 129.

"Those who are not shocked . . . understood it." Niels Bohr, as quoted by W. Heisenberg in *Physics and Beyond: Encounters and Conversations* (Harper and Row, New York, 1971), p. 206.

"It seems . . . single moment." Albert Einstein in a letter to Cornelius Lanczos, March 21, 1942, *Einstein Archive* 15, 294. Translated in H. Dukas and B. Hoffmann (eds.), *Albert Einstein: The Human Side* (Princeton University Press, Princeton, NJ, 1979), p. 68.

"One should . . . point of a needle." Wolfgang Pauli in a letter to Max Born [2], p. 223.

"Can nature . . . atomic experiments." W. Heisenberg, *Physics and Beyond: Encounters and Conversations* (Harper and Row, New York, 1971), p. 42.

Pages 66–67 "I don't like it . . . do with it." Commonly attributed to Erwin Schrödinger, referring to the probability interpretation of quantum mechanics. For example, epigraph, without citation, in John Gribbin, *In Search of Schrödinger's Cat: Quantum Physics and Reality* (Bantam Books, New York, 1984), p. v. We have been unable to find a plausible citation. This may be a variation on a comment by Schrödinger reported by Heisenberg describing Schrödinger's time spent debating Bohr in Copenhagen in September 1926: "If all this damned quantum jumping were really here to stay, I should be sorry I ever got involved with quantum theory," in Werner Heisenberg, *Physics and Beyond: Encounters and Conversations* (Harper and Row, New York, 1971), p. 75.

Papers strewn on page:

- David Z. Albert and Rivka Galchen, "A quantum threat to special relativity," *Scientific American*, March 2009, 32–39.

- Sabine Hossenfelder, "Testing superdeterministic conspiracy," *Journal of Physics: Conference Series 504*, 012018 (2014).

- Roger Colbeck and Renato Renner, "Free randomness can be amplified," *Nature Physics* 8, 450–454 (2012).

Pages 74–75 *New York Times* articles: William L. Laurence, "The week in science: Bohr and Einstein at odds," Science, *New York Times*, July 28, 1935; "Einstein attacks quantum theory; Scientist and two colleagues find it is not 'complete' even though 'correct,'" Social News-Art-Books, *New York Times*, May 4, 1935.

"Can quantum-mechanical description of physical reality be considered complete?" is the famous "EtPR" article [1] that motivated Bell's theorem.

"I cannot seriously . . . spooky action at a distance." Einstein in a letter to Max Born [2], March 3, 1947, p. 158.

"Unless one makes . . . not be possible." Einstein in [2], p. 170. From Max Born's translation of Einstein's paper "Quanten-Mechanik und Wirklichkeit,' *Dialectica* 320 (1948), reproduced as "Quantum mechanics and reality" in [2], pp. 168–173.

Pages 76–77 "Do not worry . . . all the greater." Einstein, in response to a letter from a schoolgirl, Barbara, in A. Calaprice, *Dear Professor Einstein: Albert Einstein's Letters to and from Children* (Prometheus Books, New York, 2002), p. 140: "Do not worry about your difficulties in mathematics; I can assure you that mine are still greater."

"Subtle is the Lord, but malicious he is not." Translation of "Raffiniert ist der Herrgott, aber boshaft ist Er nicht," a remark by Einstein to Princeton mathematician Oswald Veblen during a May 1922 visit to Princeton University, when Einstein gave four lectures on the theory of relativity in response to an account of some experiments by Dayton Miller, a physicist at the Case Institute of Technology in Cleveland, who claimed to have falsified the theory of relativity. See D. Brian, *Einstein: A Life* (John Wiley & Sons, New York, 1997), p. 127.

"You ask yourself . . . that outcome, no?" The reference here is to Einstein's notion of the state of an object as the "being-thus" ("So-Sein" in the original German) of the object [2], p. 170, and to the Einstein-Podolsky-Rosen criterion of reality in [1], p. 777: "If, without in any way disturbing a system, we can predict with certainty (i.e., with probability equal to unity) the value of a physical quantity, then there exists an element of physical reality corresponding to this physical quantity."

"Given your prior findings, . . . the second." This echoes the sentence in [1], p. 780: "This makes the reality of P and Q depend on the process of measurement carried out on the first system, which does not disturb the second system in any way."

"And you find yourself . . . the mere suggestion!" From Einstein's comments to Max Born on Born's book *Natural Philosophy of Cause and Chance* (Oxford University Press, Oxford, 1951), particularly on the last chapter, "Metaphysical Conclusions." The comments are reproduced after a letter from Einstein to Born dated March 18, 1948, in [2], p. 164.

Pages 78–79 "Because you must . . . not be possible." Einstein in [2], p. 170. From Max Born's translation of Einstein's paper "Quanten-Mechanik und Wirklichkeit," *Dialectica* 320 (1948), reproduced as "Quantum mechanics and reality" in [2], pp. 168–173.

"And as such . . . spooky action at a distance." Einstein in a letter to Max Born dated March 3, 1947, in [2], p. 158.

"No reasonable definition of reality . . . permit this!" From [1], p. 780.

"And if physics accepts . . . than a physicist." From Einstein's letter to Max Born dated April 29, 1924, in [2], p. 82: "I find the idea quite intolerable that an electron exposed to radiation should choose of its own free will, not only its moment to jump off, but also its direction. In that case, I would rather be a cobbler, or even an employee in a gaming house, than a physicist." Also, Einstein's letter to Schrödinger dated December 22, 1950, on p. 39 in K. Przibram (ed.), *Letters on Wave Mechanics* (Philosophical Library, New York, 1967): "If one wants to consider the quantum theory as final (in principle), then one must believe that a more complete description would be useless because there would be no laws for it. If that were so then physics could only claim the interest of shopkeepers and engineers; the whole thing would be a wretched bungle."

"And yet we cannot sulk . . . somehow incomplete." Einstein's view was that quantum mechanics is an incomplete theory, as indicated by the previous quotation.

"And like the moon, . . . closing our eyes." Attributed to Einstein by A. Pais in "Einstein and the quantum theory," *Reviews of Modern Physics* 51, 863 (1979): 'We often discussed his notions on objective reality. I recall that during one walk Einstein suddenly stopped, turned to me and asked whether I really believed that the moon exists only when I look at it."

"I, in any case, am convinced . . . does not play dice." Einstein in a letter to Max Born dated December 4, 1926, in [2], p. 9.

"Perhaps . . . God is malicious." Einstein remarked: "I have had second thoughts. Maybe God is malicious," to his assistant Valia Bargmann and Peter Gabriel Bergmann, as reported in Valia Bargmann, "Working with Einstein," in H. Woolf (ed.), *Some Strangeness in the Proportion: A Centennial Symposium to Celebrate the Achievements of Albert Einstein* (Addison-Wesley, Reading, MA, 1980), pp. 48–481.

Page 81 "Schrödinger's So-Sein Detection Services. Detecting the being-thus of ANY object." Erwin Schrödinger's views on quantum mechanics were closely aligned with Einstein's, so we think it appropriate to use Einstein's notion of the state of an object as its "being-thus," or "So-Sein" in German, for Schrödinger's device. Schrödinger's dialogue in this sequence follows the logic of his well-known cat paper [3], in which he contrasts the classical notion of the state of an object with the quantum notion.

Pages 82–83 "One can even . . . by the cat . . ." Schrödinger in [3], p.157. In the original three-part paper, the cat appears at the end of the first part, on p. 812.

The loose pages that T comments on are pp. 323 and 328 of [3].

Page 85 "Nature always knows." Schrödinger uses the expression "the system knows" in [3], p. 166: "The expression 'the system knows' will perhaps no longer carry the meaning that the answer comes forth from an instantaneous situation; it may perhaps derive from a succession of situations, that occupies a finite length of time. But even if it be so, it need not concern us so long as the system somehow brings forth the answer from within itself, with no other help than that we tell it (through the experimental arrangement) which question we would like to have answered;"

"My So-Sein machine, . . . a catalog of properties." Schrödinger refers to the quantum state as "a catalog of expectations" on numerous occasions in [3]. For example, in [3], p. 158, he writes about the quantum state: "In it is embodied the momentarily-attained sum of theoretically based future expectations, somewhat as laid down in a catalog." Later on the same page he refers to the quantum state as a "prediction-catalog." So the classical state would be a "catalog of physical properties," rather than a catalog of expectations.

"If you know . . . nature certainly knows." This was a comment by Nicolas Gisin to one of us (J) in a correspondence about quantum entanglement that highlighted for us Schrödinger's reference to "the system knows."

Pages 86–87 "I specialize . . . intractable problems." Of course, Schrödinger does not argue for classical principles in [3]. But he does contrast the classical notion of the state of an object with the quantum notion to bring out the problematic features of the quantum state as a catalog of expectations. Referring to the quantum state in [3], pp. 160–161, he points out: "Best possible knowledge of a whole does not necessarily include the same for its parts. . . . The whole is in a definite state, the parts taken individually are not."

Our Schrödinger is a classical chap who discovers entanglement through the adventure with the quoins. The real Schrödinger introduced the term "entanglement" into the literature on quantum mechanics in a seminal two-part article he wrote in response to the Einstein-Podolsky-Rosen paper [1]: "Discussion of probability relations between separated systems," *Proceedings of the Cambridge Phil-*

osophical Society 31, 555–563 (1935) and "Probability relations between separated systems," *Mathematical Proceedings of the Cambridge Philosophical Society* 32, 446–452 (1936).

"My Correlator Classique. . . . and two." The two pages show explicitly that the two possible states of the parts, in which one coin is rigged (biased) to land heads and the other tails, expresses what we *don't* have with quantum states or quoin states, in which, as Schrödinger puts it in [3], p. 160: "Maximal knowledge of a total system does not necessarily include total knowledge of all of its parts, not even when these are fully separated from each other and at the moment are not influencing each other at all." Our Schrödinger emphasizes the classical nature of his Correlator Classique (which simply biases a coin, either to land heads with certainty or tails with certainty) with the comment: "Here you see that the mechanics of the whole expresses the state of the pair of coins after they go through the correlator. This state of the whole is simply a function of its parts, the being-thus of each of the two coins which make up the pair."

Page 94 "I see you have supplied . . . each individual quoin?" This is precisely the difference between a classical correlation, with which we are all familiar, and the curious correlation of the quoins: there is no second page, because (to echo Schrödinger's comment in [3], p. 161) "the whole is in a definite state, the parts taken individually are not."

Page 97 "A hero both dead and alive!" Schrödinger [3], p. 157, referring to his thought experiment with a cat: "The ψ-function of the entire system would express this by having in it the living and the dead cat (pardon the expression) mixed or smeared out in equal parts."

The "Choose Your Own Adventure" story is our attempt to make vivid Schrödinger's insight about entanglement, specifically the peculiar relation of the parts to the whole. In [3], pp. 161–162, Schrödinger writes: "The expectation-catalog of the object has split into a conditional disjunction of expectation-catalogs—like a Baedeker that one has taken apart in the proper manner. Along with each section there is given also the probability that it proves correct—transcribed from the individual expectation-catalog of the object. But which one proves right—which section of the Baedeker should guide the ongoing journey—that can be determined only by actual inspection of the record." The *Baedeker*s were popular travel guides published by Karl Baedeker in Germany.

Page 102 Schrödinger sums up the significance of the "Choose Your Own Adventure" story with this comment: "You must then appreciate that your quoin mechanics cannot be correct because it ascribes a state to the whole which rules out any possibility of assigning definite states to its parts." In [3], p. 156, he says: ". . . if I wish to ascribe to the model at each moment a definite (merely not exactly known to me) state, or (which is the same) to all determining parts definite (merely not known to me) numerical values, then there is no supposition as to these numerical values to be imagined that would not conflict with some portion of quantum theoretical assertions."

Pages 104–105 "Our story . . . even possible." Schrödinger characterizes the quantum state (the wave function or ψ-function) in [3], p. 156, as "an imagined entity that images the blurring of all variables at every moment just as clearly and faithfully as the classical model does its sharp numerical values." He goes on to say: "Inside the nucleus, blurring doesn't bother us." The point of his thought experiment with a cat is to show [3, p. 156]: "But serious misgivings arise if one notices that the uncertainty affects macroscopically tangible and visible things, for which the term 'blurring' seems simply wrong." The cat example is set out in [3], p. 157, beginning with the remark: "One can even set up quite ridiculous cases. A cat is penned up in a steel chamber, . . ."

"You must take . . . are in effect." From [3], p.162: "Suppose the expectation-catalogs of two bodies A and B have become entangled through transient interaction. Now let the bodies be again separated. Then I can take one of them, say B, and by successive measurements bring my knowledge of it, which had become less than maximal, back up to maximal. I maintain: just as soon as I succeed in this, and not before, then first, the entanglement is immediately resolved and, second, I will also have acquired maximal knowledge of A through the measurements on B, making use of the conditional relations that were in effect."

Page 107 "Must I turn aside . . . catalog of expectations?" In [3], p.157, Schrödinger asks: "So what is left?" after having shown that the indeterminacy of quantum mechanics can't arise because we simply don't know the values of certain variables, while the cat thought experiment shows that the indeterminacy can't be an actual blurring, because "an easily executed observation provides the missing knowledge." He answers in [3], p. 157: "From this very hard dilemma the reigning doctrine [the Copenhagen interpretation] rescues itself or us by having recourse to epistemology. We are told that no distinction is to be made between the state of a natural object and what I know about it, or perhaps better, what can be known about it if I go to some trouble. Actually—so they say—there is intrinsically only awareness, observation, measurement. If through them I have procured at a given moment the best knowledge of the state of a physical object that is possibly attainable in accord with the natural laws, then I can turn aside as meaningless any further questioning about the 'actual state,' inasmuch as I am convinced that no further observation can extend my knowledge of it."

"I do not like quoin mechanics, and I am sorry I ever had anything to do with it." See the note for pp. 69–70.

Pages 110–111 "It is true that . . . seeming simplicities." Martin Gardiner, "Multiverses and blackberries," *Skeptical Inquirer* vol. 25.5, September/October 2001.

"The final strategy is . . . accept that." Sean Carroll, "Why the many-worlds formulation of quantum mechanics is probably correct," posted on June 30, 2014, to www.preposterousuniverse.com/blog/.

"[The many-worlds interpretation] . . . the picture." Scott Aaronson, "Why many-worlds is not like Copernicanism," posted on August 18, 2012, to Shtetl Optimized, https://www.scottaaronson.com/blog/.

"This universe . . . copies of itself." Bryce deWitt in B. S. DeWitt and N. Graham (eds.), *The Many-Worlds Interpretation of Quantum Mechanics* (Princeton University Press, Princeton, NJ, 1973), p. 161.

Pages 124–125 The Pauli effect. Hugh Everett is knocked unconscious by Wolfgang Pauli's dropping a roller on his head, illustrating the so-called Pauli effect: something would mysteriously go wrong with experimental equipment when Pauli entered a laboratory. From Otto Stern's interview with Res Jost: "And . . . you know, he was not allowed to enter our laboratory, because of the Pauli effect. Don't you know the famous Pauli effect?" As reported by Charles Enz in C. P. Enz, *No Time to Be Brief: A Scientific Biography of Wolfgang Pauli* (Oxford University Press, Oxford, 2002), p. 149.

Page 126–127 The GRW theory was first proposed by G. Ghirardi, A. Rimini, and T. Weber: "Unified dynamics for microscopic and macroscopic systems," *Physical Review D* 34, 470–491 (1986). Since then there have been several versions of the theory with contributions by others, notably Philip Pearle and Roderich Tumulka.

"There are subjective collapse interpretations by guys like von Neumann." John (Johann) von Neumann developed the first fully rigorous formulation of quantum mechanics in *Mathematical Foundations of Quantum Mechanics* (Princeton University Press, Princeton, NJ, 1955), a translation by Robert T. Beyer of *Mathematische Grundlagen der Quantenmechanik* (Springer, Berlin, 1932). On p. 351, he remarks: "We therefore have two fundamentally different types of intervention which can occur in a system. . . . First, the arbitrary changes by measurements which are given by the [collapse] formula . . . Second, the automatic changes which occur with the passage of time. Why do we need the special [collapse] process for the measurement? The reason is this: In the measurement we cannot observe the system S by itself, but must rather investigate the system S+M, in order to obtain (numerically) its interaction with the measuring apparatus M. . . . Moreover, it is rather arbitrary whether or not one includes the observer in M, and replaces the relation between the S state and the pointer positions in M by the relations of this state and the chemical changes in the observer's eye or even in his brain (i.e., to that which he has 'seen' or 'perceived')."

Eugene Wigner was more explicit. On the basis of a quantum-mechanical thought experiment known as the "Wigner's friend" experiment, where a friend makes a measurement and Wigner measures the friend, Wigner argued in "Remarks on the mind-body question," in I. J. Good (ed.), *The Scientist Speculates* (William Heinemann, London, 1961), pp. 171–184, that 'it is the entering of an impression into our consciousness which alters the wave function because it modifies our appraisal of the probabilities for different impressions which we expect to receive in the future. It is at this point that consciousness enters the theory unavoidably and unalterably." (pp. 175–176).

"Collapsing the Limbo of Potentialities . . . ." In *Sneaking a Look at God's Cards: Unraveling the Mysteries of Quantum Mechanics* (Princeton University Press, Princeton, NJ, 2007), p. 402, Giancarlo Ghirardi sums up Wigner's view, following his account of the "Wigner's friend" thought experiment, as follows: "It is the act of becoming conscious that makes reality pass from the limbo of potentialities to the clarity of actualities."

"Einstein's criticism . . . looking at it." Hugh Everett III, "The theory of the universal wave function," in B. S. DeWitt and N. Graham (eds.), *The Many Worlds Interpretation of Quantum Mechanics* (Princeton University Press, Princeton, NJ, 1973); pp. 3–140. Everett's remarks, recalling Einstein's 1954 Palmer lecture at Princeton, are on p. 116: "We should like now to comment on some views expressed by Einstein. Einstein's criticism of quantum theory (which is actually directed more against what we have called the 'popular' view than Bohr's interpretation) is mainly concerned with the drastic changes of state brought about by simple acts of observation (i.e., the infinitely rapid collapse of wave functions), particularly in connection with correlated systems which are widely separated so as to be mechanically uncoupled at the time of observation. At another time he put his feeling colorfully by stating that he could not believe that a mouse could bring about drastic changes in the universe simply by looking at it."

Page 131 "What about the cat, then?" This echoes Einstein's remark as reported by Everett, in the quotation on p. 130.

There is a related remark by John Bell in "Against measurement," *Physics World* 3, 33–40 (1990), p. 34: "It would seem that the theory is exclusively concerned about 'results of measurement,' and has nothing to say about anything else. What exactly qualifies some physical systems to play the role of 'measurer'? Was the wavefunction of the world waiting to jump for thousands of millions of years until a single-celled living creature appeared? Or did it have to wait a little longer, for some better qualified system . . . with a Ph.D.?"

"So you want to know . . . or the other?" This remark echoes Schrödinger's comment after his discussion of the cat thought experiment in [3], p. 157: "The ψ-function of the entire system would express this by having in it the living and the dead cat (pardon the expression) mixed or smeared out in equal parts."

Page 133 "It is the act of becoming conscious . . . the clarity of actualities." Giancarlo Ghirardi, *Sneaking a Look at God's Cards: Unraveling the Mysteries of Quantum Mechanics* (Princeton University Press, Princeton, NJ, 2007), p. 420, sums up Wigner's view following Ghirardi's account of the "Wigner's friend" thought experiment as follows: "It is the act of becoming conscious that makes reality pass from the limbo of potentialities to the clarity of actualities." This is Ghirardi's characterization of Eugene Wigner's view, not von Neumann's, but the two are sufficiently close that we've used the quote here.

"We must always . . . or the other." John von Neumann, *Mathematical Foundations of Quantum Mechanics* (Princeton University Press, Princeton, 1955), p. 420, a translation by Robert T. Beyer of

Mathematische Grundlagen der Quantenmechanik (Springer, Berlin, 1932): "That is, we must always divide the world into two parts, the one being the observed system, the other the observer. In the former, we can follow up all physical processes (in principle, at least) arbitrarily precisely. In the latter, this is meaningless. The boundary between the two is arbitrary to a very large extent. . . . That this boundary can be pushed arbitrarily deeply into the interior of the body of the actual observer . . . does not change the fact that in each method of description the boundary must be put somewhere, if the method is not to proceed vacuously, i.e., if a comparison with experiment is to be possible. Indeed, experience only makes statements of this type: an observer has made a certain (subjective) observation; and never any like this: a physical quantity has a certain value. Now quantum mechanics describes the events which occur in the observed portions of the world, so long as they do not interact with the observing portion . . . , but as soon as such an intervention occurs, i.e., a measurement, it requires the application of [the collapse] process."

Page 134 "Do you suffer from uncontrollable urges to picture an underlying reality?" Niels Bohr rejected the possibility of picturing events at the quantum level. For example, Bohr writes in *Essays 1958–1962 on Atomic Physics and Human Knowledge* (Vintage Books, New York, 1966), p. 6: "The decisive point, however, is that in this connection there is no question of reverting to a mode of description which fulfills to a higher degree the accustomed demands regarding pictorial representation of the relationship between cause and effect." Referring to the 1927 Solvay Conference in the same collection of essays, Bohr comments, on p. 89: "A main theme for the discussion was the renunciation of pictorial deterministic description implied in the new methods."

Page 135 "The world . . . by any chance observes." Sherlock Holmes to Dr. Watson in Sir Arthur Conan Doyle, *The Hound of the Baskervilles* (Dover, New York, 1994), chap. 3, p. 18.

Page 137 "There is no quantum world. . . . about nature." Niels Bohr, as quoted by Aage Petersen in "The philosophy of Niels Bohr," *Bulletin of the Atomic Scientists* 19, 8–14 (1963).

"The Heisenberg-Bohr tranquilizing philosophy . . . lie there." Albert Einstein in a letter to Schrödinger dated May 1928, in K. Przibram (ed.), *Letters on Wave Mechanics* (Philosophical Library, New York, 1967), p. 31.

Page 140 "I regard consciousness . . . from consciousness." Max Planck, as quoted in *The Observer*, January 25, 1931.

Page 142 "Ask not . . . said about a quoin." Bohr's affirmation is inspired by the comment attributed to Bohr by Aage Petersen in "The philosophy of Niels Bohr," *Bulletin of the Atomic Scientists* 19, 8–14 (1963): "It is wrong to think that the task of physics is to find out how nature is. Physics concerns what we can say about nature." The quotation is on p. 12. Also, of course, the words echo John F. Kennedy's exhortation in his inaugural address, January 20, 1961: "And so my fellow Americans: ask not what your country can do for you—ask what you can do for your country. My fellow citizens of the world: ask not what America will do for you, but what together we can do for the freedom of man."

Page 147 Einstein highlights the essential point about the theory of relativity that is relevant here in the article "What is relativity?" written for the *Times* (London), November 28, 1919: "It became clear that to speak of the simultaneity of two events had no meaning except in relation to a given coordinate system, and that the shape of measuring devices and the speed at which clocks move depend on their state of motion with respect to the coordinate system."

Page 148 Albert Einstein: "quoin mechanics is incomplete." The point of [1] was to argue that the quantum state is an incomplete description. On p. 780, the authors conclude: "While we have thus shown that the wave function does not provide a complete description of the physical reality, we left open the question of whether or not such a description exists. We believe, however, that such a theory is possible."

Niels Bohr: "How wonderful . . . making progress." As quoted on p. 196 in R. Moore, *Niels Bohr:*

The Man, His Science, & the World They Changed (Alfred A. Knopf, New York, 1966).

Niels Bohr: "Time and again . . . leads to contradictions." In "Discussion with Einstein on epistemological problems in atomic physics," in P. A. Schilpp (ed.), *Albert Einstein: Philosopher-Scientist* (Open Court, La Salle, IL, 1949), p. 221, Bohr remarks: "In any attempt of a pictorial representation of the behaviour of the photon we would, thus, meet with the difficulty: to be obliged to say, on the one hand, that the photon always chooses one of the two ways and, on the other hand, that it behaves as if it had passed both ways."

Pages 152–153 David Bohm's "Picture of Underlying Reality": "There is a 'PREFERRED' REFERENCE FRAME which reflects the REAL order of events." Referring to Bohm's theory in [4], p. 133, John Bell remarks: "It may well be that a relativistic version of the theory, while Lorentz invariant and local at the observational level, may be necessarily non-local and with a preferred frame (or aether) at the fundamental level." Giancarlo Ghirardi remarks in "Collapse Theories" in Edward N. Zalta (ed.), *The Stanford Encyclopedia of Philosophy* (Spring 2016 ed.), comparing Bohm's theory to the "dynamical collapse" theory of Ghirardi and associates: "Bohmian mechanics shows that one can explain quantum mechanics, exactly and completely, if one is willing to pay with using a preferred slicing of spacetime; our model suggests that one should be able to avoid a preferred slicing of spacetime if one is willing to pay with a certain deviation from quantum mechanics."

"We must be clear . . . creating images." Niels Bohr: "We must be clear that when it comes to atoms, language can be used only as in poetry. The poet, too, is not nearly so concerned with describing facts as with creating images and establishing mental connections." As reported by Werner Heisenberg in *Physics and Beyond: Encounters and Conversations* (Harper and Row, New York, 1971), p. 41.

"A quoin state . . . flipping." Niels Bohr in *Atomic Theory and the Description of Nature* (Cambridge University Press, Cambridge, 1961), pp. 56–57: "In any discussion of these questions, it must be kept in mind that, according to the view taken above, radiation in free space as well as isolated material particles are abstractions, their properties on the quantum theory being definable and observable only through their interaction with other systems."

Pages 154–155 Hugh Everett's "BIG PICTURE OF QUOIN MECHANICS REALITY! RELATIVITY COMPATIBLE." In [4], p. 133, John Bell points out that Everett's "many worlds" theory can be understood as Bohm's pilot-wave theory without trajectories: "It can be maintained that the de Broglie-Bohm orbits, so troublesome in this matter of locality, are not an essential part of the theory. Indeed it can be maintained that there is no need whatever to link successive configurations of the world into a continuous trajectory. Keeping the instantaneous configurations, but discarding the trajectory, is the essential feature (in my opinion) of the theory of Everett."

"Even the discovery . . . grave conceptual difficulties." Bohr often introduces a discussion of the conceptual problems of quantum mechanics by considering Einstein's insight about the relativity of simultaneity, as on p. 290 in "Causality and complementarity," *Philosophy of Science* 4, 289–298 (1937): "It was in fact the clarification of the paradoxes connected with the finite velocity of propagation of light and the judgment of events by observers in relative motion which first disclosed the arbitrariness contained even in the concept of simultaneity, and thereby created a freer attitude toward the question of space-time coördination which finds expression in the theory of relativity. As is well known, this has made possible a unified formulation of the phenomena appearing in different frames of reference, and through this brought to light the fundamental equivalence of hitherto separate physical regularities. The recognition of the essential dependence of any physical phenomenon on the system of reference of the observer, which forms the characteristic feature of relativity theory, implies, however—as especially Einstein himself has emphasized—no abandonment whatever of the assumption underlying the ideal of causality, that the behavior of a physical object relative to a given system of coördinates is uniquely determined, quite independently of whether it is observed or not."

"A challenge, implicit in . . . renounce as naïve." These comments are suggested by Bohr's response to [1] in "Discussion with Einstein on epistemological problems in atomic physics," P. A. Schilpp (ed.), *Albert Einstein: Philosopher-Scientist* (Open Court, La Salle, IL, 1949), pp. 199–241: "Indeed the finite interaction between object and measuring agencies conditioned by the very existence of the quantum of action entails—because of the impossibility of controlling the reaction of the object on the measuring instruments, if these are to serve their purpose—the necessity of a final renunciation of the classical ideal of causality and a radical revision of our attitude towards the problem of physical reality." (pp. 232–233).

Pages 156–157 "And so while . . . isn't that right, kitty?" This is a common theme of Bohr's discussions of his concept of "complementarity," as on p. 291 in "Causality and complementarity," *Philosophy of Science* 4, 289–298 (1937): "The apparently incompatible sorts of information about the behavior of the object under examination which we get by different experimental arrangements can clearly not be brought into connection with each other in the usual way, but may, as equally essential for an exhaustive account of all experience, be regarded as 'complementary' to each other."

"This exclusive . . . classical conception of causality." Bohr often talked this way, e.g., in "Causality and complementarity," p. 293: "The renunciation of the ideal of causality in atomic physics which has been forced on us is founded logically only on our not being any longer in a position to speak of the autonomous behavior of a physical object, due to the unavoidable interaction between the object and the measuring instruments which in principle cannot be taken into account, if these instruments according to their purpose shall allow the unambiguous use of the concepts necessary for the description of experience."

"That no single state . . . verifying quoin mechanics." Suggested by Bohr's comments on pp. 24–25 in "The causality problem in atomic physics," *New Theories in Physics* (International Institute of Intellectual Cooperation, Paris, 1939), pp. 11– 45: "These conditions, which include the account of the properties and manipulation of all measuring instruments essentially concerned, constitute in fact the only basis for the definition of the concepts by which the phenomenon is described. It is just in this sense that phenomena defined by different concepts, corresponding to mutually exclusive experimental arrangements, can unambiguously be regarded as complementary aspects of the whole obtainable evidence concerning the objects under investigation."

Pages 160–161 "A Snapshot of Foundational Attitudes toward Quantum Mechanics," M. Schlosshauer, J. Kofler, A. Zeilinger, *Studies in History and Philosophy of Modern Physics* 44, 222–230 (2013). The results of a poll carried out among 33 participants at the conference "Quantum Physics and the Nature of Reality," in July 2011 at the International Academy, Traunkirchen.

"An engineer . . . how they function." Nicolas Gisin, "Sundays in a Quantum Engineer's Life," in R. A. Bertlmann and A. Zeilinger (eds.), *Quantum (Un)speakables: From Bell to Quantum Information* (Springer Verlag, Berlin, 2002), pp. 199–208.

Page 166 "Thus if we accept . . . brings us to cryptography." Artur Ekert, "Less reality, more security," *Physics World* 22, 28–32 (2009), p. 5. The original quotation has EPR for [Einstein].

"The first practical application of Bell's inequality . . . detecting eavesdropping." Artur Ekert, "Less reality," p. 5.

Pages 186–187 "If you take . . . at once." From the motto at the top of Scott Aaronson's blog page, *Shtetl Optimized*.

"[This] problem is not . . . all communication problems." Sandu Popescu, on p. 266 in "Nonlocality beyond quantum mechanics," *Nature Physics* 10, 264–270 (2014). The solution to this problem is used by our heroes to break the bank in the Quantum Quasino sequence.

The three pages strewn on the page are from this article.

Page 217 "Later, when a newsman . . . a gross oversimplification." Asher Peres in "What is actually teleported," *IBM Journal of Research and Development* 48, 63–68 (2004). Asher Peres was one of the authors of the original paper on quantum teleportation: C. H. Bennett, G. Brassard, C. Crépeau, R. Jozsa, A. Peres, and W. K. Wootters, "Teleporting an unknown quantum state via dual classical and Einstein-Podolsky-Rosen channels," *Physical Review Letters* 70, 1895–1899 (1993). We use Peres's "soul" terminology to emphasize that it's the state that is teleported, not a material body.

Page 233 "Closed Timelike Curve." According to the general theory of relativity, gravity is an effect of the bending of space-time by matter and allows the possibility of "closed timelike curves"—regions of space-time where locally, at each point of the space-time trajectory, the direction is forward in time, but globally the trajectory loops backward in time, reversing the causal direction, to close in on itself. Our heroes have walked into a laboratory of experimental "bleeding edge" quantum technologies and should heed the warning.

Acknowledgments

Thanks to these next-gen potential scientists, physicists, and philosophers:

Arlo, 12, who suggested a change to the encryption section that made it way cooler (but seems to baffle PhD-level readers) and rooted out more typos than anyone else.

Anouk, 15, who tirelessly fought for rigor over humor, and substance over appearance. Thank you, the book is better for it!

Calvin, 10, who took a position on entanglement and stood his ground against his dad, and wants an Entangler of his own.

Nova, 13, who thought "this book was awesome," asked just the right questions, and was one of the first people to "get" the (non)ending.

Stephen, 12, who would like to send his friends unhackable messages and is glad he doesn't need a PhD to not understand quantum mechanics.

Solomon, 15, who feels that "one of the primary things the world could use more of would be accessible explanations of complex, important ideas in science and philosophy."

We couldn't agree more! You guys blow us away!

We owe particular thanks to Mike Dascal, who made many useful suggestions that we incorporated into early drafts; to Michel Janssen, who sent us detailed comments, including the suggestion to add a section of notes referencing the sources of the dialogue by historical figures like Einstein and Bohr; and to Clive Reis, who worked out the probabilities for various versions of our "distributed computing challenge" game in the Quantum Quasino. Thanks also to Dan Bub, Gil Bub, Kathlene Collins, Luc Janssen, Charles Midwinter, Robin Shuster, Jacqui Starkey, Dennis Whittle, and Jamie Winton, who gave us helpful input on various drafts; and to Ian Watts and Ronit Bub for their unflagging emotional support. Finally, we'd like to thank our agent, Peter Tallack; and Eric Henney, our editor at Princeton University Press, who saw the potential in a very early draft of the comic and shepherded the project to its conclusion.

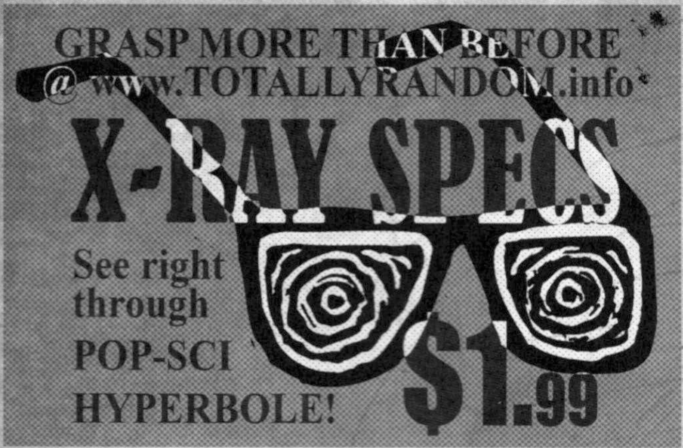
GRASP MORE THAN BEFORE
@ www.TOTALLYRANDOM.info
X-RAY SPECS
See right through POP-SCI HYPERBOLE!
$1.99

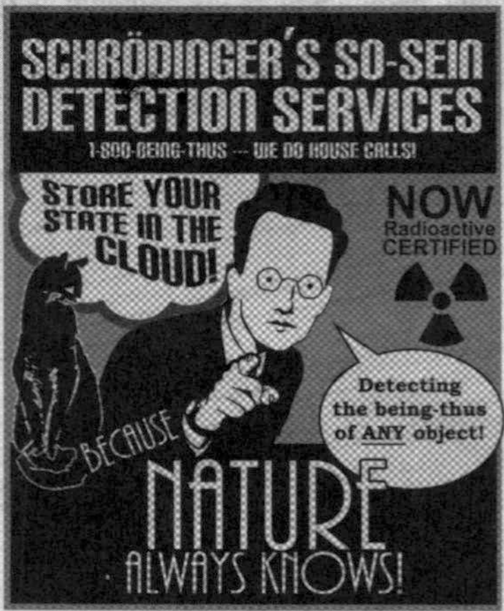
SCHRÖDINGER'S SO-SEIN DETECTION SERVICES
1-800-BEING-THUS --- WE DO HOUSE CALLS!
STORE YOUR STATE IN THE CLOUD!
NOW Radioactive CERTIFIED
Detecting the being-thus of ANY object!
BECAUSE NATURE ALWAYS KNOWS!

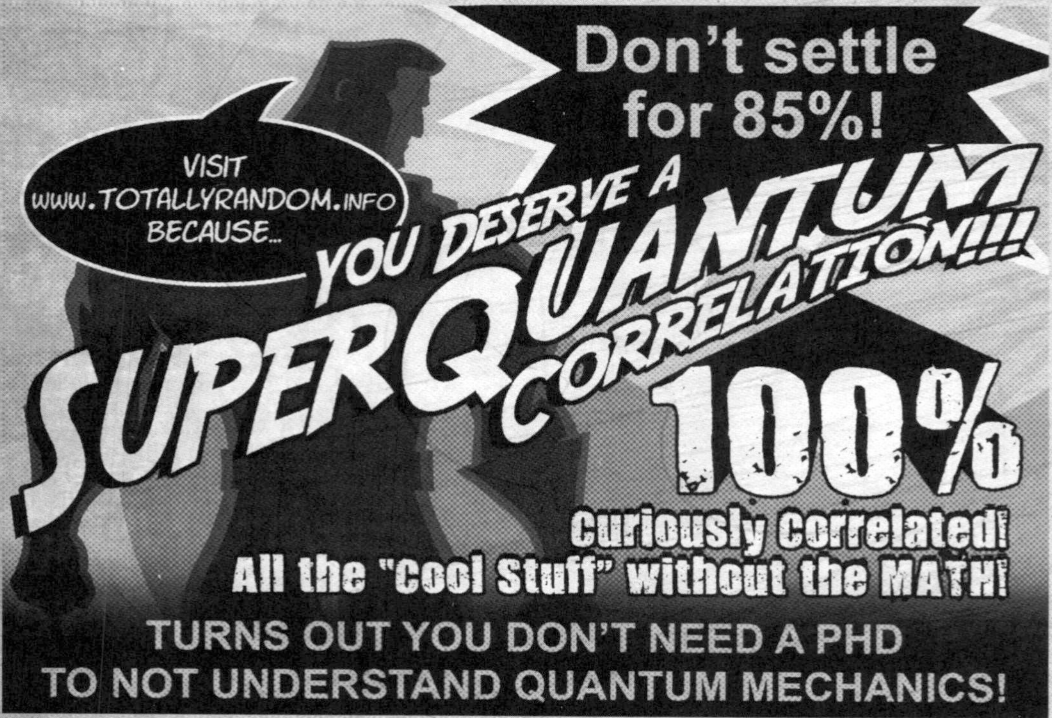
Don't settle for 85%!
VISIT WWW.TOTALLYRANDOM.INFO BECAUSE...
YOU DESERVE A SUPERQUANTUM CORRELATION!!!
100%
Curiously Correlated!
All the "Cool Stuff" without the MATH!
TURNS OUT YOU DON'T NEED A PHD
TO NOT UNDERSTAND QUANTUM MECHANICS!

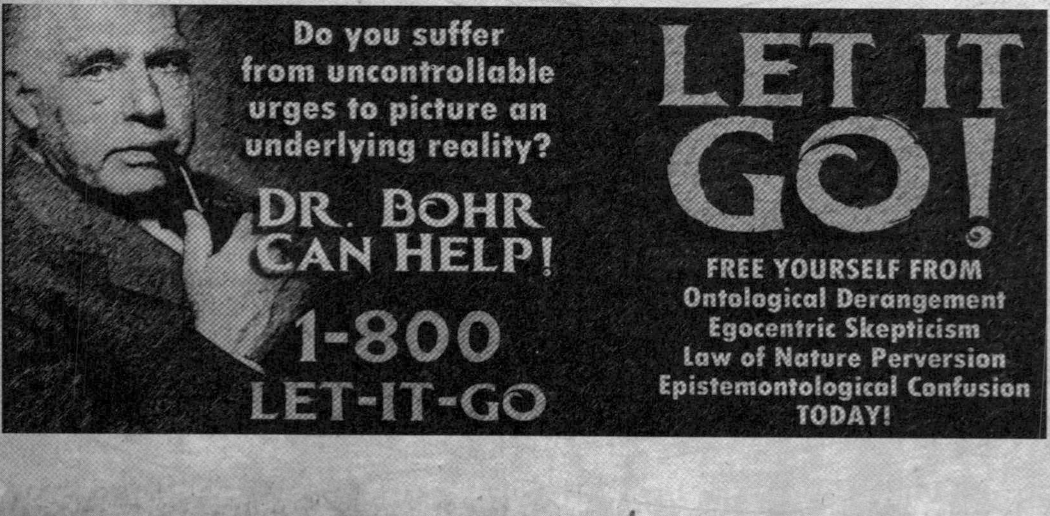
Do you suffer from uncontrollable urges to picture an underlying reality?
DR. BOHR CAN HELP!
1-800 LET-IT-GO
LET IT GO!
FREE YOURSELF FROM
Ontological Derangement
Egocentric Skepticism
Law of Nature Perversion
Epistemontological Confusion
TODAY!

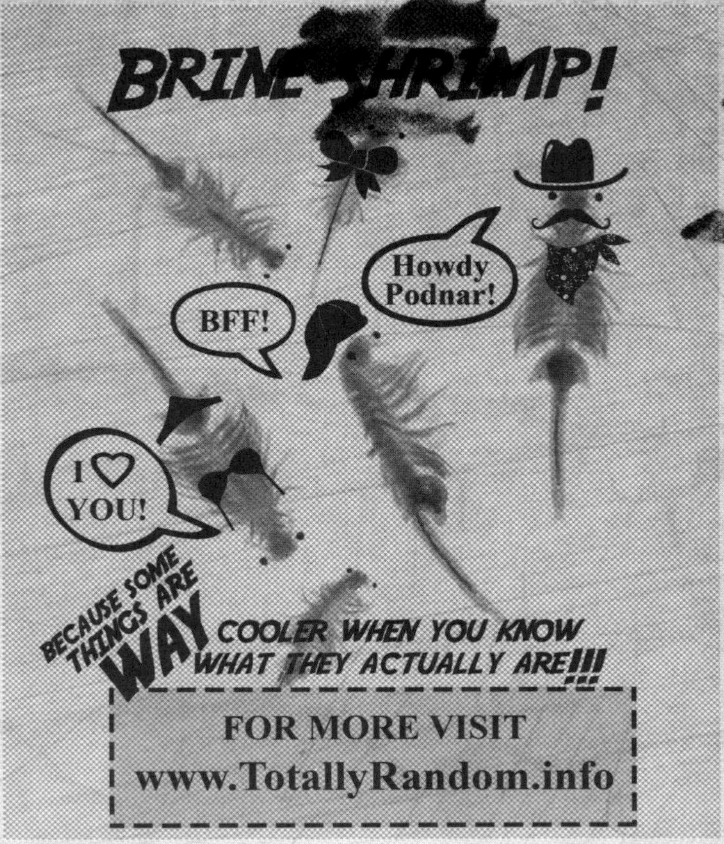
BRINE SHRIMP!
Howdy Podnar!
BFF!
I ♡ YOU!
BECAUSE SOME THINGS ARE WAY COOLER WHEN YOU KNOW WHAT THEY ACTUALLY ARE!!!
FOR MORE VISIT
www.TotallyRandom.info

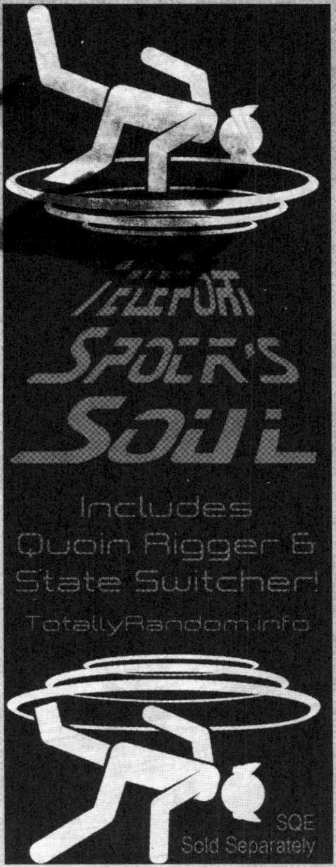
TELEPORT
SPOCK'S
SOUL
Includes
Quoin Rigger &
State Switcher!
TotallyRandom.info
SQE
Sold Separately

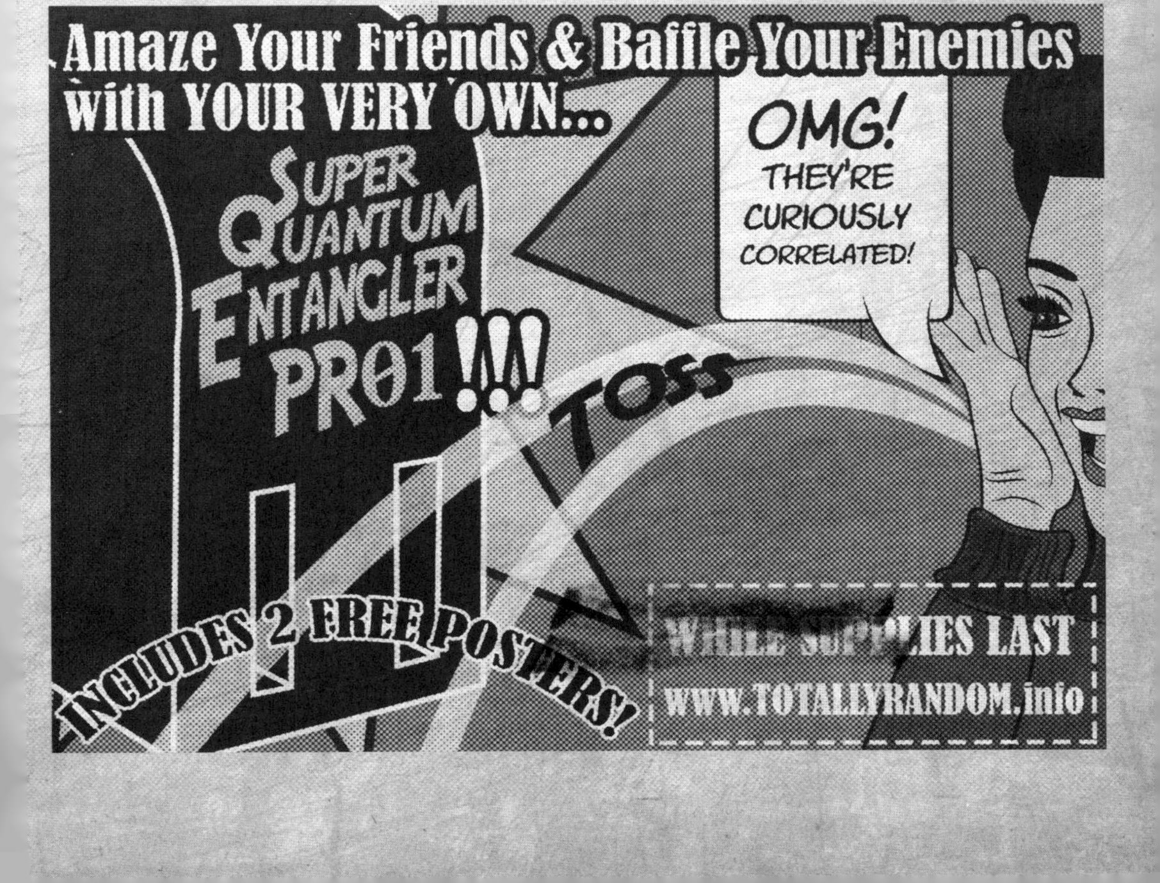
Amaze Your Friends & Baffle Your Enemies
with YOUR VERY OWN...
SUPER QUANTUM ENTANGLER PRO1!!!
OMG!
THEY'RE
CURIOUSLY
CORRELATED!
TOSS
INCLUDES 2 FREE POSTERS!
WHILE SUPPLIES LAST
WWW.TOTALLYRANDOM.info

[redacted]

[redacted]

Images Credits

From Shutterstock.com. **P. vii**: sphinx/*Nerthuz*; skeleton/ *Valentyna Chukhlyebova*; teleportation portal/*denisikym*. **Pp. viii–xii, 2–3**: sphere with lines and dots/*Login*; earth/*Egyptian Studio and NASA*; cat/*Oksana*; time warp clock/*dani3315*; human head/*Quick Shot*. **P. 3**: hand/*indigolotos*; pointing hand/*viewgene*; empty palm hand/*Evannovostro*. **Pp. 12, 14, 20, 27**: hand flipping coin/*macbrianmun*. **Pp. 12, 14, 15, 21, 32, 33**: eagle coin/*di Bronzino*. **P. 15**: Schrödinger's cat/*Rhoeo*. **P. 21**: hand thumb up/*Malyugin*. **P. 37**: fist/*Akim D*. **P. 40**: earth/*owatta*. **P. 52**: lightning/*Albert199*. **P. 65**: time warp clock/*dani3315*. **P. 67**: black cat walking away/ *bluedarkat*. **P. 83**: medicine bottle/*picturepartners*. **P. 108**: dragon/ *Valentyna Chukhlyebova*. **Pp. 125, 132, 135**: man on ground/*JOCIC*. **P. 136**: psychologist/*god.muahaha*. **Pp. 141, 148, 153**: businessman/*Sean Locke Photography*. **P. 142**: impossible geometric shape/*MSSA*. **Pp. 146–148**: train/rails/*Rashevskyi Vlacheslav*. **P. 163**: billboard/*Maxx Studio*. **P. 164**: bell/*alya haclyeva*; 3D optical illusion/*Pedro Alexandra Teixeira*; letter house/*Sedegova Alena*; isometric data wave/*aurielaki*. **P. 165**: staircase/*Aspect3D*. **P. 167**: hacker icon/*Janice Abolins*; revolver with bang flag/*MoreVector*; hacker icon/*Le Mon*; spy agent/*Sly Vectors*. **P. 168**: metal door/*Arth63*; gate with wooden doors/*Piper33*. **Pp. 169–172, 178, 182**: sphinx/*Nerthuz*; escalator/*Petr Born*. **P. 170**: robot/*stuar*. **P. 171**: knight/*FXQuadro*. **Pp. 171, 173**: helmet/*David C Azor*. **Pp. 171, 180, 181**: alarm lever/*Federico Rostagno*. **Pp. 171–175, 177**: skeletons/ *Valentyna Chukhlyebova*. **P. 175**: light burst/*Elks*. **Pp. 176, 177**: oscillating wave/ *Titima Ongkantong*. **P. 177**: 10-ton weight/*gazanfer*. **Pp. 180, 181**: 0, 1 logo/*LogotypeVector*. **P. 182**: sphinx/*Nerthuz*. **P. 183**: crypt with altar/ *Unholy Vault Designs*. **Pp. 183, 185**: rubble/*gualtiero boffi*. **P. 188**: quarter roll/*Mega Pixel*. **Pp. 189, 190**: zebra, horse in coat, dressed-up goat, Doberman Pinscher in black suit, rooster and hen couple/ *Olga Angelloz*. **Pp. 189, 191**: slot machine/*Alex Mit*. **Pp. 191–195**: rooster and hen couple/ *Olga Angeloz*. **Pp. 195–198**: fox couple/ *Olga Angeloz*. **Pp. 212–214**: rhino in tuxedo, dressed-up lion, squirrel/*Olga Angelloz*. **P. 215**: Doberman Pinscher in black suit/ *Olga Angelloz*; arch/*archideaphoto*. **P. 216**: cat and dog/*Dmitry Trubitsyn*. **P. 217**: watching TV/*Blackregis*. **Pp. 230, 231**: road sign/*Wth*. **Pp. 232, 234**: teleportation portal/*denisikm*. **P. 233**: warning sign/*IVVIVVI*. **Pp. 238, 240, 241**: people images/*grynold*. **Pp. 244, 245**: tunnel with binary language/*Carlos Castilla*. **P. 246**: X-ray glasses/*Ron and Joe*. **P. 247**: artemia plankton/*Napat*. **P. 347**: retro woman/*Evart*.

From iStock by Getty Images. **P. 79**: planetary system. **P. 134**: sketch of Bohr with pipe/ *Jacobson, Herdis & Herm*.

[redacted]

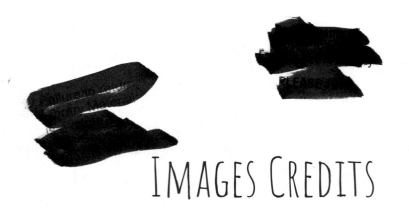

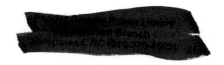